Communications in Computer and Information Science 2231

Rationale

The CCIS series is devoted to the publication of proceedings of computer science conferences. Its aim is to efficiently disseminate original research results in informatics in printed and electronic form. While the focus is on publication of peer-reviewed full papers presenting mature work, inclusion of reviewed short papers reporting on work in progress is welcome, too. Besides globally relevant meetings with internationally representative program committees guaranteeing a strict peer-reviewing and paper selection process, conferences run by societies or of high regional or national relevance are also considered for publication.

Topics

The topical scope of CCIS spans the entire spectrum of informatics ranging from foundational topics in the theory of computing to information and communications science and technology and a broad variety of interdisciplinary application fields.

Information for Volume Editors and Authors

Publication in CCIS is free of charge. No royalties are paid, however, we offer registered conference participants temporary free access to the online version of the conference proceedings on SpringerLink (http://link.springer.com) by means of an http referrer from the conference website and/or a number of complimentary printed copies, as specified in the official acceptance email of the event.

CCIS proceedings can be published in time for distribution at conferences or as post-proceedings, and delivered in the form of printed books and/or electronically as USBs and/or e-content licenses for accessing proceedings at SpringerLink. Furthermore, CCIS proceedings are included in the CCIS electronic book series hosted in the SpringerLink digital library at http://link.springer.com/bookseries/7899. Conferences publishing in CCIS are allowed to use Online Conference Service (OCS) for managing the whole proceedings lifecycle (from submission and reviewing to preparing for publication) free of charge.

Publication process

The language of publication is exclusively English. Authors publishing in CCIS have to sign the Springer CCIS copyright transfer form, however, they are free to use their material published in CCIS for substantially changed, more elaborate subsequent publications elsewhere. For the preparation of the camera-ready papers/files, authors have to strictly adhere to the Springer CCIS Authors' Instructions and are strongly encouraged to use the CCIS LaTeX style files or templates.

Abstracting/Indexing

CCIS is abstracted/indexed in DBLP, Google Scholar, EI-Compendex, Mathematical Reviews, SCImago, Scopus. CCIS volumes are also submitted for the inclusion in ISI Proceedings.

How to start

To start the evaluation of your proposal for inclusion in the CCIS series, please send an e-mail to ccis@springer.com.

Shreekumar T. · Dinesha L. · Sreeja Rajesh
Editors

Intelligent Systems in Computing and Communication

First International Conference, ISCComm 2023
Moodabidri, India, December 8–9, 2023
Proceedings, Part I

Springer

Editors
Shreekumar T.
MITE
Moodabidri, Karnataka, India

Dinesha L.
MITE
Moodabidri, Karnataka, India

Sreeja Rajesh
MITE
Moodabidri, Karnataka, India

ISSN 1865-0929 ISSN 1865-0937 (electronic)
Communications in Computer and Information Science
ISBN 978-3-031-75604-7 ISBN 978-3-031-75605-4 (eBook)
https://doi.org/10.1007/978-3-031-75605-4

This Springer imprint is published by the registered company Springer Nature Switzerland AG
The registered company address is: Gewerbestrasse 11, 6330 Cham, Switzerland

MITE

Invent Solutions

Preface

We are delighted and honored to announce the success of the First International Conference on Intelligent Systems in Computing and Communications (ISCComm 2023), held on December 8–9, 2023, at the Mangalore Institute of Technology & Engineering in Moodabidri. This conference brought together leading researchers, academicians, and industry professionals from various countries worldwide. The conference featured prominent figures from academia and industry as our conference chair, co-chair, program chairs, and track chairs, including experts from top-ranked Indian Institutes such as NITs and MITs, and prestigious foreign universities. The dedication of the Advisory, Program, and Branding & Outreach committees played a crucial role in ensuring the quality of submissions and the overall success of the conference.

We received a total of 245 technical papers across four tracks: AI & ML, IoT and Blockchain Technology, Communication Technology and Data Processing, and Image/Video Analysis and Processing. From these submissions, 53 papers ($<$22%) were selected for publication in Springer CCIS, comprising 36 long papers ($\geq$12 pages) and 17 short papers ($<$12 pages), each undergoing three technical reviews. We extend our gratitude to the staff of Springer CCIS for their continuous support and guidance.

The technical papers presented covered a wide range of application domains, including healthcare, agriculture, automobile, civil and mechanical engineering, pharmacy, cybercrime, and sports. We commend all the enthusiastic authors for submitting their innovative research works.

The conference featured keynote addresses by renowned experts such as Raghavendra Ramachandra from the Information Security & Communication department at NTNU, Norway, Shanthi Sekhar from the Training Vision Institute, Singapore, and Mahadeva Prasanna from IIT Dharwad, India.

We extend our sincere thanks to our esteemed sponsors, including Canara Bank and Mangalore Refinery & Petrochemicals Limited, for their generous support.

We believe this collection of papers will be highly beneficial for researchers and practitioners in AI & ML and their related fields.

December 2023

Shreekumar T.
Dinesha L.
Sreeja Rajesh

Organization

Chief Patron

Rajesh Chouta	MITE, India

Patron

Prashanth C. M.	MITE, India

Conference Chair

Ashok Rao	IISc, India

Conference Co-chair

Raghavendra Ramachandra	NTNU, Norway

Program Committee Chairs

Shreekumar T.	MITE, India
Dinesha L.	MITE, India
Sreeja Rajesh	MITE, India

Steering Committee

Pradeep B. S.	MITE, India
Nagesh Prabhu	NITTE University, India

Advisory Committee

Thejas G. S.	Tarleton State University, USA
Mahesha Narayana	University of the West Indies at Mona, Jamaica
Zhang Jingang T. C.	Linyi Top Network Co. Ltd., China

Vinod P.	CUSAT, India
Balakrishna R.	Raja Rajeswari College of Engineering, India
Nagaratna Hegde	Vasavi College of Engineering, India
Mohit Tahilian	NITK Suratkal, India
Sushil Kumar Pandey	NITK Suratkal, India
Suleiman Yerima	De Montfort University, UK
P. S. Hiremath	KLE Technological University, India
Brahmananda S. H.	GITAM School of Technology, India
P. Vijaya	Modern College of Business and Science, Oman
Priyamvada Singh	DigiPen Inst. of Technology, Redmond, USA
Periyasamy	NIT Tiruchirappalli, India
Venugopala P. S.	NMAMIT, India
Prakash Kumar Udupi	Middle East College, Oman

Publicity Committee

Srikrishna Shastri	MITE, India
Ravinarayana B.	MITE, India
Rashmi Samanth	MITE, India
Sunitha	MITE, India
Sunil Kumar S.	MITE, India
Uday J.	MITE, India
Jyothi S.	MITE, India
Narendra U. P.	MITE, India
Swaroop	MITE, India
Ranjith H. D.	MITE, India
Rashmi Praveen	MITE, India
Deepthi Shetty	MITE, India
Ashwin Kumar	MITE, India
Sonia Peal K. P.	MITE, India
Vasudha	MITE, India
Shivaprasad T. K.	MITE, India
Bindu Madhavi J.	MITE, India

Main Technical Committee

Nawali Noushath	University of Technology and Applied Sciences, Sohar, Oman
Raghavendra Ramachandra	NTNU, Norway
Sreeja Rajesh	MITE, India

Nagesh Prabhu	NITTE University, India
Roshan Fernandes	NITTE University, India

Technical Program Committee Chairs

Sunil Jacob	SOTI Aerospace, Canada
Muhammed Siddique	Majan University College, Oman
Jacintha Menezes	Majan University College, Oman
Saira Jacob	SOTI, Canada
Ramalingam	EC-Council, Oman
Manju Jose	Majan University College, Oman
Sannidhan M. S.	NMAMIT, India
Syed Ziaur Rahman	Majan University College, Oman
Shiva Prakash M.	Dr. AIT, India
Srinath Acharya	SJEC, India
Radhakrishna Doddamane	NMAMIT, India
V. Sowmya	Amrita Vishwa Vidyapeetham, Coimbatore, India
Sudharshan K. M.	REVA University, India
Geetha	NITK Suratkal, India
Karthik	KSSEM, India
Narendra V. G.	Manipal Institute of Technology, India
S. G. Totad	KLE Technological University, India
Ravi	GAT, India
Guruprasad M.	Graphic Era University, Dehradun, India
J. D. Pujari	SDM College of Engineering, India
Shreeknath Acharya	St. Joseph Engineering College, India
Ashwini	NMAMIT, India
Varun G. Menon	SCMS School of Engineering & Technology, India
Varghese Paul	Rajagiri School of Engineering and Technology, India
Mahesh T. R.	Jain University, India
Krishna Prakash	Manipal Institute of Technology, India
Ramakrishna	Manipal Institute of Technology, India
Regitha M. R.	Sacred Heart College, Thevara, India
Srikrishna Shastry	MITE, India
Ramalingam	MITE, India
Kiran Kumar	MITE, India
Babu Rao	MITE, India
Maryjo M. George	MITE, India

Additional Reviewers

Ahamed Shafeeq B. M.
Ajanthaa Lakkshmanan
Akash Gandhar
Anitha N.
Anoop B. K.
Arunadevi
Arunakumari B. N.
Ashwini
B. Perumal
Babitha Ganesh Kulal
Basappa Bharamappa Kodada
Bhargavi K.
Biswajit R. Bhowmik
Chandan Kumar Behera
Dhanesha R.
Divya Mohan
Gaurav Prasad
Geeta R. Bharamagoudar
Geetha
Geetha Mary A.
Guruprasad M.
Haraprasad Mondal
Harshalatha Y.
Hemalatha N.
J. D. Pujari
Jacintha Menezes
Jose Alex Mathew
Jyoti Metan
K. A. Varunkumar
K. Annapoorneshwari Shetty
Kallinatha Hd
Kanakam Siva Rama Prasad
Kanmani
Karthik
Krisha M.
Krishna Prakash
Krishna Prakasha K.
Kshama S. B.
Kuldeep BabanVayadande
Latha H. N.
Laxmi Gulappagol
Madhusudhan G.
Mahesh T. R.
Malathi S. Y.
Manju Jose
Manjula
Manjula Gururaj Rao H.
Mohammed Siddique
Mohan Naik R.
Muhammed Siddique
Nandana Anantha Prabhu
Nandini Prasad K. S.
Narendra V. G.
Neelima K.
Nguyen Quang Vinh
Nimisha Patel
P. Vijaya
Padmanayana
Pavan Kunchur
Poovammal
Prabha Niranjan
Pradeep Kanchan
Pradeep V.
Prajna M. R.
Prakash S. Hiremath
Pramod H. B.
Prashant Narayankar
Praveen Gujjar J.
Praveen J.
Premjith B.
Priyadarshini Patil
Puneet Mittal
Pushpa R.
RadhakrishnaDoddamane
RafidhaRehiman K. A.
Rajan Patel
Rajashekhar B. Somasagar
Rajitha M. R.
Raju K.
Ramakrishna
Ramalingam
Ramesh G.
Ramesh N. Koppar
Ramesh S. Wadawadagi
Rashmi Laxmikant Malghan
Rashmi Naveen

Rashmi P. Karchi
Ratnababu Pilli
Ravi
Ravishankara Kulamarva
Roshan Fernandes
Rudresh Deepak Shirwaikar
S. Gopal Krishna Patro
S. Sethu Selvi
S. G. Totad
S. Kannadhasan
Sai Venkatramana Prasada G. S.
Sainithin Artham
Saira Jacob
Sanjay Thakur
Sanjeevakumar M. Hatture
Sanket S. Kulkarni
Sannidhan M. S.
Sarojadevi H.
Savitha C. K.
Savitha G.
Shashank M. Gowda
Shashank Shetty
Shekhappa Ankaliki
Shiva Prakash M.
Shiva Tyagi
Shreeknath Acharya
Sindhu C.
Sowmya V.
Srikanta Dash
Srinath Acharya
Srinivasa K.
Srinivasa Rao Burri
Sudharshan K. M.
Sudheendra P.
Sudheer Shetty
Suja C. Nair
Sumangala Biradar
Sumathi
Sunil Jacob
Sunil Kumar Nahak
Suresha D.
Syed Ziaur Rahman
T. Ganesh Kumar
Thangam S.
Urmila T. S.
Varghese Paul
Varun G. Menon
Vasudeva
Veena Devi Shastrimath V.
Vidya A.
Vidya Rao
Vidyalakshmi K.
Vimuktha E. Salis
Viswavardhan Reddy Karna

Contents – Part I

Contents – Part II

Artificial Intelligence and Machine Learning

A Novel Approach for Skin Infections Classification Using Transfer Learning

Dhruv Singhal[1], Deepanshu Verma[1](✉), Ankita Nainwal[1], B. Dhananjaya[2], and Garima Sharma[1]

[1] Department of Computer Science and Engineering, Graphic Era Deemed to be University, Dehradun, Uttarakhand, India
ankitanainwal1424@gmail.com

[2] Department of EEE, NMAM Institute of Technology-Affiliated to Nitte (Deemed to be University), Nitte, India

Abstract. Skin infections in today's era are proven to be very deadly if not treated at an early stage. They can gradually results in skin cancer. Doctors use a computer-assisted diagnostic support system to help detect these at an early stage so that the patient's life can be saved. The research in these areas is essential for the development of effective treatments, prevention strategies, and improved patient outcomes in the field of skin infections. In this work, a deep learning model is developed to identify skin infections at a precision of near about 92%. The model is evaluated using a dataset that includes a very large number of samples of mainly 7 types of skin cancers that are classified as dangerous. By using CNN with transfer learning techniques, the class imbalance and complexity in the dataset are reduced, and the model performs better than most of the deep learning techniques in terms of accuracy and computational cost.

Keywords: Deep Learning · Convolutional Neural Network · Transfer Learning · MobileNet · skin infection classification

1 Introduction

Skin diseases in the current era in humans are increasing at a rate of 5.3% every year. Skin diseases generally spread in the body when infections develop in human blood and circulate in the whole body in vessels. Thus, early detection of skin cancer is very important for the proper diagnosis of the disease. When there is an uncontrolled growth of irregular skin cells, they tend to develop tumors in the skin which finally lead to cancers. Skin cancer like cell basal and cell carcinoma are the two most common. Till now, five to six million cases of skin infection that have led to cancer have been found in the UN alone and we can think if we think of the whole world what can the number be. These infections nowadays comprise mainly seven types of skin cancers which can be further classified into many sub- divisions. The treatment and prognoses of these infections depend on various types and stages of the disease. In the traditional clinical method preliminary screening is used by dermatologists to diagnose skin cancer, specialized

T. Shreekumar et al. (Eds.): ISCComm 2023, CCIS 2231, pp. 3–13, 2025.
https://doi.org/10.1007/978-3-031-75605-4_1

expertise is required for this. Diagnosis is critical in determining the most effective treatment and recovery of the disease. So in these cases, early detection is the key to managing increasing skin infections. Hence regular skin checks and medical attention is required for suspicious lesion, So this study plays an important role.

Skin cancer can be detected using a number of measures, starting with a visual inspection of suspicious tumors, followed by dermoscopy and biopsy. Dermoscopy improved the detection rate by 50% with an accuracy range of 75% to 84%. However, an accurate diagnosis still depends heavily on the skill of the doctor. Computer-aided diagnostics can help medical professionals analyze dermoscopy procedures when expertise or availability is limited. Artificial intelligence is a new area of study that is similar to the way we have been making improvements to our lives in the last few years. Machine learning (ML) [1] or deep learning methods help us to do things faster and more efficiently, and recent studies have shown that they can help us to predict cancer more accurately. Deep learning [2] is one of the most important aspects of artificial intelligence, and it is used to do things like to identify images and classify skin diseases. Some computer programs that use deep learning are as accurate as professional dermatologists.

To develop a deep learning model for diagnosing skin cancer, the first step is to identify the location of a lesion. After that, the computer system will use various techniques to detect and measure the characteristics of the lesion. Based on that information, the system will create a classification system to help doctors make a diagnosis. This way can help doctors as well as on an individual level to find out the type of diseases before it gets too late. Using this the health benefit also goes higher as all people will get access and can take proper steps regarding the diagnosed skin disease. Generally, image classification involves a model being trained to be able to recognize different categories of skin infections based on their classification [3]. In today's era, there are numerous ML and deep learning algorithms for the same. Deep learning is one of the most important aspects of artificial intelligence, and it is used to do things like to identify images and classify skin diseases Some computer programs that use deep learning are as accurate as professional dermatologists It has been developed using a large range of skin samples approximately ten to fifteen thousand for testing, training, validation, and building the model respectively.

2 Literature Survey

The doctors have been studying how to automatically classify skin lesions for a long time. However, most of this research has been focused on classifying tumors, rather than a variety of diseases. In past recent years, a growing interest in using ML and deep learning techniques has been considered for improving classification. Studies in the same field is done by many researchers in recent years and many of them proposed work that plays an important role. A neural network detection method combined with rational Gaussian curve fitting was utilised in the research by M.I. Rajab, M.S. Woolfsom, and SP Morgan [4]. Five independent machine learning techniques—GBDT, DT, CART, RF, and SVM—were combined with neural networks and support vector machines by T Parvathaneni Naga Srinivasu and Jalluri Gnana Siva Sai [5]. Abderrahim Bourouis and Ali Zerdazi [6] employed the feed-forward neural network and the Multi-Layer Perceptron (MLP) method. With a feedforward back propagation neural network and the

K-nearest neighbour technique, Mohmad Elgamar achieved high accuracy [7]. Images of skin infections were categorised using an ANN classifier for artificial intelligence and digital image processing by Aswin R.B. and J. Abdul Jaleel [8]. Manjusha K.K. and K. Shankararayan employed data mining techniques in conjunction with the Naive Bayesian theorem to find skin lesions [9]. Joanna Jaworek-Korjakowska uses a Support Vector Machine with k-fold cross-validation to provide findings with good sensitivity and specificity [10]. In order to use the K Neural Network with a hierarchical representation of the data, Manish Kumar and Rajiv Kumar used KNN [11]. In order to outperform earlier studies, P. Kavimathi combined Otsu's segmentation approach with comparative analysis of classifiers [12]. K-means clustering was used in the research by Mohd Anas and Ram Kailash Gupta [13] using six distinct color-texture feature extractors. In order to identify photos using GLCM, Uzma Bano Ansar and Tanuja Sarode employed a Support Vector Machine and achieved comparable accuracy results [14]. According to a comparative analysis of all algorithms, Seema Kolkur, Dr. D.R. Kalbande, and Dr. Vidya Kharkar classified skin lesions using a range of algorithms, including Multilayer Perceptron (MLP), Nave Bayes, and Decision tree algorithms [15]. K Neural Network using KNN employed for the computation of Euclidean distance was used by YA Gerhana, MA Ramdhani (2018), Zulfikar, and AH to locate related regions [16]. In order to identify pictures using GLCM and classify them using SVM, John Mitchell's study [17] extracted texture feature information. ResNet50, DenseNet 121, and MobileNet with CNN [18] models were used in the study by Anabik Pal, Saunak Ray, and Utpal Garain, which successfully detected skin cancer [19]. Convolutional neural networks were utilised by Saurav Kumar Patnaik to compare InceptionV3, Inception ResnetV2, and MobileNet as the best accuracy architecture [20]. In order to categorise skin illnesses into three groups, V. Pugazhenthi and Sagar K. Naik suggested a method using the GLCM (Grey Level Co-occurrence Matrix), decision trees, and image quality evaluation techniques [20]. To categorise skin illnesses, Nazia Hameed and Fozia Hameed presented a method using CNN-SVM, support vector machines (SVM), and quadratic support vector machines (QSVM) [21]. The suggested research by Samuel Akyeramfo-Sam and Acheampong Addo Philip employed smoothing, aggregation normalisation, and attribute creation approaches in a convolutional neural network (CNN) classifier for the classification of skin diseases [23]. By categorising the aim as distinct shots as 1-shot, 3-shot, and 5- shot classification, Kushagra Mahajan and Monika Sharma applied meta-learning, especially Reptile and Prototypical Network, for illness identification [23]. In Alaa Ahmed Abbas' method, skin lesion edges were segmented using the Canny edge detector [26]. The assessment of 27 research articles for the proposed project by Jamal Mustafa Al-Tuwaijari and Saja Salim Mohammed led to the conclusion that image processing methods and CNN generate excellent accuracy for skin disease classification [27]. K-means clustering and a colour gradient technique are used by V Rajesh and Ruth Ramya Kalangi to separate skin lesion pictures from the Xiangya-Derm dataset [28]. In their paper, Fiaz Majeed and Ansar Siddique suggested a method that combines features-based classification with GLCM and CNN-based segmentationFor the categorization of skin diseases, Prof. Amiya Bhaumik and Dr. Sandeep Poddar employ Multi Support Vector Machines (MSVM). The suggested method included strategies for dataset selection, preprocessing, and classification [28]. Convolutional Neural Networks (CNN) were utilised by Ebrahim Mohammed Senan to classify skin conditions and extract features using GLCM and CNN [29]. The work of Mostafiz Ahammed and Md. Al Mamun made use of an impainting technique and black-hat algorithm transformation [30]. Convolutional Neural Networks (CNN) were utilised by Mrs. S. Hemavathi and colleagues to categorise skin disorders using feature extraction and segmentation [31]. In order to categorise skin illnesses, Javed Rashid and Noor Samand employ a transfer learning model with a trained MobileNetV2 model [32]. This states that study in early detection of various skin disease types can be proven helpful for society.

3 Methodology

Artificial intelligence is mostly based on self -a learning algorithm which acts as the building block. They can continue to evolve as new information is received, which can help resolve any issues that may arise. The methodology applied in this model is stated. Firstly the whole dataset has been cleaned, and images have been pre-processed, and augmented (since the data is less, so data augmentation using ImageDataGenerator was applied). ImageDataGenerator generates augmentation of images in a real-time scenario when the model is in the process of being trained. An individual who wants to develop can use the application use of any random transformations on each image that need to be trained as it is passed to the model) and finally the dataset is split into testing and training sets. After this the CNN model is built, it is designed to detect features as well as patterns in images. A well- defined CNN light weighted architecture consists of main 3 layers first convolutional layers, second pooling layers, and thirdly fully- connected layers (Fig. 1).

The first layer of convolutional helps to extract image features and the pooling layer reduces its dimensionality the third i.e fully connected layer helps to classify these features. Then a pre-trained CNN-Model is chosen as the base model for ap- plying the transfer learning technique then this pre-trained model is loaded and its layers get freeze now this new model is being got trained on the split training set and performance is evaluated.

Dataset: The used dataset HAM 10000 is a collection of images of human skin diseases, used for the study of different types of skin cancer. The dataset was divided into three category sets: training, testing, and validation. In this way, there was no difference in the obtained results between the different sets. Few definitions used in this methodology are-:

A. One-Hot Encoding

   It is used to enhance the accuracy score of the results of classification and improve the features of the input images. This is done by using a CNN approach on a large-scale image dataset.

B. Image Rescaling

   All the images of a dataset that are original are available in a much larger resolution of 6000 by 4000 pixels. However, for increasing the speed of the processing process, the dataset is reduced to a much smaller size of 256 by 256 pixels.

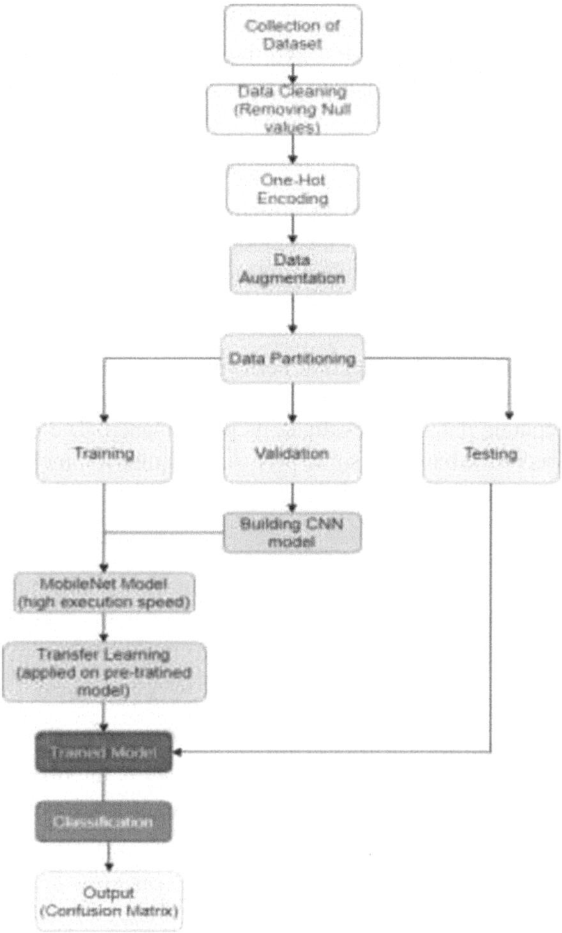

Fig. 1. Proposed Skin Infection Classification System

C. Extracting Features

Classifying skin lesions can be difficult, because the features that are important for one disease may not be important for another. This research is trying to find features that are common to many skin diseases so that the classification process is simplified.

D. Training Deep learning model and then performing Validation and Testing

The dataset was split into three main parts: training used to train the model, testing, and validation for testing and building respectively. The set of datasets used for training was used to train the MobileNet CNN-based model, whereas the validation and testing sets were utilized to measure the model's productivity.

E. MobileNet Architecture

In this study, the challenge of skin infection mainly 7 types of cancer classification was addressed using the deep transfer learning MobileNet architecture. The thinking of building this model was influenced by several factors, including the small size of the dataset used for training after splitting the big dataset. MobileNet, being a small but expressive model, helped to mitigate this effect significantly. Overall, the MobileNet model proved to be a promising solution for tackling the challenge of melanoma classification.

F. Transfer Learning

Transfer learning is described as a method that can be used to improve skin disease detection by using a pre-trained model of CNN which has been trained on a large dataset of general images. This model can then be finely organized on a smaller dataset of skin disease images, which allows it to learn specific features that are relevant to skin disease detection.

G. Measures for Evaluating Classification Performance

Following the completion of the training process, the proposed method underwent testing on a separate dataset to evaluate its performance. The effectiveness of the architecture was tested based on some performance metrics, including accuracy, F1 score, precision, and recall. These performance metrics will now be discussed in greater detail.

H. Classification Accuracy

Classification accuracy can be described as the ratio of the number of correctly predicted instances to the overall number of instances in the dataset. In other words, it measures the percentage of accurate predictions made by the classification model.

$$\text{Classification Accuracy} = \text{Ratio of}(\text{TP} + \text{TN})\&(\text{FN} + \text{FP} + \text{TN} + \text{TP}) \quad (1)$$

I. Precision Value (Exactness)

When measuring something repeatedly, accuracy is important, but if the thing being measured is very different each time, accuracy is not very useful. Precision is a measure of how consistent a measurement is.

$$\text{Precision} = \text{TP}/(\text{TP} + \text{FP}) \quad (2)$$

J. Recall

The recall is a statistic that tells us how well the system can predict which samples belong to which group.

$$\text{Recall} = \text{TP divided by}(\text{TP} + \text{FN})\text{ i.e. TP}/(\text{FN} + \text{TP}) \quad (3)$$

K. F1Value

The f1 value is a measure of how well you remember information and how precise your information is.

$$\begin{aligned} &\text{F1Value} = [(\text{Precision} * \text{Recall}) * 2]/(\text{Y}) \\ &[\text{Y} = (\text{Precision} + \text{Recall})]. \end{aligned} \quad (4)$$

4 Results

The In this study the dataset taken as mentioned above in different sections is HAM10000, and before moving to build any model, some work has been done on the dataset that is mentioned further. The one-hot encoding i.e. image preprocessing is done to resize the image and show the result of 7 main types of skin infections (Fig. 2).

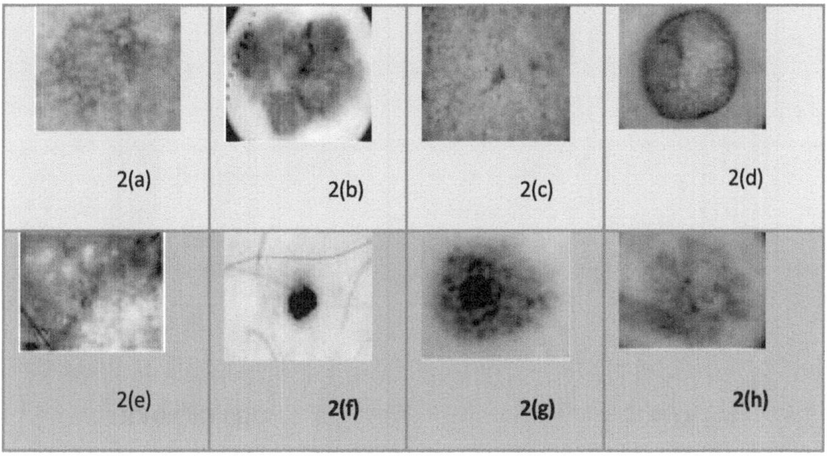

Fig. 2. 08 types of skin cancer images

After the above mentioned using exploratory data analysis, we identified and corrected errors, to detect outliers and patterns within the dataset based on different parameters, result is attached above (Fig. 3). Here the explanation of Fig. 3 is demonstrated after analyzing the result obtained from the output graphs i.e. Fig. 3(a) depicts that skin infections are found to be maximum in humans around the age of 45. Figure 3(b) denotes that men as compared to women have higher skin infections. Figure 3(c) says about the maximum presence of skin infections on back body parts. Figure 3(d) depicts that melanocytic nevi are the most common whereas dermatofibroma is the least. Now the CNN model has been built without using transfer learning, as it is known to be ideal for image classification. It is a repeated network of mainly 5 layers i.e. Convolution, pooling, dropout, flatten, and dense. These layers are responsible to detect higher-level features, to reduce the spatial dimensionality, and to take the final decision on the outcome.

After running a certain number of epochs and finding out their respective accuracy, a summarized study in the form of a table is stated further (Table 1).

Then, the confusion matrix has been plotted to evaluate the performance of this CNN model without transfer learning by breaking down predictions into four categories i.e. true positives, false positives, true negatives, and false negatives (Fig. 4).

In the same model after applying transfer learning with CNN (MobileNet), the mean recall obtained is near about 0.865, the accuracy to about 92%. Transfer learning significantly reduces the time and resources that are required to train the CNN model and is used to improve the accuracy of the CNN model by using the features learned by a pertrained model on a large dataset such as MobileNet. The below figure shows the training and validation accuracy of the model that is being denoted by the blue and yellow color lines with an accuracy of near about 95 and 92% respectively (Fig. 5).

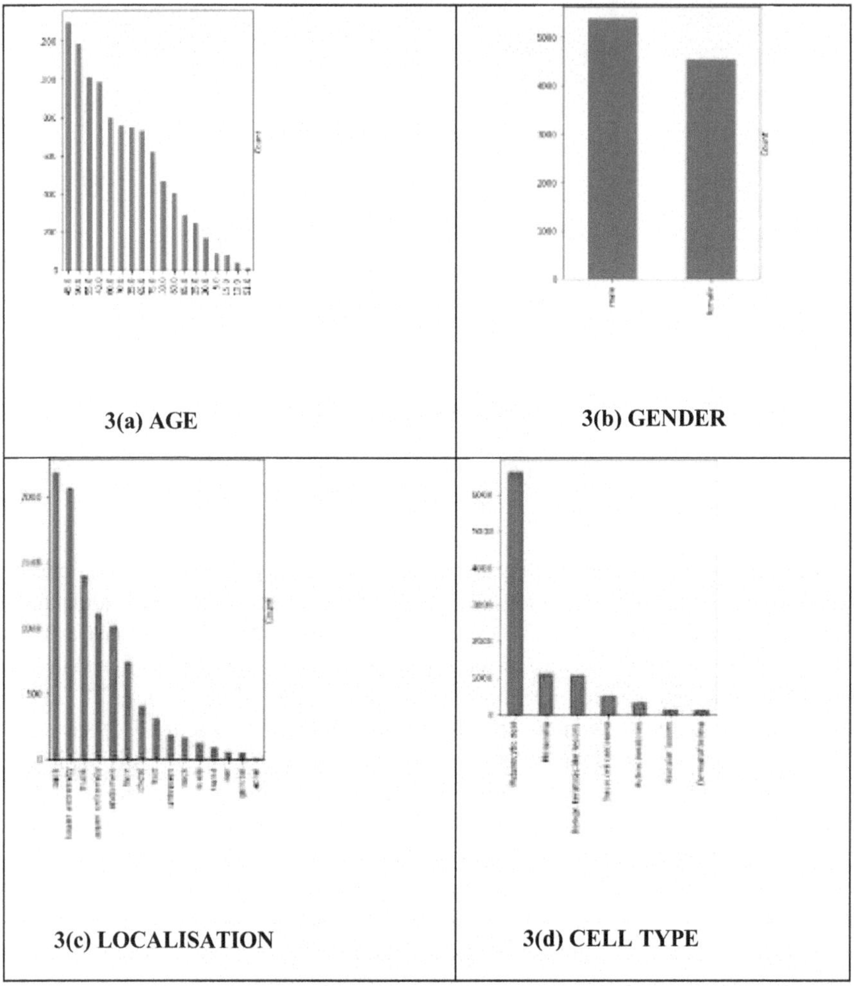

Fig. 3. Univariate analysis of dataset

Table 1. Comparative study of the accuracy after a certain number of epochs after building the CNN model without transfer learning

Number of epochs	Accuracy(%)
10	72.87
20	73.60
30	74.90
40	76.5
50	77.2
60	78.67

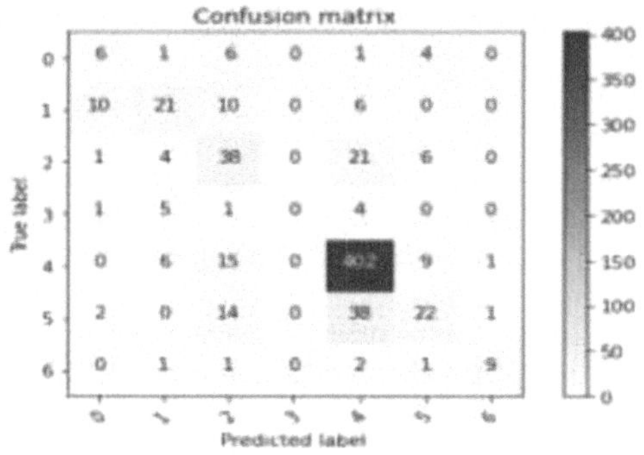

Fig. 4. Confusion matrix

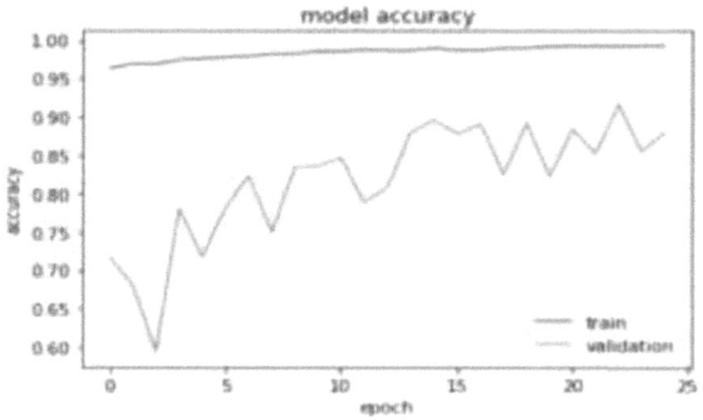

Fig. 5. Accuracy plot of the developed CNN model when used with transfer learning

5 Future Scope and Conclusion

This paper has suggested a CNN using a transfer learning model-based approach to recognize seven main skin infections that are Melanocytic nevi, Melanoma, Benign keratosis-like lesions, Basal cell carcinoma, Actinic keratosis, Vascular lesions, Dermatofibroma. This model has a very promising future scope in the medical field especially dermatology. One possible use in our view for the skin detection model is that we can expand its scope from binary to multiclass classification, in which the model can identify different skin infections with high accuracy and it can cover almost all types of skin cancer diseases. And the second use of this model, if integrated with time-changing technology, is that it can be merged with mobile applications or web-based platforms making it more accessible to dermatologists and patients. This study concludes that a model which is based on CNN with transfer learning has great potential giving more accuracy than most of the studies ever done. After analyzing near about 30 research papers in this field and applying the CNN architecture implementation, the accuracy score comes to around 78%-80%. Now the implementation of transfer learning using MobileNet significantly increases the accuracy value to 92%. Hence, the use of deep learning techniques can help in the early detection, treatment, and management of skin diseases.

References

1. Asha, P., et al.: Role of machine learning in attaining environmental sustainability. Energy Rep. **8**, 863–871 (2022)
2. Rajagopal, A., et al.: A deep learning model based on multi-objective particle swarm optimization for scene classification in unmanned aerial vehicles. IEEE Access **8**, 35383–135393 (2020)
3. Rawat, V., Singh, D.P., Singh, N., Kumar, P., Goyal, T.: A comparative study of various skin cancer using deep learning techniques. In: 2022 International Conference on Computational Intelligence and Sustainable Engineering Solutions (CISES), pp. 505–511 (2022)
4. Rajab, M.I., Woolfson, M.S., Morgan, S.P.: Application of region-based segmentation and neural n-etwork edge detection to skin lesions. Comput. Med. Imaging Graph. **28**, 1–2 (2004)
5. Srinivasu, P.N., SivaSai, J.G., Ijaz, M.F., Bhoi, A.K., Kim, W., Kang, J.J.: Classification of skin disease using deep learning neural networks with MobileNet V2 and LSTM. Sensors **21**(8), 2852 (2021)
6. Bourouis, A., Zerdazi, A., Feham, M., Bouchachia, A.: M-health: skin disease analysis system using a smartphone's camera. Procedia Comput. Sci. **19**, 1116–1120 (2013)
7. Jaleel, J.A., Salim, S., Aswin, R.B.: Computer aided detection 01 skin cancer. International Conference on Circuits, Power and Computing Technologies, pp. 2395–0056 (2013)
8. Jaworek-Korjakowska, J., Kłeczek, P.: Automatic classification of specific melanocytic lesions using artificial intelligence. BioMed Research International (2016)
9. Kumar, M., Kumar, R.: An intelligent system to diagnosis the skin disease. ARPN JEAS **11**(19), 11368–11373 (2016)
10. Kavimathi, P.: Comparative analyses of classifiers for diagnosis of skin cancer using dermoscopic images. Indian J. Sci. Technol. **9**, 43 (2016)
11. Whiteman, D., Green, A.: Melanoma and sunburn. Cancer Causes Control **5**, 564–572 (1994)
12. Younis, H., Bhatti, M.H., Azeem, M.: Classification of skin cancer dermoscopy images using transfer learning. 2019 15th International Conference on Emerging Technologies (ICET), pp. 1–4. IEEE (2019)
13. Chintawar, V., Sanghavi, J.: A review on computer-aided melanoma skin cancer detection using image processing. Easy Chair Preprint, p. 584 (2018)
14. Suleiman, M., Akio, K.: An SVM-based diagnosis of melanoma using only useful image features. In: 2018 International Workshop on Advanced Image Technology (IWAIT), IEEE (2018)
15. Kolkur, M.S., Kalbande, D.R., Kharkar, V.: Machine learning approaches to multiclass human skin disease detection. International J. Computational Intelligence Res. **14**(1), 1–12 (2018)
16. Sonali, J.: Analysis and detection of skin cancer. IOSR Journal of Electronics and Communication Eng. **9**(4), 50–54 (2014)
17. Goel, R., Singh, S.: Skin cancer detection using GLCM matrix analysis and SVM classifier. International Journal of Applied Engineering and Technology ISSN (2015)
18. Khari, M., Garg, A.K., Crespo, R.G., Verdú, E.: Gesture recognition of RGB and RGB- D static images using convolutional neural networks. Int. J. Interact. Multim. Artif. Intell. **5**(7), 22–27 (2019)
19. Gerhana, Y.A., Zulfikar, W.B., Ramdani, A.H., Ramdhani, M.A.: Implementation of nearest neighbor using HSV to identify skin disease. In: IOP Conference Series: Materials Science and Eng. **288**(1), 012153 (2018)

20. Wei, L.S., Gan, Q., Ji, T.: Skin disease recognition method based on image color and texture features. Computational and Mathematical Methods in Medicine (2018)
21. Pal, A., Ray, S., Garain, U.: Skin disease identification from dermoscopy images using deep convolutional neural network (2018)
22. Hameed, N., Hameed, F., Shabut, A., Khan, S., Cirstea, S., Hossain, A.: An intelligent computer-aided scheme for classifying multiple skin lesions. Computers **8**(3), 62 (2019)
23. Akyeramfo-Sam, S., Philip, A.A., Yeboah, D., Nartey, N.C., Nti, I.K.: A web-based skin disease diagnosis using convolutional neural networks. International Journal of Information Technology and Computer Science **11**(11), 54–60 (2019)
24. Sriwong, K., Bunrit, S., Kerdprasop, K., Kerdprasop, N.: Dermatological classification using deep learning of skin image and patient background knowledge. International Journal of Machine Learning and Computing, **9**(6), 862–867 (2019)
25. Mahajan, K., Sharma, M., Vig, L.: Meta-dermdiagnosis: few-shot skin disease identification using meta-learning. In: Proceedings of the IEEE/CVF Conference on Computer Vision and Pattern Recognition Workshops, pp. 730–731 (2020)
26. Abbas, A.A., Abu-Almash, F.S.: Skin lesion border detection based on optimal statistical model using optimized colour channel. J. Autonomous Intelligence **3**(1), 18–26 (2020)
27. Mohammed, S.S., Al-Tuwaijari, J.M.: Skin disease classification system based on machine learning technique: a survey. IOP Conference Series: Materials Science and Eng. **1076**(1), 012045 (2021)
28. Inthiyaz, S., et al.: Skin disease detection using deep learning. Adv. Eng. Softw. **175**, 103361 (2023)
29. Manzoor, K., et al.: A lightweight approach for skin lesion detection through optimal features fusion. Computers, Materials & Continua **70**(1), 1617–1630 (2022)
30. Bandyopadhyay, S., Bhaumik, A., Poddar, S.: Skin Disease Detection: Machine Learning vs Deep Learning (2021)
31. Ahammed, M., Al Mamun, M., Uddin, M.S.: A machine learning approach for skin disease detection and classification using image segmentation. Healthcare Analytics **2**, 100122 (2022)
32. Rashid, J., et al.: Skin cancer disease detection using transfer learning technique. Applied Sciences **12**(11), 5714 (2022)

A Unified Image Repository for Cucurbits Diseases and Pests

Rithika Adapa[1], Raksha[1], N. Hemalatha[1(✉)], Prathamesh Pai[2], K. M. Sreekumar[3], P. K. Sajeesh[3], and Sainamole Kurian[3]

[1] St. Aloysius Institute of Management and Information Technology, Mangalore, India
rakshamohan729@gmail.com, hemalatha@staloysius.ac.in
[2] Lcode Technologies Pvt. Ltd., Mangalore, India
[3] College of Agriculture, Padanakkad, Nileshwar, India
{sreekumar.km,sajeesh.pk,sainamole.k}@kau.in

Abstract. This study offers a user-friendly digital platform that hosts an integrated database of images encompassing diseases and pests that harm cucurbits plants. The database consists of two crucial aspects in cucurbits plants – pests and diseases. It allows users access to a vast image repository, encouraging research collaboration and facilitating quick responses for cucurbits crop protection. Our platform makes use of the strength of shared knowledge and cutting-edge technology to serve as a valuable resource for the agricultural community. The creation of this extensive image repository of cucurbits pests and diseases are described in detail in this study, with an emphasis on its importance for improving agricultural disease and pest control.

Keywords: Cucurbits diseases · Pest identification · Collaborative platform · Crop protection

1 Introduction

The ability of agriculture to feed the population of the world, maintain economic stability, and provide critical nutrition is crucial. Cucurbits are one of the many crops that are grown all over the world and are a basic food. In addition to being a favourite in salads and sandwiches, these fresh, crisp vegetables also make a substantial contribution to the agricultural economy. Cucurbits are more valuable economically and agriculturally than their culinary purposes. Their cultivation provides significant income, sustains livelihoods, and guarantees a steady supply of food.

The most serious issue that cucumber farmers deal with is the attack of plant diseases and destructive pests. These microscopic and macroscopic foes cause significant economic losses by lowering crop yields, putting the safety of the food supply at risk, and making it more expensive for farmers to operate.

Agriculture has undergone a revolution in this era of technological development, driven by innovation. To address persistent issues, modern agriculture increasingly relies on technology. Particularly notable among technological aspects are computer vision and

T. Shreekumar et al. (Eds.): ISCComm 2023, CCIS 2231, pp. 14–25, 2025.
https://doi.org/10.1007/978-3-031-75605-4_2

image analysis. These innovative techniques have enormous potential for resolving the ongoing difficulties that cucumber producers encounter.

In order to better understand cucurbits diseases and pests that cause destruction on cucurbits crops, we plan to build a comprehensive and meticulously curated image database. The major reason for creating this image database is that there were no sufficient number of images available in the browser related to the pests and diseases that affect cucurbits plants in India. So, we have made obtained some of the images from the College of Agriculture, Nileshwar. This image database is intended to be a strong tool that can support agricultural stakeholders, researchers, and farmers alike. The webpage also provides the user with relevant information about each pest and disease that has been included in the paper. With early intervention made possible and agricultural output eventually increased, it promises to revolutionise disease and pest diagnosis.

This paper is described with the following sections – Sect. 2 describes the different pests and diseases which are included in our database. Section 3 describes the methodology used for creating database. Results of our work is described in Sect. 4 and finally paper ends with a conclusion.

2 Literature Review

In their article from 2021, Kumar *et al.*, discuss how important it is to have complete datasets in the agriculture field, especially for machine learning initiatives. The authors carefully choose a picture dataset of fruits of various grades, photographed from various angles, and supplemented by appropriate weight data, focusing on the 'Ruby' cultivar of pomegranate. The effectiveness of this dataset, which consists of 12 folders that each reflect a different quality grade, has previously been established by the authors' earlier work. The dataset's value for data science engineers and machine learning experts active in precision agriculture is emphasized in the article, along with its potential for a variety of machine vision applications in the industry [1].

In the agriculture sector, there is a critical need for quick and precise fruit classification based on quality criteria, which will be addressed by Meshram and Patil in 2021. The authors introduce FruitNet, an image dataset of Indian fruits including apple, banana, guava, lime, orange, and pomegranate, acknowledging the value of a well-structured dataset for creating reliable machine-learning models. The collection, which consists of about 19,500 processed photos, is divided into folders that depict fruits of good, bad, and mixed quality. Meshram and Patil highlight the dataset's potential in computer vision and deep learning applications by showing its usefulness for developing, evaluating, and validating fruit classification and identification models [2].

In 2013, Gutman *et al.*, emphasized the need for improved biomedical data integration, concentrating on the confluence of morphological observations, patient outcomes, and genetic events within TCGA. They introduced the Cancer Digital Slide Archive (CDSA), a freely available digital platform that includes more than 20,000 complete slides depicting various cancer types. The CDSA supports URL-based picture sharing and selective annotation sharing based on important metadata, enabling seamless visualization and integration of pathology images with clinical and genetic data. Given the increased accessibility of whole-slide scanners, the study emphasizes the growing value of digital pathology imaging in developing precision medicine and cancer research [3].

Forero *et al.*, addressed the lack of accessible datasets for seedling research by presenting a unique LiDAR-based platform for 3D plant phenotyping. Their platform offers synchronized, high-precision point clouds and photos of 362 maize seedlings, capturing a heterogeneous dataset. Raw and processed data with manually assigned plant labels are included in the database. A Random Forest classifier successfully identifies seedling components using morphological characteristics with an astonishing 89.41% accuracy, demonstrating the platform's effectiveness [4].

Avraham *et al.,* in their article describes the community-driven Plant Ontology Database, a resource that offers controlled language and annotations for information on the anatomy, morphology, and developmental stages of plants. The database has aided developments in plant genomes and systems biology and is crucial for standardizing data on plants across many research programs [5].

3 Pests and Diseases

This section provides detailed information on common pests and diseases affecting cucurbits plants which are included in our image repository.

3.1 Pests

This section describes Aphids, Red Pumpkin Beetle, Fruitfly, Whitefly, and Serpentine Leafminers along with their impact on cucurbits crops.

3.1.1 Aphids

Numerous aphid species, such as the destructive melon aphid (*Aphis gossypii*) and green peach aphid, have been found to feed on cucurbits. They use their narrow mouthpart to puncture the delicate plant part and suck plant fluids from it [6]. Aphid infestations result in a number of symptoms, such as diminished plant vigour and growth, as well as yellowing, browning, curling, and withering of the leaves. These symptoms ultimately lead to low economic yields and occasionally plant mortality. The first indication of an aphid infestation is the downward curling and tearing of the leaves. Aphids feeding on fruit and flower buds cause the development of malformed blooms or fruits. Honeydew of Aphids serves as a growing medium for sooty molds (fungi), which eventually obstruct light and disrupt photosynthesis on leaves and other plant parts. Additionally, aphids spread several viruses that harm all types of cucurbits, leading to a high rate of crop failure and substantial financial losses [7].

3.1.2 Red Pumpkin Beetle

Aulacophora foveicollis, frequently referred to as the Red Pumpkin Beetle, is a major and destructive pest of cucurbits in India. When creepers are extremely young, the adults eat the cotyledons, blossoms, and leaves, which results in holes. The early-sown cucurbits are so severely injured that they might need to be resown [8]. The beetles begin to consume shortly after seed germination and slow down seedling growth because of extensive foliar damage. This beetle damages crops in both stages as grubs and adults, with the grubs that reside underground damaging the roots of crops. The roots, underground stems, and occasionally the fruits that contact the ground are all points of entry for the grubs. In other cases, the entire field needs to be resowed because the adult beetles are mostly accountable for completely defoliating the plant sections above ground [9].

3.1.3 Fruitfly

On a variety of cultivated plants, specifically fruits, fruit flies are a common sight. According to reports, there are 4000 different species of fruit flies in the world, and thirty five percent of these pests affect commercially grown produce with significant economic value. The Cucurbitaceae family, which includes watermelon and cucumber, is the primary host group for the fruit fly species *Bactrocera cucurbitae* [10]. Melon flies, also known as *Bactrocera cucurbitae* (Coquillett), are a serious hazard to cucurbits, fruits, and vegetables. More than 80 hosts have been reported for melon flies. This insect is regarded as one of the worst for cucurbit crops, such as cucumber [11]. Their attacks not only lower yields but also have an impact on fruit quality, making farming operations unprofitable [12]. The insect severely damages cucurbitaceous fruits, either entirely or partially, rendering them unfit for human consumption. Due to the ability of the female fly to place its eggs in the fruit tissues, the maggots cause damage to the fruits. The maggots cause rotting of the fruit after they feed on the fruit flesh. The fruit fly also lays its eggs in delicate plant tissues like seedling terminals, unopened blooms, immature stems, and roots. The plant might die as a result of this [13].

3.1.4 Whitefly

One of the most significant pests affecting cucumber plants during their three growth stages, seedling, blooming, and fruiting, is the whitefly, *Bemisia tabaci* (Genn.). Chemical management of this insect is challenging because of its habit, which involves adult feeding, mating, oviposition, and larval development taking place on the abaxial (lower) portion of the leaves [14].The whitefly, *Bemisia tabaci*, weakens Cucurbitaceae plants by sucking the sap, infusing toxins into the vascular system of the plant, covering the leaf with honeydew, which promotes the formation of soots mold fungi, and triggering physiological problems with the leaf (squash silverleaf) [15]. Sessile whitefly nymphs and adults usually appear on the undersides of leaves. The yield is decreased by nymphal honeydew production and the growth of sooty molds. Chemical management in the field is challenging due to the protective habitat of the species on the underside of host plant leaves, the extremely rapid rate of population growth, and other factors [16].

3.1.5 Serpentine Leafminers

One of the main pests of cucumber grown in greenhouses all over the world is the serpentine leaf miner, *Liriomyza trifolii*. Leafminers develop as pupae in the soil over the winter, emerging as adult flies in the spring and summer [17]. The upper surface of the leaf is largely covered with white stipples caused by feeding and egg-laying punctures, which can enable bacterial and fungal illnesses to penetrate the plant. However, feeding of the larva and tunnelling through the mesophyll layer of the leaf, results in blisters, blotchy mines, or serpentine tunnelling, is what harms the plant. The mine gets larger as the larva grows. Mines may contain faeces that are visible and can taint leaves meant for human consumption. Photosynthesis is hindered by severe feeding damage, and leaves may dry out and fall off, which lowers plant quality and productivity. The most severe infestations can damage huge portions of the leaf and typically appear late in the season. Damage may be more severe in warm climates and greenhouse farming [15].

3.2 Diseases

This section describes Alternaria Leaf Spot, Powdery Mildew, Downy Mildew, Cercospora Leaf Spot and Viral diseases along with their impact on cucurbits crops.

3.2.1 Alternaria Leaf Spot

Alternaria Leaf Spot, caused by the fungus *Alternaria cucumerina*, is a common cucurbits disease Alternaria Leaf Spot is characterized by the appearance of tiny, water-soaked lesions on cucumber leaves. Over time, these lesions develop into brown dots with recognizable yellow haloes. The illness causes serious risks to the production of cucumbers in addition to affecting the overall appearance of the plant. Defoliation, a frequent side effect of Alternaria Leaf Spot, can result in decreased crop output and fruit quality. Additionally, when the illness worsens, the different spots may mix, severely harming the foliage. Extreme defoliation further weakens the damaged plants, impairing photosynthesis and, as a result, reducing productivity [18].

3.2.2 Powdery Mildew

Powdery Mildew is a common disease that may be identified by the characteristic look of white, powdery growth on cucurbits leaves, stems, and even fruits. Powdery Mildew is brought on by the fungus *Podosphaera xanthii*. This disease poses a serious threat to cucumber plants since it alters their physiological functions as well as their aesthetic appeal. Powdery mildew can impede photosynthesis, decreasing the capacity of the plant to produce energy and ultimately having an effect on fruit growth. Additionally, it may result in early fruit senescence, which would reduce crop yields even more. Powdery mildew can be controlled using a variety of techniques, including as choosing resistant cucumber types, making sure that growing sites have adequate airflow and ventilation, and using potent fungicides like sulfur, potassium bicarbonate, and neem oil. These actions are essential to reducing the negative impacts of this fungus illness and maintaining the health of the cucumber crop [19].

3.2.3 Downy Mildew

Pseudoperonospora cubensis, an oomycete infection, causes this damaging disease on cucurbits plants. On cucurbits leaves, it appears as angular, pale-yellow lesions with downy, grayish-white growth on the underside. Fruit and foliage can both be severely harmed by this disease. Spreading lesions result in leaf defoliation, which reduces photosynthesis, slows growth, and reduces yields. Under conditions of high humidity and moderate temperatures, it spreads quickly. Strategies include using timely fungicides, resistant cultivars, and good air circulation. To treat this illness and safeguard cucumber output, preventative measures and early detection are essential [20].

3.2.4 Cercospora Leaf Spot

The fungus *Cercospora cucumerina* is responsible for Cercospora Leaf Spot, which manifests as round to angular lesions on cucurbits leaves with dark brown borders and grey centers. The centers of the lesions may become lighter in colour as the condition worsens, giving them a distinctive appearance. In extreme circumstances, these spots may combine, seriously harming the leaf. The vigour of the plant and productivity may suffer greatly as a result of infected leaves turning prematurely yellow, defoliating, and having less ability for photosynthetic activity. Crop rotation, selecting resistant types, and timely fungicide spraying are only a few examples of management techniques that are effective [20].

3.2.5 Viral Diseases

Cucurbits crops are vulnerable to a number of viral infections, including Papaya Ringspot Virus (PRSV), Cucumber Green Mottle Mosaic Virus (CGMMV), Zucchini Yellow Mosaic Virus (ZYMV), and Cucumber Mosaic Virus (CMV). These viruses can produce a range of symptoms, including yellowing of the foliage, reduced growth, deformed fruit, and mosaic patterns on leaves. Many of these viruses are frequently transmitted by aphids, underscoring the importance of efficient vector management strategies. In order to control viral illnesses in cucumber agriculture, it is also crucial to practice crop rotation, use seeds free of disease, and plant resistant cucumber types. For sustaining healthy cucumber crops and guaranteeing optimal yields, effective disease prevention and control are essential.

4 Materials and Methods

Creating a database and uploading images through a web server involves various steps, including creating a database, building a web server, creating a user interface for image uploads, and establishing a link between the web server and the database. The initial stage of the project is selecting a database management system (DBMS). We can use well-known DBMS alternatives like MySQL, PostgreSQL, SQLite, or MongoDB for database creation. We should establish a new database to store image data using the command-line interface of DBMS or a graphical user interface (like phpMyAdmin for MySQL). In addition to creating a database, to write server-side code a programming language must be chosen. PHP, Python (with frameworks like Django or Flask), Node.js, Ruby, etc. are preferred choices. Developing a web application is also necessary. Server-side code is used to manage HTTP requests and communicate with the database to create a web application. This code will manage image uploads and save relevant information to the database, such as file locations and metadata.

Additionally, it is crucial to create an HTML form for image uploads. A web page with a form was created to enable image upload by users. To handle file uploads, the form requires to have an < input type = “file” > element. Next, server-side code handles the file uploads. In other words, the server-side code will receive the uploaded file when a user submits the form. The file is processed and saved to a specified location on the server. Relevant information (such as file path, metadata, etc.) to the database is added after the file is saved. This will help in accessing and showing images using this information. Finally, application is deployed using a server or hosting provider.

4.1 Data Collection

Images of cucurbits pests and diseases were gathered from the College of Agriculture, Nileshwar, Kerala.

The data collection largely comprised of image files in commonly used formats such as JPEG or PNG. Each picture file represents the image of pests or diseases that commonly affect the Cucurbits plants and is labelled with the associated pest name or disease name. In this paper, we have mainly focused on five different types of pests–Aphids, Pumpkin Beetle, Whitefly, Serpentine leafminer, and Fruitfly (Figs. 1, 2, 3, 4 and 5) and five different types of diseases –Alternaria leaf spot, Cercospora leaf Spot, Powdery mildew, Downy mildew, and viral diseases (Figs. 6, 7, 8, 9 and 10).

The dataset collected for this project consists a total of 66 pest images, showcasing five different pests including aphid images, pumpkin beetle images, fruit fly images, serpentine leafminer images, and whitefly images. Additionally, the dataset includes 329 images depicting various cucurbits diseases including Alternaria leaf spot images, Cercospora leaf spot images, Downy mildew images, Powdery mildew images and viral diseases images. These disease images are a vital component of the dataset, facilitating a comprehensive study of cucurbits health.

Fig. 1. Aphids

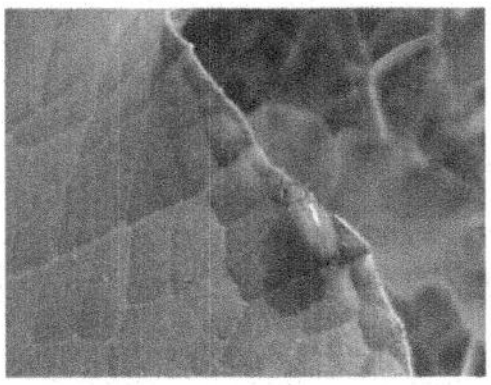

Fig. 2. Pumpkin Beetle

Fig. 3. Whitefly

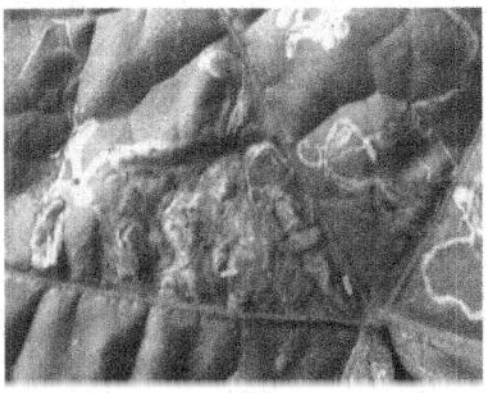

Fig. 4. Serpentine leafminer

Fig. 5. Fruitfly

Fig. 6. Alternaria leaf spot

Fig. 7. Cercospora leaf spot

Fig. 8. Powdery mildew

Fig. 9. Downy mildew

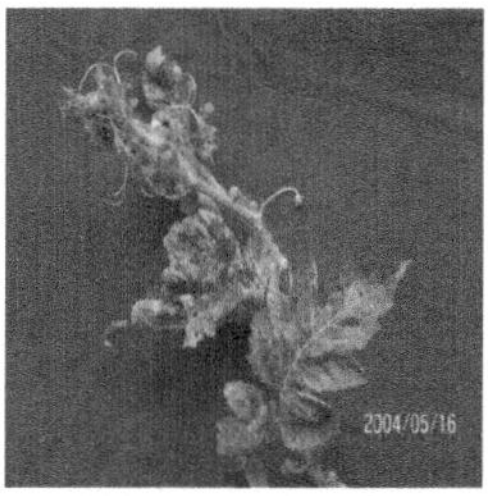

Fig. 10. Mosiac disease

4.2 SQL Server

MySQL Server stands as one of the most widely used open-source relational database management systems globally. The MySQL database, created by Oracle, is renowned for its efficiency, dependability, and simplicity. It supports various applications, from web-based content management systems to expansive business solutions, and excels at organizing structured data. MySQL uses the industry-standard SQL language to query and manipulate data, providing database administrators and developers with a comfortable user interface. It can be easily customized and extended due to its modular architecture, which in turn allows it to be flexible to various workloads. In addition, MySQL is renowned for its high-performance capabilities, efficient storage engines, and robust transactional support. For companies and developers looking for a dependable and affordable database solution, MySQL continues to be a top choice because of its thriving community and comprehensive documentation.

4.3 PHP

Hypertext Preprocessor, or PHP, is a well-known server-side programming language that is essential to the creation of websites. Initially designed for the development of dynamic web pages, PHP has developed into a flexible and popular scripting language. By easily integrating with HTML, it enables site designers to insert PHP code directly into web pages. PHP is an essential tool for creating interactive websites and online apps because of its ability to handle forms, connect with databases, and produce dynamic content. Its large function and framework library speed up development and offer options that range from data manipulation to encryption.

4.4 Workflow Diagram

Figure 11 describes the workflow of how the database was developed for cucurbits pests and diseases.

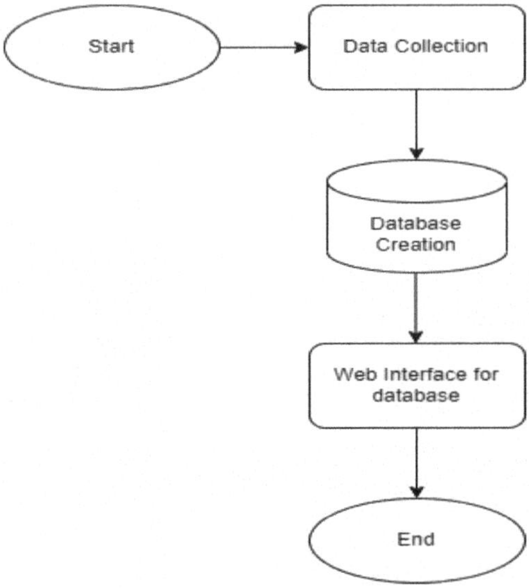

Fig. 11. Workflow Diagram

5 Results

5.1 Database Creation

A database of images was developed using HTML and PHP to provide other researchers access the image data. To develop and manage the picture database, a database management system such as MySQL from XAMPP was used. We have established a new database to store image data using the graphical user interface, phpMyAdmin. We have selected PHP, Python programming languages to write the server-side code. The database stores relevant information about the photos, such as file names. To retrieve and update picture data, the database can be queried and edited using SQL.

5.2 Web Server Creation

The main page provides a brief description of the cucurbits plant, highlighting significance of cucurbits in agriculture.

Below the description, there are ten card bodies representing five different pests and five different diseases that affect cucurbits plants shown in Fig. 12. Each card body displays an image representing the pest or disease and along with it a button with the name of the pest or disease.

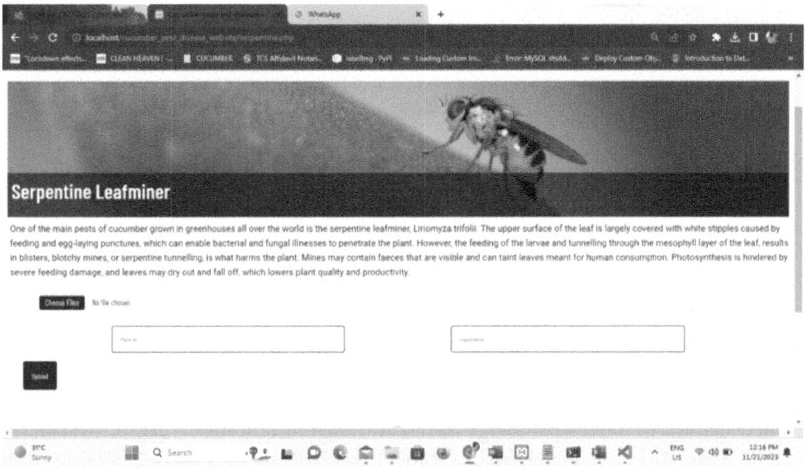

Fig. 12. Image representing webpage of a pest

The user is sent to a specific homepage for each pest or disease by clicking on any of the card bodies on the home page. For ease of identification, each pest or disease webpage has a header or title that includes the name of the pest or disease along with a small description of the pest or disease. Additionally, the page also allows users to download the images. The website design provides users with an interactive and user-friendly interface to discover information regarding pests and diseases.

6 Conclusion

In conclusion, this study has produced an essential resource for the agricultural industry: a comprehensive image database containing diseases and pests that affect cucurbits plants. The potential of this resource to completely alter cucurbits farming methods is what makes it so important. It provides farmers, academicians, and stakeholders with a powerful tool for quick and accurate identification by providing a database of photos. As a result, early intervention and improved crop protection are made possible, ultimately resulting in higher agricultural output and financial stability. Looking ahead, the future of this project shows potential in several areas.

In the future, we would like to expand the database to include a wider range of diseases and pests. Also, since this paper is a part of a project, attempt will be made to update the database periodically as time progresses. This will help the researchers use these images for their academic purposes. The creation of user-friendly interfaces will also be crucial for maximising the utility of the database. These tactical improvements ensure sustainability of cucurbits farming and resilience in the face of changing agricultural difficulties, laying the framework for revolutionary change.

Acknowledgements. This work is a part of research project funded by St Aloysius college management numbered No: SAC/MRP/2022–23/02. Also would like to acknowledge the expert advice rendered by Professors Dr. K M Sreekumar, Dr Saina Mol and Dr. Sajeesh PK, Asst Professor, College of Agriculture, Nileshwar during the data collection for developing this database.

References

1. Kumar, R.A., Rajpurohit, V.S., Gaikwad, N.N.: Image dataset of pomegranate fruits (Punica granatum) for various machine vision applications. Data Brief. **26**(37), 107249 (2021). https://doi.org/10.1016/j.dib.2021.107249. PMID: 34307808; PMCID: PMC8258780.
2. Meshram, V., Patil, K.: FruitNet: Indian fruits image dataset with quality for machine learning applications. Data Brief. **40**, 107686 (2021). https://doi.org/10.1016/j.dib.2021.107686. PMID: 34917715; PMCID: PMC8668825.
3. Gutman, D.A., et al.: Cancer digital slide archive: an informatics resource to support integrated in silico analysis of TCGA pathology data. J Am Med Inform Assoc. **20**(6), 1091–8 (2013). https://doi.org/10.1136/amiajnl-2012-001469. Epub 2013 Jul 25. PMID: 23893318; PMCID: PMC3822112
4. Forero, M.G., Murcia, H.F., Méndez, D., Betancourt-Lozano, J.: LiDAR platform for acquisition of 3D plant phenotyping database. Plants (Basel). **11**(17), 2199 (2022). https://doi.org/10.3390/plants11172199. PMID: 36079580; PMCID: PMC9459957.
5. Avraham, S., et al.: The plant ontology database: a community resource for plant structure and developmental stages controlled vocabulary and annotations. Nucleic Acids Research **36**(suppl_1), D449–54 (2008)
6. Das, A., Chakraborti, S.: Biointensive integrated pest management of cucurbitious crops. In: Biointensive Integrated Pest Management for Horticultural Crops, pp. 149–178). CRC Press (2021)
7. Barbercheck, M.E.: Biology and management of aphids in organic cucurbit production systems. Organic Vegetable Production Systems, Insect Management in Organic Farming Systems. eOrganic article (2014)
8. Ratnakar, V., Vijayaraghavendra, R.S., Reddy, S., Padmasri, A.: Efficacy of certain insecticides to red pumpkin beetle, aulacophorafoevicolis on cucumber, Cucumis sativus. Progressive Research-An International J. Society for Scientific Dev. **2**, 478–480 (2016)
9. Khursheed, S., Raj, D., Ganie, N.A.: Population dynamics of red pumpkin beetle on cucumber in mid-hill Himalayas. J. Appl. Hortic. **15**(2), 5–12 (2013)
10. Apriyadi, R., Saputra, H.M., Sintia, S., Andini, D.E.: The diversity of fruit fly (Diptera: Tephritidae) on combination of attractant and different trap height in cucumber field (Cucumis sativus L.). InIOP Conference Series: Earth and Environmental Science, **926**(1, p. 012059). IOP Publishing (2021)
11. Devi, Y.K., Thakur, P., Ibrahim, M.M.: Destructive fruit fly species in cucumber and their management. Indian Farmer. **5**(02), 196–200 (2018)
12. Sharma, S.K., Kumar, R.: Management of fruit fly (Bactrocera spp.) in cucumber (Cucumis sativus Linn.) grown organically. J. Biopesticides **9**(1), 73 (2016)
13. Kate, A.O., Bharodia, R.K., Joshi, M.D., Pardeshi, A.M., Makadia, R.R.: Estimation on yield losses in cucumber due to fruit fly, Bactroceracucurbitae (Coquillet). International J. Plant Prot. **2**(2), 276–277 (2009)
14. Mohamed, M.A.: Impact of planting dates, spaces and varieties on infestation of cucumber plants with whitefly, Bemisiatabaci (Genn.). The Journal of Basic & Applied Zoology **65**(1), 17–20 (2012)
15. Alston, D.G., Mull, A.: Leafminers of Vegetable Crops (Order: Diptera; Family: Agromyzidae)

16. Abubakar, M., Koul, B., Chandrashekar, K., Raut, A., Yadav, D.: Whitefly (Bemisiatabaci) management (WFM) strategies for sustainable agriculture: a review. Agriculture **12**(9), 1317 (2022)
17. Vignesh, M., Patil, R.K., Biradar, M.S.: Pseudomonas fluorescens induced resistance against Liriomyzatrifolii (Burgess) on parthenocarpic cucumber. J. Pharmacognosy and Phytochemistry **8**(6), 1021–1024 (2019)
18. Hassan, R.A., et al.: In-vitro and in-vivo evaluation of different fungicides against leaf blight causing fungus Alternaria cucumerina in bitter gourd. J. Saudi Soc. Agric. Sci. **21**(3), 208–215 (2022)
19. Trueman, C.L., Role, S.E.: Powdery mildew of cucurbits. In: Diseases of Fruits and Vegetables, pp. 463–471. Springer, Cham (2019)
20. Hausbeck, M.K., Lamour, K.H.: Phytophthora capsici on vegetable crops: research progress and management challenges. Plant Dis. **88**(12), 1292–1303 (2004)

Towards Hands-Free Computing: AI Virtual Mouse Interface Powered by Gestures

A. Menaka Gowda, M. Krisha(✉), Shrividya Shashidhara, and B. N. Lakshmi

Department of Computer Science Engineering, BMS Institute of Technology and Management, Avalahalli, Bangalore, Karnataka 560064, India
krishamanoj7@gmail.com, lakshmibn@bmsit.in

Abstract. This research presents an innovative virtual mouse control system driven by hand gestures, enhancing human-computer interaction. Using Python and libraries like OpenCV, MediaPipe, and PyAutoGUI, the system tracks and interprets hand gestures in real-time, enabling intuitive control of computers. Users can effortlessly manipulate cursor movement, perform clicks, scroll content, and adjust settings like volume and brightness through natural hand motions. Key features encompass gesture recognition, precise cursor tracking, multi-hand support, and diverse gesture compatibility (e.g., fists, pinch, V-gestures), catering to varied user preferences. This system advances human-computer interaction by providing a hands-free, efficient control method, particularly beneficial when conventional input devices are impractical. Its responsiveness and versatility suggest applications in accessibility technology, virtual reality, and interactive presentations.In summary, this paper introduces an efficient virtual mouse control system that leverages hand gestures, revolutionizing computer interaction. Its performance and adaptability make it a valuable addition to gesture-based interfaces, promising novel possibilities in human-computer interaction.

Keywords: Human-computer interaction · Python · OpenCV · MediaPipe · PyAutoGUI · Gesture recognition · Cursor tracking · Multi-hand support · Accessibility technology · Virtual reality · Interactive presentations · Gesture-based interfaces

1 Introduction

In the ever-evolving landscape of technology, the virtual mouse project stands as a testament to our commitment to innovation and refinement. It has become an instrumental tool, forging new paths in state-of-the-art on-screen cursor control.At its core, the virtual mouse leverages the power of computer vision technology to recognize specific hand gestures made by users through a simple webcam. This recognition process involves detailed feature extraction, where the system studies finger positions, curves, and distances. Once recognized, the virtual mouse is trained to perform corresponding actions using sophisticated algorithms and decision trees, continually evolving for greater efficiency.The virtual mouse's paramount mission is to enhance accessibility, empowering

T. Shreekumar et al. (Eds.): ISCComm 2023, CCIS 2231, pp. 26–49, 2025.
https://doi.org/10.1007/978-3-031-75605-4_3

users to perform tasks without relying on a physical mouse. It's a game-changer for professionals requiring precision and offers relief to those spending extended periods at computers. While traditional mice remain compatible across various operating systems, the virtual mouse thrives in augmented reality, medical environments, and even gaming, adding excitement to user experiences.Developed with Python libraries like OpenCV for image processing, Mediapipe for hand tracking, and pyAutoGUI for system control, the virtual mouse orchestrates cursor movements elegantly. It uses classes like "Gest" for gesture enumeration and "Hlabel" for hand distinction. Key commands like scrolling, clicking, and system control respond flawlessly to recognized gestures through the "controller" class. Looking ahead, the virtual mouse project envisions voice recognition-based virtual mice and sensor integration, promising a new chapter in innovation. It continues to enchant and empower, playing a pivotal role in human-computer interaction's advancing tapestry.

2 Literature Survey

The evolution of computer mice from Douglas Engelbart's wooden box in 1964 to today's advanced virtual mouse systems marks a fascinating journey through the annals of technology history. In today's technology landscape, we've witnessed the birth of an ultra-portable Bluetooth alternative to traditional mice, driven by the rapid growth of AI and virtual technology. This innovation has led to the emergence of what we now call the "Virtual Mouse."

Li Wensheng et al. [8] has spoken on how machine vision can be implemented using virtual mouse in their research paper. It offers software-based mouse control, particularly valuable in scenarios where traditional mouse are impractical or limiting. One notable method for achieving this is tracking of the fingertip, which proves computationally cost-effective and highly efficient. With the help of a histogram that contains colour tips in both RGB and HSV values we are able to track down the finger tips. However, the limitations include rate of tracking can be difficult at times, the coordinates of finger tips can be distorted. Therefore, the activation of the gestures might take some time.

Aditi Khandagale et al. [4] has given us an insight on Virtual Mouse based on AI in their research paper. The system is developed using Python, OpenCV, MediaPipe, Pyput, and PyAutoGUI. It combines hand gestures and movement to capture the speed of hand motion, resulting in smoother trajectories for mouse pointer control. The system offers an array of functions, including signaling for stopping and various other gestures. The limitation being requiring a high-sensitivity microphone to capture voice commands effectively, particularly in noisy open environments.

Rajat Dudhapachare et al. [6] has implemented voice assistance in their Virtual Mouse System, which is being spoken about in the paper. The mouse cursor actions likeclicking, scrolling, and pointing, as well as controlling brightness and volume happens with the help of web cam or a camera that is already inbuilt. It mainly works on how the video is captured and the images are being processed. Additionally, libraries like sapi5 and pyttsx3 are employed for voice interaction. However, a limitation in this setup is that captured frames of images need to undergo processing to isolate hands and identify the pixels of the skin, adding some complexity to the system.

Manav Ranawat et al. [2] has mentioned on how the mouse events play an important role in recognising the gestures of the hand in their research paper. The key points of this paper includes the images being captured, detection of the hand, to recognize the gesture and finally call an event accordingly. RGB is being converted to HSV images. A CNN model is implemented to achieve the above. The limitations include involvement with complex datasets and working with various models.

Vantukala VishnuTeja Reddy et.al [7] has mentioned on how we can control the gestures using finger tips that are coloured in their research paper. This paper talks on more of mathematical aspects such as subtraction of background, to identify the colour and the circle. Images are converted to frames here. In the background subtraction, the unnecessary elements are neglected such as the shadows formed. Every image is scanned for the colour based on which the shape of the objects are noted. The raw image is converted to final image for processing. The limitation of this virtual mouse is that the accuracy varies with respect to the brightness.

The journey from wooden boxes with wheels to sophisticated virtual mice exemplifies the continuous innovation in human-computer interaction, opening doors to new possibilities and improving accessibility in the digital realm.

Dinh-Son Tran et al. [1] has spoken on how the images such as RGB-D and the detection of fingertip can power up a virtual mouse efficiency.Here first they detect the regions of interest and the palm area of the hand.Whereas,the next few steps includes using of an algorithm that traces the border for detecting the contour coordinates of the hand and the usage of K-cosine algorithm for the detection of fingertip. Then finally these coordinates are being mapped to the images that shall trigger the necessary operations of a mouse.

Changbo Hu et al. [2] has given us an insight on how virtual mouse acts an a device of input for recognizing the gestures of the hand. Their main focus was on capturing the hand motion using an algorithm called the condensation algorithm. There is a multi-model approach being suggested in the development of the virtual mouse.Many statistical procedures are carried out to bring out an effective way to track the hand gestures such an sampling and mean methods.

Roshnee Matlani et al. [3] has spoken of how laptops being controlled using gestures have gained a lot of attention. They make use of leap motion to track the gestures of the hand. They have specified that using a normal laptop in very difficult at times, when the environment isn't well lit. Therefore with the simple hand motions complex mouse functions become really easy to operate. There is no need of any extra electronic device for the implementation. The laptop's web cam is more than enough to capture the necessary images and to process them. Therefore this way the mouse operations can be performed.

C Saritha et al. [24] has spoken on the usage of the simple webcam for mouse operations. They have used finger tips as the detection factor and have implemented their algorithms for functions like left-click, right-click etc. They have spoken on the importance of the virtual mouse in this new era. They have spoken of how the difficulties of using a traditional mouse can be overcome by using their detection algorithms. However they have a limited set of functions that can be performed.

Aarti Moraikar et al. [25] have come up with an idea of implementing hand gesture mainly for disabled children. The have created a gesture recognition model using CNN. The have included various set of options to be performed by the user like clicking, selection etc. However their model gives a delay of approximately 10 s which shall affect the overall performance.

Hrithik Joshi et al. [26] have bought up the idea of human-machine interaction in their research paper. They have focused on how cumbersome touchpads can be replaced. Therefore with the cost effective solution of using gesture control and machine learning they have developed an interface to make mouse control operations way easier. However, the application is limit to basic functionality like scrolling, selecting and changing slides only.

T. P. Jacob et.al [14] have presented a virtual mouse mainly using OpenCV. They have trained their model using machine learning before the final implementation. The open source python platform, Anaconda has been used for writing the main functions. They have used color detecting techniques too. OpenCV is mainly used for video capture here. This way the virtual mouse is being made use for a set of applications.

However our paper includes a wide range of mouse control features like switching tabs, right-clicking, left-clicking, allows us to take screenshots. The system works fine even in a not so well lit environment. The system detects multiple hands all allows only one hand to be detected for operational use. The functions are immediately triggered as soon the system is set up for use.The finger tip detection algorithm is made really precise to enhance its capabilities. Therefore this way there is no time delay in the functionality.

3 Methodology

In the following section, we will delve into the methodology behind our AI-powered Virtual Mouse System, providing a comprehensive overview of its system architecture and key components. We will explore how the system utilizes gesture recognition to offer precise cursor control and a range of actions triggered by specific hand gestures. Additionally, we will discuss how the system supports multi-hand interactions and fine-grained control, highlighting its potential applications in various domains.

3.1 System Architecture

The Webcam/Camera as the Initial Unit of Processing:
To support the capturing of frames, our own laptop's webcam will be sufficient enough. To increase the performance we can use a high definition camera that shall provide a wider range. By using the Python computer vision library OpenCV, the video capture object is created and the web camera will start capturing video. Using the WebCam frames are captured that continue until the termination of the system.

Screen Coordinates Initialization:
The PyautoGUI function is used to map the screen coordinates. The screen is the spatial area available for hand movement. A larger screen area ensures higher vertical and horizontal scrolling speed. The ideal size of the screen would be a 12-inch display.

Finger Control Points Enabling Hand Tracking:
For every finger there is a minimum of three control points plotted, each point that is involved in the gesture recognition is treated as a special set. We calculate the distance between the control points associated with the fingers and the hand result together gives the necessary gesture that the user wants to perform. The finger state is set to 1 if the finger is open else 0.

Handling Fluctuations Due to Noise:
Sometimes between performing the necessary operation, there comes a lot of unwanted noise that disrupts the necessary operation from being performed. We have intelligently implemented a library of Python called pycaw that performs Audio Control. This ensures that before the recognized gesture is sent to the controller for implementation, the noise is filtered out.

Gesture Recognition and Mouse Functions:
A function called controller executes commands according to detected gestures. There are flags maintained to perform the executions such as if the pinch control flag is raised then the necessary adjusting of brightness or volume action is performed. Finally, the Gesture Controller function is the execution that allows smooth movement of the operation. The various gestures performed are scrolling, clicking, double-clicking, grab and drag, and pinch control.

Figure 1 is the representation of various functional blocks in the proposed system where a web cam is used to initiate the system to start functioning. With the help of the webcam, we capture the images that is used for further processing. Hand Detector makes sure to recognize a single hand to avoid having multiple hands in the image captured. There is initializing of the screen coordinates using bitmap feature. Based on if the finger is up or down and the necessary control points we are able to simulate various gestures, s uch as a 'Mid Gesture' enables left-clicking.

3.2 Implementation

The implementation of the gesture-based virtual mouse control system is a culmination of software libraries and modules, primarily leveraging Python and several key libraries. This section provides an overview of the practical execution of the system.

Software Framework and Libraries
The system's implementation relies on the following essential software components:

Python:

Python serves as the primary programming language for developing the virtual mouse control system. Its versatility and extensive library support make it an ideal choice for this project.

OpenCV (Open Source Computer Vision Library):

OpenCV is pivotal for image processing and computer vision tasks. It provides the foundation for detecting and tracking hand landmarks and interpreting gestures.

Mediapipe:

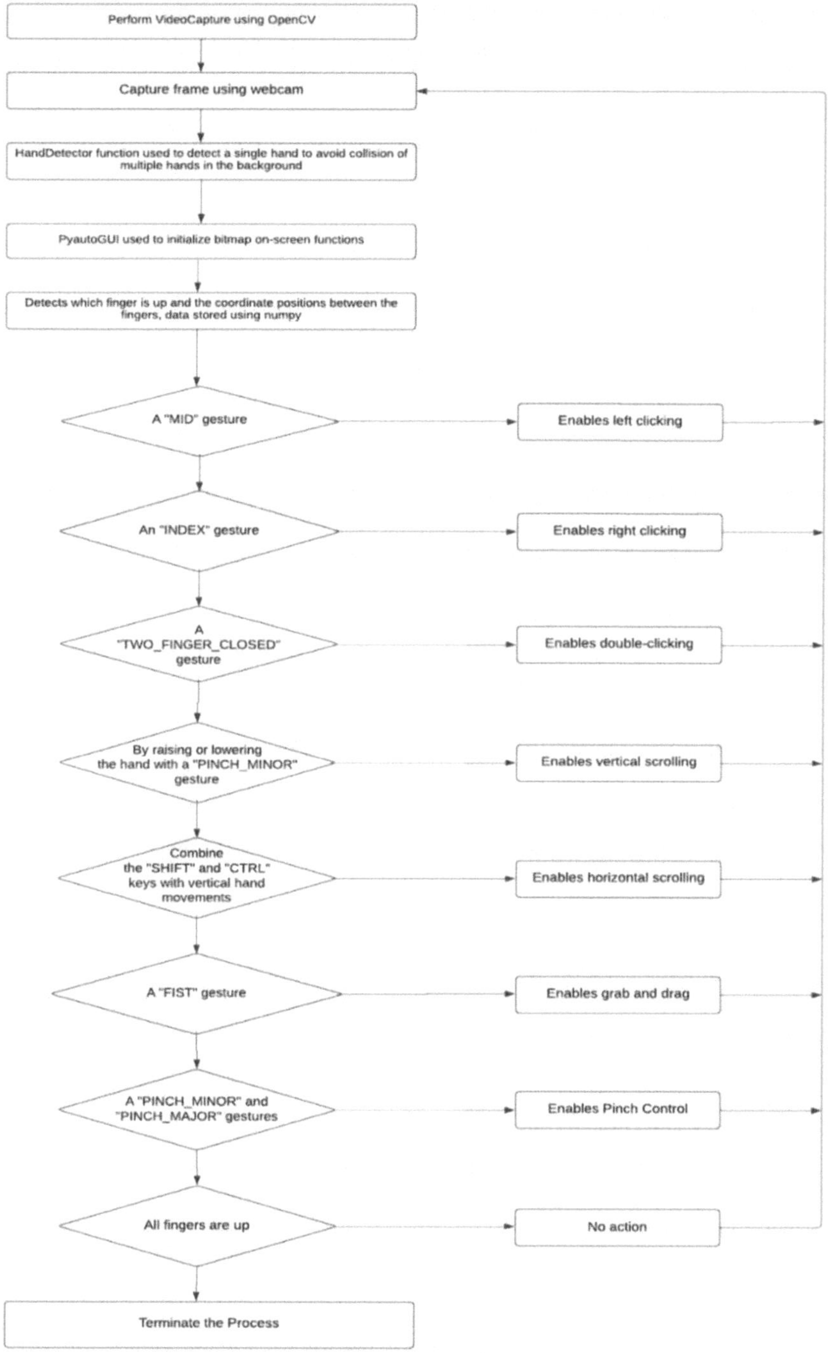

Fig. 1. Circuitry Blueprint of the AI Powered Virtual Mouse

Mediapipe offers efficient hand tracking capabilities, including hand landmark identification. It plays a central role in recognizing and interpreting gestures.

PyAutoGUI:

PyAutoGUI empowers the system with the ability to control cursor movements and execute actions triggered by hand gestures. It enhances the system's interaction with the operating system.

Gesture Recognition and Control Logic

Hand Landmark Detection:

The system detects 21 key landmarks on each hand using the Mediapipe library. Of particular interest is the tip of the index finger, which acts as a reference point for gesture interpretation. The positions of these landmarks are continuously tracked to facilitate real-time recognition.it is explained in detail in further sections.

Gesture Classification Algorithm:

Upon landmark detection, the system employs a custom gesture classification algorithm to interpret specific hand gestures. This algorithm relies on the relative positions and movements of fingers, especially the thumb and four fingers. It classifies each finger as either open or closed, representing their states as binary values as explained in detail in further sections.

Gesture Encoding:

Based on the binary states of the fingers, the system generates a unique binary encoding for each recognized gesture. These encodings provide a clear and unambiguous means of identifying different gestures. Key gestures and their respective encodings have been predefined.it is explained in detail in further sections.

Real-Time Responsiveness:

To ensure a seamless user experience, the system incorporates techniques to minimize latency. This involves optimizing gesture recognition algorithms, hardware integration, and the use of low-latency communication protocols.

Hardware Requirements

The system's implementation demands minimal hardware requirements. A standard webcam, integrated or external, is the primary hardware component, serving as the input device for detecting hand gestures. This ensures the system's compatibility with a wide range of computing setups.

Execution and User Interaction

The system is initiated by creating an instance of the Gesture Controller class, invoking the start() method. This sets in motion the recognition of hand gestures, cursor control, and execution of actions, all of which are responsive to recognized gestures. Users interact with the system through natural hand movements, providing an intuitive and engaging interface.

In summary, the implementation of the gesture-based virtual mouse control system is founded on a robust combination of software libraries and modules. It facilitates real-time hand gesture recognition, cursor control, and interaction with the operating system, offering users a seamless and engaging means of interfacing with their computers. The system's simplicity and minimal hardware requirements make it accessible and adaptable to a wide range of computing environments.

3.3 Gesture Recognition

Gesture recognition is foundation to our virtual mouse control system. We begin by detecting key hand landmarks, particularly focusing on the index finger's tip for intuitive cursor control.

Our gesture classification algorithm consists of many stages like Finger State Determination, Gesture Encoding, Handling Fluctuations as discussed below.

Finger State Determination: Here, we analyse the relative positions and movements of the fingers, especially the thumb and the other four fingers, to ascertain whether they are open or closed. Each finger's state is represented as a binary value (open or closed).The finger state determination is captured as the input via cameras. The primary focus is to capture data related to angles and textures of the palms and fingers so as to accommodate the necessary triggers for each individual user. This is a vital and crucial step to be carried out with precision and accuracy in order to initiate and execute predefined commands that the user wishes.

Gesture Encoding: During this stage, based on finger states, we create a gesture encoding that represents the hand gesture. This encoding is a binary value capturing the state of each finger, enabling us to distinguish various gestures. This also avoids the risk of misinterpretation or being unable to distinguish between various gestures. The precision and accuracy mentioned above during the finger state determination is achieved through binary encoding. Another benefit is the fact that gesture encoding proves to be compatible with several applications, including the ability to be integrated with hardware.

Handling Fluctuations: To mitigate noise and ensure gesture stability, here, we implement a mechanism to manage fluctuations in hand movements. This involves tracking the previous gesture and counting frames to confirm the stability of the detected gesture. The algorithm deployed above is trained to recognise false positives and refrain from triggering the wrong mouse actions. Gesture validation stems from error correction and recovery strategies as well by being able to maintain accurate cursor positioning and appropriate user interaction. Gesture encoding helps mapping between binary triggers and tasks.

Table 1 provides a concise mapping of action triggers associated with their respective binary codes to represent a streamlined flow of optimised task execution. Each binary code corresponds to specific actions which enhances automation in various applications. By decoding these binary triggers, users can effortlessly interact with a broad range of systems, devices, and software. Because of the fact that the complexity of action assignments are simplified, the virtual mouse interface proves as a valuable tool for developers and users alike.To maintain robust gesture recognition, we use a frame-counting mechanism. A gesture is confirmed if it persists for a set number of consecutive frames, preventing false positives and ensuring accurate system responses.

Table 1. Mapping Binary Triggers to Tasks

Sl. no	Binary Digit Code	Represented Finger Gesture	Description of Trigger
1	00000	Fist	The closed fist gesture enables the user to move objects across the screen via grab and drop mechanism
2	00001	Pinky Finger Extended	This gesture is best suitable for formatting purposes such as highlighting a text as well as increasing or decreasing the font size
3	00010	Ring Finger Extended	Extending the ring finger away from the rest of the fingers will enable a user to proceed to the next page while bringing the ring finger closer to the other fingers will help the user to navigate back to the previous page
4	00100	Middle Finger Extended	This gesture is enables the user to click the left button of the mouse
5	00111	Last Three Fingers Extended	This gestured is a customised command that is helpful for switching tools or tabs on the system
6	01000	Index Finger Extended	Extending the index finger will allow a user to enable the right click of the mouse.
7	01100	First Two Fingers Extended	When the first two fingers are extended, the user can bring them closer to trigger a double click action
8	01111	Last Four Fingers Extended	When a user extends the last four fingers, keyboard shortcuts are activated. This enables the user to take a screenshot
9	10000	Thumb Extended	An extended thumb gesture increments the volume whereas retracting the thumb reduces the volume

3.4 Cursor Control and Actions

In our gesture-controlled project, precise cursor control using hand gestures is a core functionality. This section outlines the methods and techniques employed for cursor control and the corresponding actions triggered by specific gestures.

Gesture-Based Cursor Positioning
We utilize the Mediapipe library to detect and track hand movements in real-time, aiming for accurate cursor positioning on the computer screen. The following techniques are employed to achieve gesture based cursor positioning.

Hand Landmark Tracking: The Mediapipe library provides hand landmark tracking, identifying 21 key landmarks on each hand. By focusing on specific landmarks, such as the index finger's tip, we determine the cursor's position. Mapping hand coordinates to screen pixel coordinates ensures precise and responsive cursor movement.

Dampening for Smooth Cursor Movement: To maintain smooth cursor movement, we implement dampening. Dampening adjusts cursor speed based on hand movement speed. Slow hand movements result in reduced cursor speed for precise control, while rapid hand movements accelerate cursor navigation.

Gesture-Triggered Actions
Our gesture-controlled system also triggers various actions based on detected hand gestures, enhancing user interaction and productivity:

Clicking: The "MID" gesture serves as a left-click action, mimicking traditional mouse clicks for easy interaction with icons, buttons, and applications.

Right-Click: An "INDEX" gesture triggers a right-click action, providing access to context menus for context-specific options and settings.

Double-Click: Recognizing a "TWO_FINGER_CLOSED" gesture allows users to perform double-click actions, simplifying tasks like opening files or folders.

Scrolling: The system supports vertical and horizontal scrolling gestures. Vertical scrolling mimics a mouse scroll wheel, and horizontal scrolling is activated with "SHIFT" and "CTRL" keys, enhancing navigation of documents and web pages.

Grab and Drag: The "FIST" gesture simulates click-and-drag operations, useful for moving windows or objects on the screen.

Pinch Control: "PINCH_MINOR" and "PINCH_MAJOR" gestures provide fine-grained control over system brightness and volume. Users can adjust settings intuitively, improving user comfort in varying environments.

These techniques ensure precise cursor control and enable seamless interaction with the computer interface, enhancing overall user experience.

3.5 Multi-hand Support

There exists custom logic in order to handle multiple hands in the system. As discussed earlier, first it is necessary to initialize and configure the system by importing necessary libraries and integrating configurations. Landmark detections for hand tracking are made easier with the Media Pipe library. Parallelly, camera initialization is set up using the GestureController class to capture video frames. Once the hands are captured in the frame, The HandRecog class recognizes the signs and gestures of individual hands. This is the classification phase wherein once the motions are captured, the dominant hand is recognized. It is necessary to note that despite the distinctions that are made to recognize the dominant hand, the same logic is applied to all objects observed in the motion frame. Further user interactions are processed by the GestureController class which is a loop for capturing, detecting, and generating results based on the type of input given by the hands and fingers. Once the system proves to satisfy all the necessary requirements of the user, the camera is released and the window is closed by pressing the Enter key.

Although the virtual mouse weighs more in proportion to radiating benefits, recognizing gestures from various hands can be challenging. The accuracy of the landmark detection may not be precise due to the fact that there is a vast amount of variability in the sizes and shapes of the hands of various users. Complementary to this, different objects in the background may also be fed as the input to the camera, thus affecting the system's ability to track the right landmark. Fortunately, there is a way to overcome these potential problems by adapting robust algorithms that can identify and adapt to a unique set of user's hands. An alternative approach is to implement a more diversified training set that personalizes user-specific profiles, thus enhancing the compatibility as well as the overall efficiency of the virtual mouse system.

Moreover, the multi-hand support in HCI (Human-Computer Interaction) mirrors real- world actions as users are able to manipulate digital content with their bare hands. Multiple actions can be performed simultaneously without the need to run complex commands. In summary, it improves user experience, technology is made user-friendly and the chances of crafting new possibilities for collaborative circumstances is an elevated probability favouring a promising outlook.

3.6 Fine-Grained Control

The accuracy rate that the virtual mouse tends to achieve to satisfy user requirements is unattainable through methods involving the traditional mouse. Since the cursor moves across the screen pixel by pixel based on the gestures that the training set recognizes, graphic designing, documentation, and 3D modelling applications are heightened. This high-tech tool allows the manipulation of minute adjustments, thus offering fine-grain control as a game changer in scenarios where user interaction is ideal.

Tasks such as digital illustrations or paintings make the best use of fine-grained control since the virtual mouse preserves the integrity of intricate details where every stroke or pixel matters. Delicate surgical simulations profit from this system as the virtual mouse offers a level of control necessary to analyse medical images and real-time cases. Gamers often appreciate fine-grained control, especially in games that require accurate aiming and precise timing. Moreover, the virtual mouse expands its benefits for users with disabilities. In summary, stakeholders are offered promising precision in various domains with a high degree of control, which eventually enlivens creativity and quality of outcomes.

3.7 Rationale Begind Performance Metrics

The selection of performance metrics in the methodology is crucial to assess the effectiveness and efficiency of the AI-powered Virtual Mouse System. These metrics provide a quantitative basis for evaluating the system's performance, identifying strengths, and areas for improvement. Here are the key rationales behind the selection of performance metrics:

Gesture Recognition Accuracy: This metric measures how accurately the system recognizes and classifies hand gestures. The rationale for selecting this metric is to ensure that the system correctly interprets user gestures, minimizing false positives and false negatives. High accuracy is essential for user satisfaction and system reliability.

Cursor Positioning Precision: The accuracy of cursor positioning on the screen is critical for the system's usability. This metric assesses how closely the system can map hand gestures to screen coordinates. It is important for tasks that require fine-grained control, such as graphic design or medical simulations.

Latency: Latency measures the delay between performing a gesture and the system's response. Low latency is crucial to provide a responsive and real-time user experience. High latency can lead to frustration and reduced usability. This metric ensures that the system can keep up with the user's actions.

Noise Handling: This metric evaluates the system's ability to filter out unwanted noise and distinguish between intentional gestures and unintended movements. Robust noise handling is important for maintaining the accuracy of gesture recognition in real-world environments.

Multi-Hand Support: As the system claims to support multi-hand interactions, assessing its ability to handle multiple hands is essential. Metrics related to the recognition of gestures from multiple hands and their coordination are important to ensure a seamless user experience in collaborative scenarios.

Hardware Resource Utilization: It's important to measure the system's resource usage, including CPU and memory consumption. Efficient resource utilization ensures that the system can run on a wide range of hardware configurations without causing performance bottlenecks.

User Experience and Satisfaction: Collecting user feedback and satisfaction scores can provide valuable insights into the system's overall performance. This qualitative metric helps in understanding the user's perspective and identifying areas for improvement.

Task Execution Speed: For actions triggered by gestures, such as clicking, scrolling, or adjusting brightness, measuring the time it takes to complete these tasks is important. This metric assesses the system's efficiency in performing actions.

Error Rate: Error rate measures the frequency of incorrect interpretations or misclassifications of gestures. It helps in identifying areas where the system may need improvements in gesture recognition algorithms.

Compatibility: Assessing the system's compatibility with different applications and use cases is crucial. Metrics related to the ease of integration with various software and hardware configurations ensure the system's versatility and adaptability.

In summary, the rationale behind selecting these performance metrics is to comprehensively evaluate the AI-powered Virtual Mouse System's functionality, accuracy, efficiency, and user-friendliness. These metrics collectively help in assessing the system's capabilities and its potential for a wide range of applications and user scenarios.

4 Application and Use Cases

The development of a gesture-based virtual mouse control system has a wide range of practical applications, each showcasing its potential impact:

4.1 Accessibility Technology: Gesture control serves as a game-changer in accessibility technology, offering individuals with physical disabilities a means to navigate the digital world. It simplifies tasks like web browsing and software interaction, making digital environments more inclusive and user-friendly.

4.2 VR and AR: In the realm of virtual and augmented reality, hand gestures unlock immersive experiences. Users can interact with virtual environments, manipulate objects, and control characters with natural hand movements, enhancing gaming, educational simulations, and professional applications.

4.3 Gaming Gesture-based control revolutionizes the gaming industry by allowing players to use their hand gestures for in-game actions. This adds a new level of immersion and interactivity, offering gamers a more intuitive way to control characters, make decisions, and engage with virtual worlds.

4.4 Presentations and Conferences: Presenters benefit from gesture-based control when delivering presentations. They can seamlessly navigate slides and multimedia content, making their presentations more engaging and dynamic. This technology enhances audience interaction and provides a futuristic touch to presentations and conferences.

4.5 Remote Control: Gesture-based remote control extends convenience and accessibility to various devices. Users can control smart TVs, home automation systems, drones, and more with intuitive hand gestures, eliminating the need for physical remote controls and simplifying device management.

4.6 Medical and Healthcare: In healthcare settings, touchless control of medical images and equipment is invaluable. It reduces the risk of contamination, streamlines medical procedures, and allows for precise manipulation of images and data, ultimately enhancing efficiency and patient care.

4.7 Industrial and Manufacturing: Gesture based control finds application in industrial environments, where workers can interact with machinery and equipment using natural hand gestures. This technology improves productivity, safety, and efficiency in manufacturing and industrial settings by eliminating the need for physical controls.

4.8 Design and 3D Modeling: Professionals in design and 3D modeling leverage precise gesture-based control to expedite their work. It allows for intricate adjustments and fine-grained control over design elements, benefiting industries like graphic design, architecture, and animation.

4.9 Education and Training: Gesture-based control enhances interactive learning experiences. Students can engage with educational software and simulations using natural hand movements, making learning more immersive and enjoyable. It's particularly beneficial in virtual labs and training scenarios..

4.10 Simulations and Simulators: Training simulations leverage gesture-based control to create realistic and immersive training environments. This technology prepares individuals for real-world scenarios, whether it's pilot training, medical simulations, or military exercises, by allowing them to interact with the simulation naturally and intuitively.

These diverse applications underscore the adaptability and potential impact of gesture- based virtual mouse control, positioning it as an advancement in Human-Computer Interaction across multiple industries.

5 Evaluation and Performance

A system isn't complete without testing its performance. Therefore, we got 5 people to try the virtual mouse. Each one of them tried the AI-powered mouse on various backgrounds as one tried to slide through the presentations using the scrolling function, the other tried to drag and drop a file into a folder and the rest performed pause and play actions of videos using the clicking function. Likewise, every single one of them had a different outlook on it. Having about 6 h of testing sessions and discussing their opinions on it, we finally concluded that the scrolling feature has the most accuracy and precision and the grab and drag feature is slightly difficult to enable. It was tried out on various environmental conditions such as low lighting, and farther distance from the camera. The amount of time taken by the system to recognize the features and to enable the control functions was also noted.

Figure 2 depicts the representation of mouse control gestures through a line graph where the horizontal axis consists of the various available mouse gestures. The vertical axis consists of how precisely the gestures are operating in the system. The left click having a value of 150 is definitely more precise and easier to operate than the right click that has a value of only 145.We observe that the highest precision is shown by the vertical scrolling gesture having value of 158.The pinch control has a precision value of about 130 that lies in between the value of horizontal scrolling and grab and drag. The grab and drag is the least precise of value about 120.

After the evaluation comes the final performance of the mouse in the market. Looking at the performance rate over the years, the rate increases concerning the set of new features added. The implementation of characteristics like Colour Identification, fingertip tracking, Hand Detection, Gesture recognition, Human-Computer Interaction, and Volume and brightness control has increased our performance rate. Such an AI- powered virtual mouse shall perform extremely well if implemented in hospitals. We all

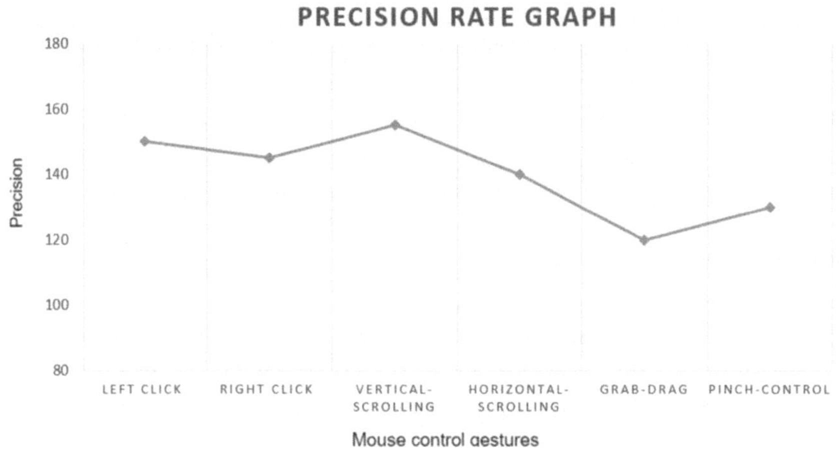

Fig. 2. The line graph represents the set of mouse gestures available and their ex- perimental precision rates.

know germs spread through physical contact, by creating a mouse that is completely contactless, we shall also be creating a germ-free environment. It is really easy to train our model and therefore can be easily implemented. With all such enhancements, the system is to be seemingly perfect (Fig. 3).

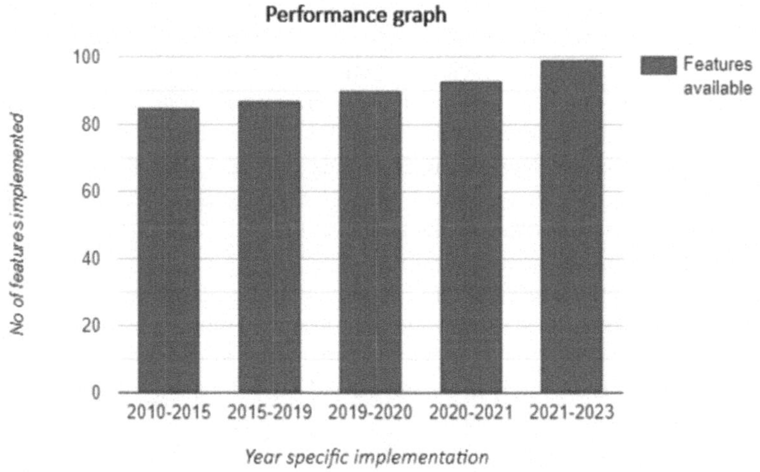

Fig. 3. The bar graph represents the additional features being added every year and the performance of the system accordingly.

In the bar graph given above we can see that the technology is consistently growing and the number of features available in a virtual mouse system in today's world is about 99 percentile. With the help of advancing innovations in the field of Artificial Intelligence there are features like voice assistance being included with the activation of mouse features through a virtual mouse. The graph symbolizes that there is at least 2–3% advancements done every year. During 2010–2015 the percentage was around 82 but now it has drastically increased up to 99%. In 2015–2019 there was a 2% increase from the previous year. In the mid of 2019 the bar rose up to 85%. In 2020–2021 the range was recorded to be 93%. Currently the virtual mouse system is made available to all by including the simplest features as the pre- requisites therefore having 99% stability (Figs. 4, 5, 6, 7, 8, 9, 10, 11, 12 and 13).

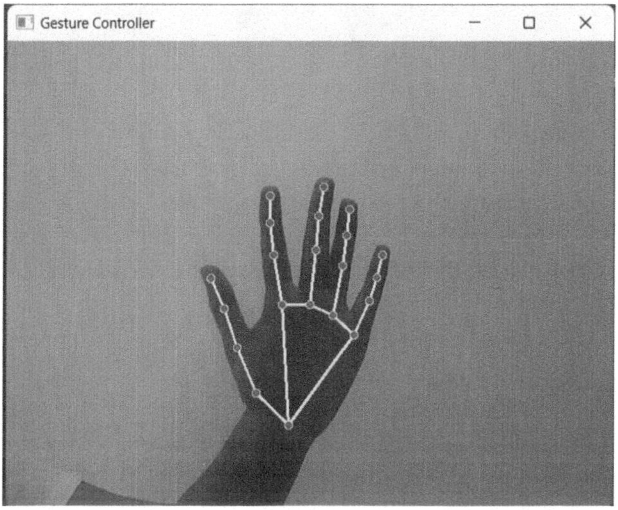

Fig. 4. Palm is held up against the screen. This gesture is meant for recognition of each finger by facilitating necessary coordinate points and structures which is useful to deploy for further gesture recognition.

6 Challenges and Future Directions

This section highlights the challenges encountered during the development and deployment of the virtual mouse control system and proposes avenues for future research and enhancement.

6.1 Challenges

Noise and Environmental Factors

The system faces challenges related to varying lighting conditions, background clutter, and noisy environments, which can hinder hand gesture recognition. To mitigate this, future research should focus on improving noise filtering and adaptability using advanced computer vision techniques and noise reduction algorithms.

Fig. 5. Representation of a closed fist gesture. Can be used to select objects to be moved across various positions on the screen.

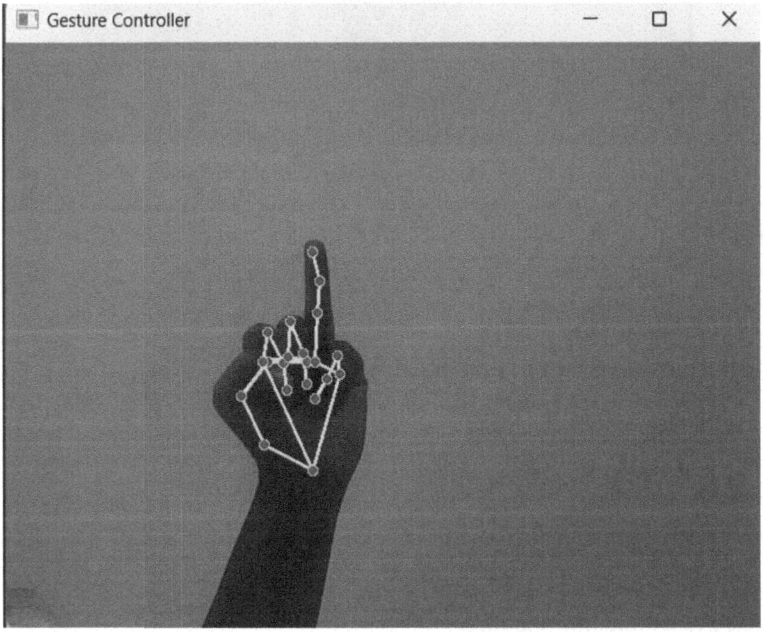

Fig. 6. Representation of gesture when ring finger is extended. This gesture when captured by the system has the ability to comprehend that the user is trying to navigate to the next or previous pages.

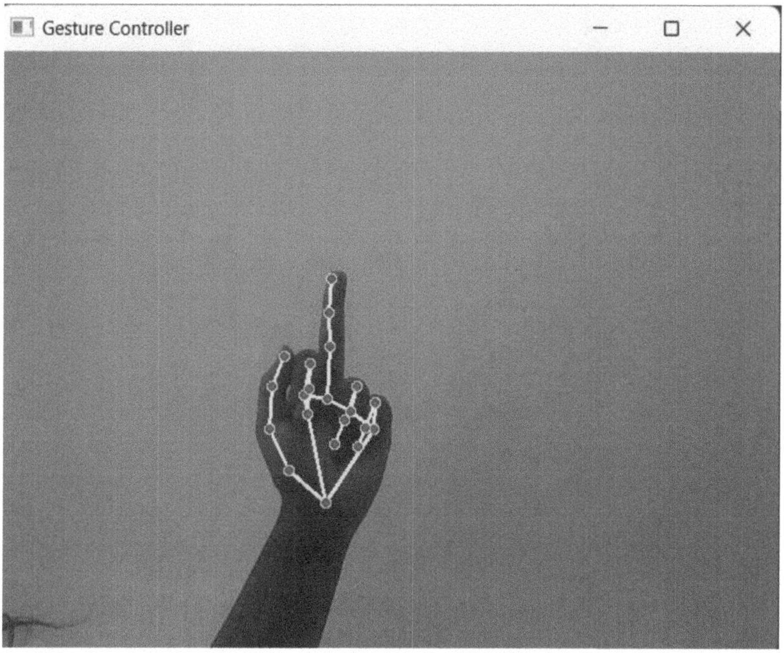

Fig. 7. Extension of middle finger gesture captured by the webcam. Performs the functionality of the left click of a mouse button.

Multi-hand Coordination

Coordinating gestures from multiple hands, especially in two-handed scenarios, is complex. Distinguishing between gestures from different hands and ensuring smooth interaction poses challenges. Potential solutions involve investigating methods for seamless multi-hand gesture recognition and enabling collaborative gestures.

Gesture Variability

Natural variations in hand gestures among users, including hand size, shape, and orientation, can affect the system's recognition of predefined gestures. To address this, research should explore adaptive gesture recognition models, potentially using personalized gesture models to enhance accuracy.

Real-Time Responsiveness

Achieving real-time responsiveness with minimal latency is crucial for user satisfaction. Delays in cursor control or gesture recognition can lead to frustration. To mitigate this, research should focus on latency reduction through optimization techniques for algorithms, hardware integration, and communication protocols.

Fig. 8. Recognition of pinky finger extension which is primarily meant to satisfy custom text requirements which includes highlighting, formatting and adjusting the size of the text.

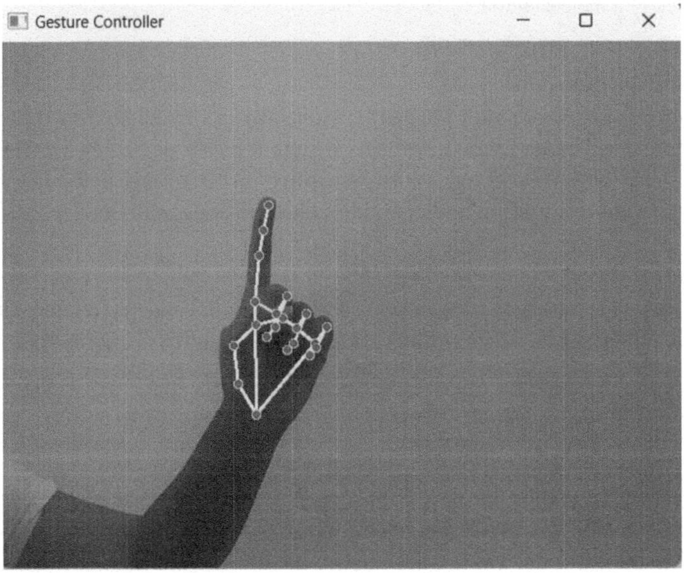

Fig. 9. Detection of index finger by the webcam as input gesture provided by the user. This gesture is used to perform the action of right clicking in a mouse.

Fig. 10. Extension of the first two fingers as gesture input by the user. This triggers the action of double clicking in a mouse.

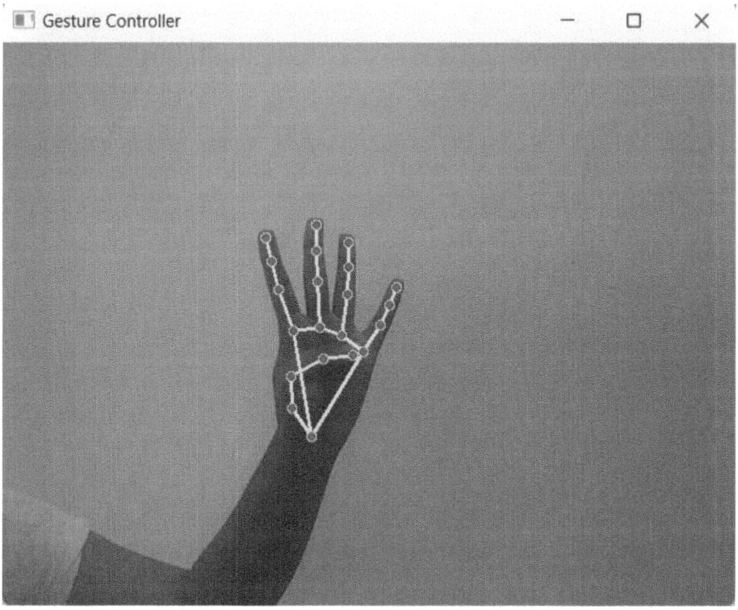

Fig. 11. Recognition of last four extension of fingers from the user. This triggers the AI mouse to take a screenshot or make use of keyboard shortcuts.

Fig. 12. Thumb gesture detected by the webcam. Deployed to customize the volume requirements.

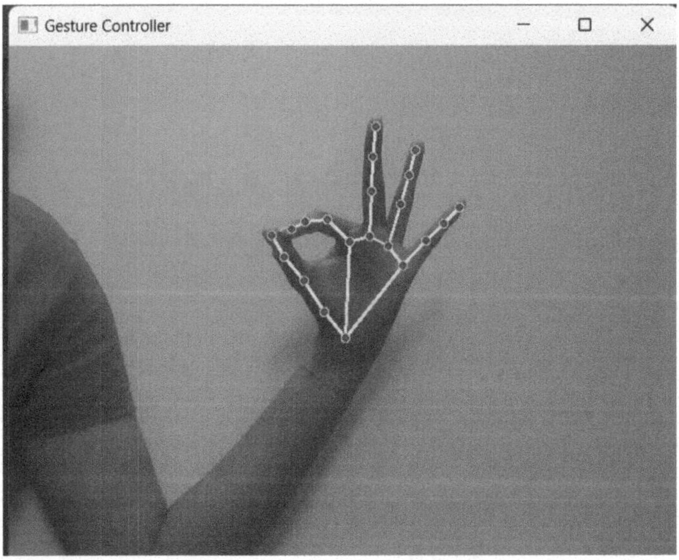

Fig. 13. The extention of the last three fingers which triggers the action to switch between tabs on the system.

6.2 Future Directions

Machine Learning Advancements

Leveraging advancements in machine learning, including convolutional neural networks (CNNs) and recurrent neural networks (RNNs), can enhance gesture recognition accuracy, making the system more robust.

User-Centric Adaptation:

Developing adaptive systems that learn and adapt to individual user preferences and variations can lead to improved accuracy and user satisfaction in gesture recognition.

3D Gesture Recognition:

Extending the system to recognize three-dimensional (3D) gestures by incorporating depth sensors or stereoscopic cameras can provide users with more immersive interactions in digital environments.

Accessibility and Inclusivity:

Exploring applications in accessibility technology to benefit individuals with mobility impairments or disabilities can open up new possibilities for inclusive Human- Computer Interaction (HCI).

Integration with Virtual Reality:

Investigating integration with virtual reality (VR) environments can enable users to interact with VR applications and simulations using natural hand gestures, enhancing immersion and interactivity.

Collaboration and Multi-user Environments:

Extending the system to support collaborative gestures in multi-user environments and researching ways to enable multiple users to interact simultaneously within digital spaces can foster new levels of cooperation.

In conclusion, addressing these challenges and pursuing these future research directions has the potential to revolutionize Human-Computer Interaction (HCI) by providing more intuitive and immersive interactions in digital environments.

7 Conclusions

In conclusion, the development of a gesture-based virtual mouse control system represents a significant leap forward in the realm of Human-Computer Interaction (HCI). This innovative system, driven by hand gestures and powered by Python, OpenCV, MediaPipe, and PyAutoGUI, offers a wide array of practical applications and potential benefits across various industries. Through meticulous gesture recognition, precise cursor control, and support for multiple hands, this system opens doors to improved accessibility, immersive experiences in virtual and augmented reality, and enhanced productivity in fields ranging from gaming to industrial settings. The performance evaluation of the system demonstrates its capability to recognize gestures accurately, particularly excelling in tasks such as scrolling. However, challenges related to noise, multi-hand coordination, gesture variability, and real-time responsiveness must be addressed to further enhance the system's robustness and user satisfaction. In summary, the gesture-based virtual mouse control system is a testament to the relentless pursuit of innovation in technology. Its potential to improve accessibility, transform user experiences, and streamline various industries underscores its significance in the ever-evolving landscape of Human-Computer Interaction. All of these features are majorly devoted to enhance the cognitive thinking and intelligent behaviour. They mainly focus on building an interactive student-mentor environment.

8 Future Scope

The implementation of an AI-powered virtual mouse project using a webcam is just the beginning. There are numerous promising future scopes for this technology. These include expanding hardware compatibility with devices like depth-sensing cameras, refining gesture recognition for improved accuracy, integrating voice control, adapting the technology for mobile devices, exploring applications in healthcare, education, and gaming, and enhancing accessibility features for individuals with disabilities. There exists the potential to revolutionize human-computer interaction and offer a wide range of applications across various industries, making digital interactions more intuitive and efficient. Looking ahead, future research directions in machine learning advancements, user-centric adaptation, 3D gesture recognition, accessibility, and integration with virtual reality hold the promise of revolutionizing HCI and ushering in a new era of intuitive and immersive digital interactions. The simplicity of implementation and usage definitely makes this particular system an impeccable innovation.

References

1. Tran, D.-S., Ho, N.-H., Yang, H.-J., Kim, S.-H., Lee, G.S.: Real-time virtual mouse system using RGB-D images and fingertip detection. Department of Artificial Intelligence Convergence, Chonnam National University, Gwangju, pp. 500–757, South Korea (2020)
2. Hu, C., Liang, L., Ma, S., Lu, H.: Virtual mouse—inputting device by hand gesture tracking and recognition. Natural Laboratory of Pattern Recognition, Institute of Automation, Chineese Academy of Sciences, P.O. Box 2728, Beijing 10080, China
3. Matlani, R., Dadlani, R., Dumbre, S., Mishra, S., Tewari, A.: Virtual mouse using hand gestures. In: Proceedings of the 2021 International Conference on Technological Advancements and Innovations (ICTAI)
4. Khandagale, A., Thakkar, N., Patil, S., Jadhav, V., Nehete, C.: Jarvis-AI based virtual mouse. In: Proceeding of the International Conference on Signal Processing, Computation, Electronics, Power and Telecommunication (IConSCEPT). IEEE (2023)
5. Dudhapachare, R., Rajadhyaksha, M., Lakhani, N., Shankarmani, R.: Hand gesture recognition based virtual mouse events. In: Proceeding of 2nd International Conference for Emerging Technology (INCET) Belgaum, India (2021)
6. Dudhapachare, R., Awatade, M., Kakde, P., Vaidya N., Kapgate, M., Nakhate, R.: Voice guided, gesture controlled virtual mouse. In: Proceeding of 4th International Conference for Emerging Technology (INCET) Belgaum, India (2023)

7. Reddy, V.V., Dhyanchand, T., Krishna, G.V., Maheshwaram, S.: Virtual mouse control using colored finger tips and hand gesture recognition. IEEE-HYDCON, Hyderabad India (2020)
8. Li, W., Deng, C., Lv, Y.: Implementation of virtual mouse based on machine vision.. The International Conference on Apperceiving Computing and Intelligence Analysis Proceeding (2010)
9. Henzen, A., Nohama, P.: Adaptable virtual keyboard and mouse for people with special needs. FTC Future Technologies Conference , San Francisco, United States (2016)
10. Han, M., Chen, J., Li, L., Chang, Y.: Visual hand gesture recognition with convolution neural network. In: 17th IEEE/ACIS International Conference on Software Engineering Artificial Intelligence Networking and Parallel/Distributed Computing (SNPD), pp. 287–291 (2016)
11. Dogan, R.O., Dogan, H., Kose, C.: Virtual mouse control with hand gesture information extraction and tracking. In: 23nd Signal Processing and Communications Applications Conference (SIU) (2015)
12. Murthy, R.S.G., Jadon, R.S.A.: Review of vision based hand gestures recognition. Int. J. Inf. Technol. **2**(2), 405–410 (2009)
13. Basha, S.J., Sowmya, L.S.L., Rumiya, S., Afroz, S.: Virtual mouse with hand gestures using AI. In: 4th International Conference on Electronics and Sustainable Communication Systems (ICESC) (2019)
14. Varun, K.S., Puneeth, I., Jacob, T.P.: Virtual mouse implementation using open CV. In: 3rd International Conference on Trends in Electronics and Informatics (ICOEI), pp. 435–438 (2019)
15. Patil, M., Dudhane, S.U., Gandhi, M.B.: Cursor control system using hand gesture recognition. In: International Journal of Advanced Research in Computer and Communication Eng. **2**(5) (2013)
16. Zhou, Y., Jiang, G., Lin, Y.: A novel finger and hand pose estimation technique for real-time hand gesture recognition. Pattern Recognition **49**, 102114 (2016)
17. Abhilash, S.S., Thomas, L., Wilson, N., Chaithanya, C.: Virtual mouse using hand gesture. International Research Journal of Engineering and Technology (IRJET) **5**(4) (2018)
18. Bharath Reddy, K., et al.: Artificial intelligence based virtual mouse. International Journal of Computer Science and Mobile Comput. **11**(1), 25–35 (2022)
19. Angel, N.P.S.: Real time static & dynamic hand gesture recognition. International Journal of Scientific & Engineering Res. **4**(3) (2013)
20. Vinay, K.P.: Cursor control using hand gestures. International Journal of Critical Accounting, 0975–8887 (2016)
21. Thomas, L.: Virtual mouse using hand gesture. International Research Journal of Engineering and Technology (IRJET) **5**(4) (2018)
22. Subhash Chandra, N., Venu, T., Srikanth, P.: A real time static & dynamic hand gesture recognition system. International Journal of Engineering Inventions **4**(12), 93–98 (2015). e-ISSN: 2278–7461, p-ISSN: 2319–6491
23. Hasan, M.M., Mishra, P.K.: Hand gesture modeling and recognition using geometric features: a review. Canadian Journal on Image Processing and Computer Vision **3**(1) (2012)
24. Saritha, C., Phanitha, Y., Krishna, A.R.: Virtual Mouse Operations Using Webcam (2023)
25. Morajkar, A., James, A.M., Bagwe, M., James, A.S., Pavate, A.: Hand Gesture and Voice-Controlled Mouse for Physically Challenged using Computer Vision (2023)
26. Joshi, H., Waybhase, N., Litoriya, R., Mangal, D.: Design of a Virtual Mouse using Gesture Recognition and Machine Learning (2022)
27. Reddy, V.V., Dhyanchand, T., Krishna, G.V., Maheshwaram, S.: Virtual Mouse Control Using Colored Finger Tips and Hand Gesture Recognition (2020)

SmartAgro: Precision Yield Prediction, Crop Insights, and Real-Time Dashboard

Parth Mody(✉), Ankur Singh(✉), Param Sejpal(✉), Anurag Singh(✉), and Archana Nanade

Mukesh Patel School of Technology Management and Engineering, NMIMS, Mumbai, India
modyparth7@gmail.com, archana.nanade@nmims.edu

Abstract. The agricultural sector is a vital component of India's economy, which has led the country to become the second-largest global agricultural producer. Despite this achievement, India's crop yields per hectare remain notably low, resulting in challenges for farmers and financial losses. To address this issue, our research project proposes a comprehensive set of tools to empower Indian farmers with informed decision-making in crop production. The proposed tools include (1) Crop production prediction - employing various regression models to predict crop production in different regions of India. (2) Crop recommendation system - utilizing multiclass classification models to provide personalized crop suggestions to farmers. (3) Business Intelligence (BI) dashboard enabling farmers to visualize and explore crop price data from diverse markets over multiple years. These tools aim to improve crop planning and management, offer valuable insights into price trends, and enable farmers to select crops tailored to their specific requirements and conditions. Ultimately, our project seeks to enhance crop production and food security, thereby revolutionizing India's agricultural sector and contributing positively to the lives of millions of farmers.

Keywords: Agriculture in India · BI dashboard · Crop yield Prediction · Decision-making · Dashboard · Recommendation system

1 Introduction

The economy of India leans heavily on the agricultural industry, rendering a noteworthy input of over 17% to the nation's Gross Domestic Product and engaging more than 60% of the labor force [1]. In 2022, this sector's contribution to India's GDP reached over $310 billion [2]. As the fundamental pillar of the Indian economy, farming assumes a critical function in guaranteeing sustenance security, occupation, and country advancement. Nonetheless, Indian cultivators frequently confront hindrances in settling on judicious choices concerning yield arranging and management, resulting in considerable economic setbacks for both the individual farmers and the entire nation.

Global yield statistics reveal that India has yet to reach its full agricultural potential, with its performance being average compared to other nations [3]. To address this, agricultural practices in India necessitate technological advancements. Hence, this research

T. Shreekumar et al. (Eds.): ISCComm 2023, CCIS 2231, pp. 50–71, 2025.
https://doi.org/10.1007/978-3-031-75605-4_4

project aims to develop a comprehensive set of tools to empower Indian farmers with data-driven decision-making capabilities in crop production.

We have designed three key tools to solve the above-mentioned problem which are as follows:

Crop Production Prediction. This tool accurately predicts crop production for different regions of India. To achieve this, we employed various regression models, including Multiple Linear Regression, Extra Tree Regressor, XGBoost, and the innovative Stack-Net Regressor. These models leverage a broad range of variables, encompassing state, crop, season, and numerous other factors. The prediction model will not only aid farmers in making better decisions but also enable policymakers to make informed choices regarding crop planning and management.

Crop Recommendation Model. We have utilized multiple classification models like SVM, DT, LR, and AdaBoost, on variables such as NPK, rainfall, and other relevant parameters. By processing these inputs, the model recommends the most suitable crop for individual farmers. This personalized crop recommendation approach will help farmers make optimal choices in their crop selection.

Market Analysis Dashboard. We have created a Business Intelligence (BI) dashboard that visualizes the current daily prices of various crops from different markets (Mandis) across India. The dashboard has been fabricated utilizing information collections encompassing numerous years, thus allowing for effortless exploration of both state-specific and agricultural commodity price information sets. This effectively eases the evaluation of pricing trends over time, presented in an unambiguous and approachable format. The objective of the instrument panel is to furnish cultivators with a comprehension of agro-commodity price fluctuations and patterns. This enables them to make well-informed determinations regarding the timing and locations for vending their yields.

In its entirety, this undertaking harbors tremendous capacity to augment agricultural yield and guarantee sustenance safeguarding in India. By enhancing crop yields, minimizing financial losses, and promoting sustainability among Indian farmers, these tools represent a significant step towards advancing agricultural practices and achieving a more prosperous future for the Indian agricultural sector.

2 Literature Review

The subsequent Table 1 illustrates the crucial data that elucidates the research endeavor that has been executed within this particular realm.

In the context of India's agricultural sector, characterized by low crop yields and challenges faced by farmers, the need for technological interventions is imperative. Our research addresses this need by proposing a holistic approach that combines advanced regression and classification models, leveraging the latest machine learning techniques, to empower Indian farmers with tools for informed decision-making in crop production which is further discussed in sect. 3.

Table 1. Literature Review

Authors	Research Focus	Methods / Models Used	Latest Technology /Methodology	Key Findings/Remarks
Saiteja, SaiShivanz, Sathya Kande, and Nageswara [4]	Crop yield prediction in India	K-Nearest Neighbors (KNN), remote sensing technology, machine learning	Integration of remote sensing technology and machine learning for improved accuracy	Accurate crop yield prediction in India using KNN and remote sensing technology
Nagini and Kanth [5]	Agricultural yield growth prediction	Linear, multiple linear, and non-linear models, stepwise regression	Exploration of multiple models including traditional and statistical techniques	Explored various predictive models and influential variables for yield prediction
Sreemathy J and Prasath N [6]	Precision agriculture crop recommendation system	BiLSTM-MERNN algorithm (deep learning)	Utilized deep learning techniques for precise crop recommendations	Used historical agricultural data and climatic conditions for crop suggestions
Shabari Shedthi et al. [7]	Crop recommendation system	Machine learning techniques	Utilized machine learning for practical crop recommendations	Matches user inputs with appropriate crops based on soil conditions. Multi-lingual interface and nutrient recommendations
C Sagana et al. [8]	Agricultural guidance for Indian farmers	Decision Trees (DT), Support Vector Machine (SVM), Naive Bayes (NB), and other algorithms	Leveraged various machine learning algorithms	Demonstrated the effectiveness of various machine learning algorithms in agriculture
Aayush Kumar et al. [9]	Intelligent Crop Recommendation System	Regression techniques, ensemble models	Employed regression techniques and ensemble models	Suggests crops based on soil characteristics and environmental factors, explored precision agriculture
Kampalli Ramu and Priydarsini K [11]	Crop yield prediction under complex conditions	Random Forest (RF), Support Vector Machine (SVM)	Enhanced predictive accuracy using RF and SVM	Used RF and SVM to enhance predictive accuracy under complex conditions
Anusha Ashok Deshmukh et al. [12]	Crop yield prediction with focus on genotype-environment interaction	Hybrid models, diverse metrics for evaluation	Utilized hybrid models and diverse metrics	Evaluated hybrid models and genotype-environment interaction with diverse metrics
Richa Kumari Karn and A. Suresh [13]	Wheat harvest prediction in Punjab, India	Hybrid machine learning models, genetic algorithms	Optimized model hyperparameters and genetic algorithms	Optimized model hyperparameters for increased prediction efficiency in wheat harvest

(continued)

Table 1. (*continued*)

Authors	Research Focus	Methods / Models Used	Latest Technology /Methodology	Key Findings/Remarks
Ankita Sharma et al. [14]	Yield prediction and crop suggestion	Random Forest, Decision Tree regressors	Achieved remarkable accuracy rates through Random Forest and Decision Tree	Achieved remarkable accuracy rates through Random Forest and Decision Tree regressors
Sivanandam K, Prasanth M, Naveen B, and Saran S [15]	Crop recommendation model	K-Nearest Neighbors (KNN) algorithm	Leveraged KNN algorithm for precise crop suggestions	Demonstrated potential of KNN algorithm in making precise crop suggestions based on key features
Katarya [16]	Influence of machine learning on precision agriculture	Comparative analysis of different algorithms	Comparative analysis of various machine learning algorithms	Analyzed various algorithms, including KNN and Neural Networks, with a focus on Random Forest's effectiveness
R. Kavitha, M. Kavitha, and R. Srinivasan [17]	Crop recommendation using classification algorithms	Support Vector Machine (SVM), Random Forest Classifier (RFC), Logistic Regression (LR)	Utilized SVM, RFC, and LR for crop type prediction	Evaluated classification algorithms in predicting crop types based on various factors
N. Sai et al. [18]	Crop yield prediction for Indian agriculture	Regression methodologies (Kernel Ridge, Lasso, ENet), Stacked Regression	Utilized regression methods and Stacked Regression	Introduced Stacked Regression for enhanced prediction accuracy. Acknowledged limitations in data
Aruna [19]	Comparative analysis of crop selection and yield prediction methods	Random Forest Regression, Multivariate Regression, Support Vector Machine	Leveraged Random Forest, Multivariate Regression, and SVM	Achieved notable accuracy rates but lacked comprehensive information on data preprocessing and validation techniques
Muthoni [20]	Maize yield prediction in Southern Africa	Machine learning, regional focus	Utilized machine learning for maize yield prediction	Demonstrated reasonable accuracy, but regional focus limited generalizability
Venkatesh. A and Saravanan. M. S [21]	Crop yield prediction models evaluation	Lasso regression, polynomial regression	Employed Lasso and polynomial regression	Compared Lasso and polynomial regression models, highlighting Lasso's superior accuracy

(*continued*)

Table 1. *(continued)*

Authors	Research Focus	Methods / Models Used	Latest Technology /Methodology	Key Findings/Remarks
Swarnim Rai et al. [22]	Crop yield prediction using regression models	Linear Regression, Random Forest, Decision Tree, LASSO	Utilized Linear Regression, Random Forest, Decision Tree, and LASSO	Random Forest model achieved high accuracy rates (94%). Suggested inclusion of more error performance metrics
Anupama Aravatagimath et al. [10]	Facilitating communication in the agricultural supply chain	K-means algorithm to analyze consumer purchase data	Utilized K-means algorithm for analyzing consumer purchase data	Enabled direct communication between farmers and consumers
Shilpa Rani, Sandeep Kumar, Venkata Subbamma. T, Arpit Jain, A. Swathi, and Ramakrishna Kumar M. V. N. M [23]	Agricultural commodity price estimation	Regression-based method, user interface	Employed regression-based method and user interface	Estimated commodity prices based on various inputs, limited to predicting prices for a specific year

3 Methodology and Implementation

The flowchart Fig. 1 outlines a typical machine learning pipeline starting with data collection and preprocessing, followed by exploratory data analysis to understand the patterns and relationships. Based on these insights, relevant features are selected and ML algorithms chosen based on data characteristics. The models are trained on the prepared data, evaluated using appropriate metrics, and hyperparameters tuned to optimize performance. If the models achieve satisfactory results on held-out data, they can be deployed; otherwise, the process iterates with different features or algorithms. The key aspects covered are data preparation, feature engineering, model selection, evaluation, tuning, and deployment once the models are performant. The flow is explained thoroughly further.

3.1 Yield Prediction

Our research pipeline began with data collection and preprocessing, involving the retrieval of a dataset related to crop production in India. Categorical data were encoded, continuous variables were scaled, and missing values were managed. Exploratory Data Analysis (EDA) provided key insights, highlighting trends followed by feature engineering; and selecting relevant attributes for modeling. Multiple regression models were used since the prediction of continuous target features is needed. The model evaluation relied on metrics such as MAD, MSD, and R2, with the data split into training and testing subsets. This rigorous approach ensured the robustness of the models in predicting crop yield for practical agricultural scenarios which is explained in detail as follows.

Data Collection and Preprocessing. It involved retrieving dataset from Kaggle's "India Agriculture Crop Production [24]", comprising about 345,408 instances. The dataset exhibited variables such as province, territory, and cultivation, among others. Categorical 'state' data were encoded, and continuous variables ('Year', 'Area Units', 'Production

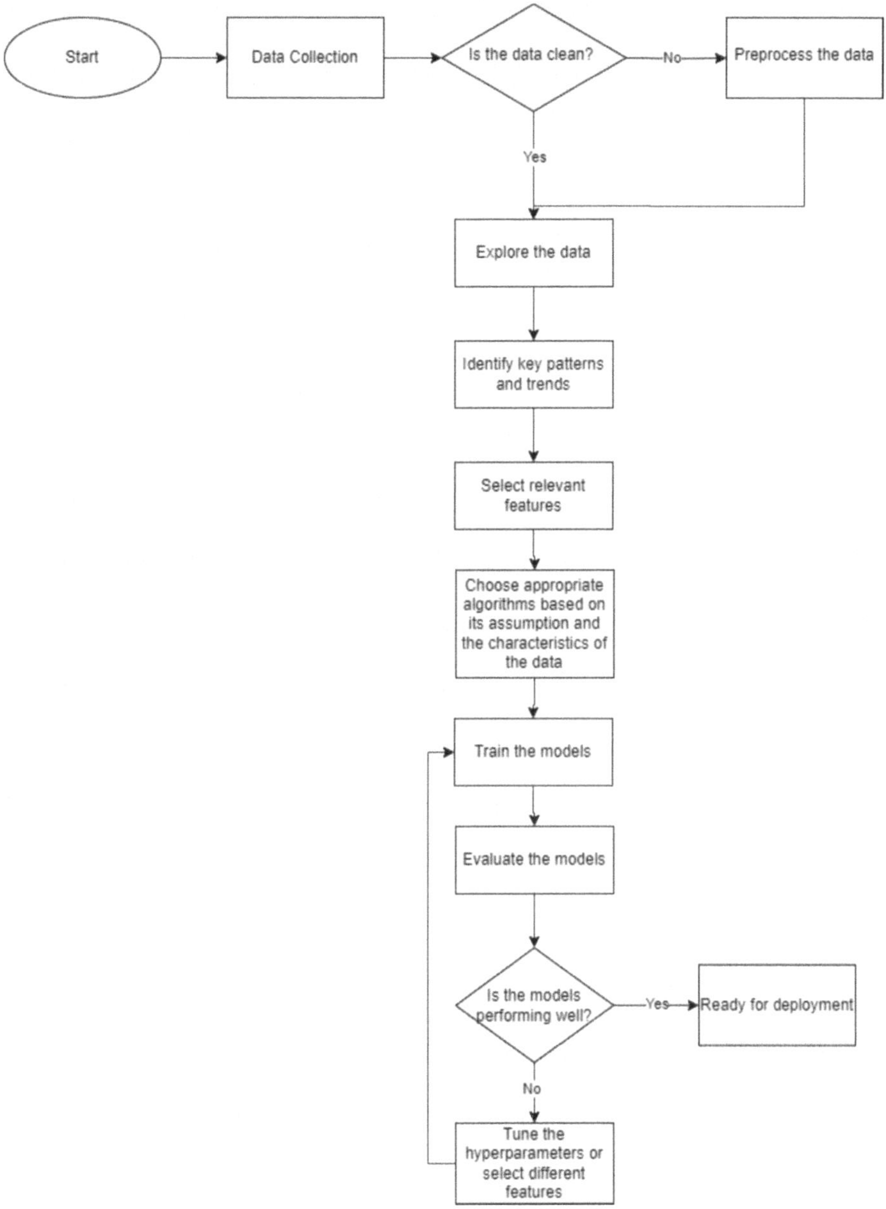

Fig. 1. Flowchart of our proposed approach

Units') underwent scaling. Special attention was given to managing missing values within the 'Area', 'Production', and 'Yield' columns. Notably, a logarithmic transformation was applied to 'yield' to improve linearity and reinforce resistance against outlier influences.

Exploratory Data Analysis (EDA). It was carried out with the objective of acquiring a profound comprehension of the yield prediction dataset. Several observations as shown in Fig. 2 emerged from our analysis, shedding light on key patterns and trends. Notably, when examining the relationship between production units and crop years, we observed that the crop yield was highest in 2011–12, and a decrease in yield was noticeable in 2020–21, potentially attributed to the impact of the COVID-19 pandemic. Furthermore, our investigation into production across different seasons revealed that crops categorized as "Whole Year" consistently exhibited the highest production levels. These insights offer valuable context, contributing to the foundation of our subsequent modeling decisions.

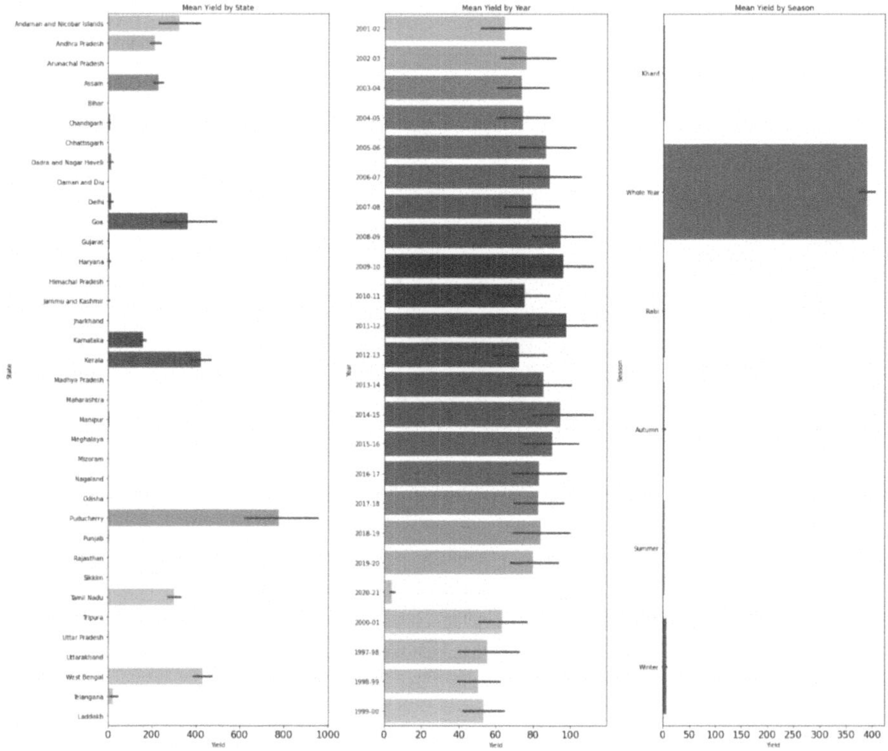

Fig. 2. Plot for yield prediction based on attributes

Feature Engineering. It involved selecting pertinent features from the initial dataset columns such as province, territory, and cultivation, among others. In the process of model formulation, the features of Province, Territory, Crop, Periodicity, Region, and Harvest were elected as the focal points. The exclusion of the Year column was driven by the project's non-requirement for a time series model, given the presence of a dedicated dashboard to address temporal aspects. The exclusion of the Year column was driven by the project's non-requirement for a time series model, given the presence of a dedicated dashboard to address temporal aspects. Additionally, 'Area Units' and 'Production Units' were omitted due to their consistent values (e.g., Hectare and Tonnes), which minimally impacted the model. Feature selection utilized a correlation matrix and data distribution chart to identify influential variables. These analyses guided the selection of features that maximized the model's predictive capability, resulting in a refined feature set contributing significantly to model performance.

Model Selection. It initially involved the use of linear regression (MLR) for modeling, but it couldn't handle complex non-linear relationships and yielded unsatisfactory results. Therefore, ensemble algorithms were chosen based on their ability to generalize well and address the complexities within the data. This transition to ensemble algorithms significantly improved predictive performance, capturing intricate patterns and non-linear interactions in the dataset.

Model Architectures and Training Procedures. In this subsection, we shall elucidate the architectural blueprint and instructional course for every single one of the models employed within our study.

Linear Regression (MLR). A linear regression model was initially considered but later discarded due to its inability to capture complex non-linear relationships within our dataset.

The architecture of this model is presented below: -

$$Y = \beta_0 + \beta_1 X_1 \ldots + \beta_k X_k + \varepsilon \tag{1}$$

Y is the predicted yield (dependent variable).
β_0, β_1, and β_k are the coefficient values.
X_1, X_k are the sets of features (independent variables).
ε is the error term for e.g.: - random noise.
Training Algorithm:

1. Gather the training dataset, comprising features (independent variables) and the target variable (dependent variable).
2. Initialize the linear regression model.
3. Train the model using the training dataset to estimate the coefficient values (β_0, β_1,.. β_k) that minimize the error term (ε).
4. Once trained, the model is ready to make predictions.

Testing Algorithm:

5. Gather the test dataset with the same features used during training.
6. Apply the linear regression Eq. (1)
7. Produce the final prediction using the trained model.

Ensemble Model. To address the complexities within the data and enhance predictive performance, ensemble algorithms were selected. These models demonstrated exceptional capability in capturing intricate patterns and non-linear interactions within the dataset. The following ensemble algorithms were utilized: -

Extra Tree Regressor. This ensemble model employs randomized feature selection, effectively minimizing overfitting. The training and testing algorithms are presented below:

Training Algorithm:

1. Randomly select subsets of features and instances.
2. Build multiple decision trees with these subsets.
3. Aggregate predictions to produce the result.

Testing Algorithm:

4. Pass test data through each tree.
5. Aggregate the results to produce the final prediction.

XGBoost (Extreme Gradient Boosting). Another ensemble model is known for handling non-linear complexities. Its training and testing algorithms are as follows:
Training Algorithm:

1. Initialize with a base prediction.
2. Iteratively add decision trees to minimize errors.
3. Aggregate predictions to produce the result.

Testing Algorithm:

4. Pass test data through each tree.
5. Aggregate the results to produce the final prediction.

PyStackNet: It is a Python implementation of StackNet, originally developed in Java. It resembles a feedforward neural network but uses stacked generalization.
Training Algorithm:

1. Base Model Initialization: Start with a base prediction.
2. Iterative Model Building: Progressively add supervised machine learning models to the ensemble to minimize errors.
3. Aggregation: Aggregate predictions from individual models in each layer.
4. Result Generation: Produce the final prediction based on aggregated results.

Testing Algorithm:

5. Passing Test Data: Feed the test data through each layer of the ensemble.
6. Model Execution: Execute each model within the ensemble.
7. Aggregating Predictions: Aggregate the predictions produced by each layer.
8. Final Prediction: Combine the aggregated results to generate the final prediction for the test data.

Regression Model Implementation. It employed the use of two robust algorithms Extra Tree Regressor, and XGBoost were adeptly harnessed. Extra Tree Regressor's ensemble nature and randomized feature selection proved effective in minimizing overfitting. XGBoost's ensemble of weak learners showcased its prowess in handling non-linear complexities, leading to enhanced regression performance. Each of these algorithms exhibited exceptional performance in predicting yield. Alongside these two algorithms StackNet Regressor was employed which was made possible by utilizing the pystacknet [30] library. The integration of the StackNet Regressor significantly enhances our regression model's performance and suitability for crop yield prediction. StackNet's iterative, layered construction, contrasting traditional back propagation, enables effective multi-level ensemble modeling. Each layer employs diverse supervised machine learning algorithms, fine-tuned for specific tasks like GradientBoostingRegressor, LGBMRegressor, CatBoostRegressor, and RandomForestRegressor. StackNet's architecture and versatility make it a powerful tool for precise and robust crop yield forecasting. The assessment procedure encompassed measures like Mean Squared Deviation (MSD), Mean Absolute Deviation (MAD), and Coefficient of Determination (R2) to holistically evaluate the prognostic precision and appropriateness of the frameworks.

Hyperparameters for Stacknet Regressor. The hyperparameters for this model is shown in Table 2.

Table 2. Hyperparameters for Stacknet Regressor.

Regression Model	Hyperparameters
Level 0 Gradient Boosting	n_estimators: 100 learning_rate: 0.1 max_depth: 5 random_state: 0
Level 0 LGBMRegressor	boosting_type: 'gbdt' objective: 'regression' learning_rate: 0.1 num_leaves: 31 random_state: 0
Level 0 CatBoostRegressor	learning_rate: 0.1 depth: 6 iterations: 100 loss_function: 'RMSE' random_seed: 0
Level 1 RandomForestRegressor	n_estimators: 250 max_depth: 5 max_features: 'sqrt' random_state: 24

Training and Testing. It was executed with precision. The corpus was segregated into a training subset, comprising 80% of the information, and a testing subset, representing the residual 20%. Throughout the training process, models assimilated intricate relationships between features and yield, leveraging the training subset. Notably, data partitioning was carried out using the train_test_split function, ensuring an objective evaluation by assessing model performance on the independent testing set. This arduous examination entailed the calculation of efficacy parameters, comprising of average mean deviation (MSD), mean absolute deviation (MAD), and coefficient of determination (R2) measurements, providing comprehensive insights into the model's predictive prowess within real-world agricultural scenarios.

Performance Evaluation. It involved the use of essential metrics mean absolute deviation (MAD), mean squared deviation (MSD), and coefficient of determination (R2).

$$\mathrm{MAD} = (1/n)\Sigma\left|\mathrm{Yi} - \widehat{\mathrm{Y}}\mathrm{i}\right| \tag{2}$$

where:

MAD is the Mean Absolute Deviation.
n is the number of data points.
Yi is the observed (actual) value.
$\hat{Y}$i is the predicted (model) value.
Σ represents the summation symbol over all data points.

$$\mathrm{MSD} = (1/n)\Sigma\left(\mathrm{Yi} - \widehat{\mathrm{Y}}\mathrm{i}\right)^2 \tag{3}$$

where:
MSD is the Mean Squared Deviation.
n is the number of data points.
Yi is the observed (actual) value.
$\hat{Y}$i is the predicted (model) value.
Σ represents the summation symbol over all data points.

$$R^2 = 1 - \left(\Sigma\left(\mathrm{Yi} - \widehat{\mathrm{Y}}\mathrm{i}\right)^2\right)/\left(\Sigma\left(\mathrm{Yi} - \bar{\mathrm{Y}}\right)^2\right) \tag{4}$$

where:
R^2 is the Coefficient of Determination.
Yi is the observed (actual) value.
$\hat{Y}$i is the predicted (model) value.
$\bar{Y}$ is the mean of the observed values.
Σ represents the summation symbol over all data points.

These metrics provide a concise snapshot of the model's predictive accuracy and overall goodness of fit. A comprehensive discussion and interpretation of these metrics, along with their implications for the model's effectiveness, will be expounded upon in the forthcoming "Results and Discussion" section.

3.2 Crop Recommendation

Our research process began with data collection from a Kaggle dataset containing essential features for crop recommendation. Minimal preprocessing was conducted, including categorical encoding and feature scaling to optimize the dataset for model development. Exploratory Data Analysis unveiled trends highlighting the interplay of environmental factors in crop preferences. Feature engineering selected pertinent attributes for modeling. Multiple algorithms, including Logistic Regression, Support Vector Machine, Decision Tree, and AdaBoost, were employed for classification, and rigorous hyperparameter tuning was carried out to mitigate overfitting. The training and validation phases ensured robust model performance. Performance evaluation relied on the accuracy, precision, recall, and F1 score, providing a comprehensive assessment of the models' effectiveness in crop recommendation which is explained as follows.

Data Collection and Preprocessing. It involved using a dataset retrieved from Kaggle named "Crop_recommendation [25]". It contains approximately 2201 instances and comprises essential features including 'nitrogen', 'phosphorus', 'potassium', 'humidity', 'pH', and 'rainfall'. With a well-structured dataset, minimal preprocessing was needed. The dependent variable, 'crop_type', encompassed categorical data representing diverse crop types (e.g., rice, maize, chickpea). To facilitate model training and prediction, categorical encoding was executed utilizing the label encoder technique, which mapped the categorical values to numeric values and features were scaled for equitable model training. Scaling ensured consistent influence among features and improved model performance. These steps optimized the dataset for effective crop recommendation model development and evaluation.

Exploratory Data Analysis (EDA). For crop recommendation dataset revealed enlightening trends which are evident in Fig. 3. That underscore the collective impact of environmental factors on crop preferences. For instance, regions characterized by higher humidity levels were found to be conducive to the growth of moisture-loving crops such as coconut, watermelon, and muskmelon. Similarly, areas with elevated nitrogen levels witnessed thriving yields of nitrogen- hungry crops like cotton, pomegranate, and coffee. These examples illustrate how a blend of factors, rather than isolated influences, shapes crop choices. By analyzing the combined effects of nitrogen, rainfall, potassium, phosphorus, pH, humidity, and temperature levels, we gained a comprehensive perspective on crop growth dynamics. This holistic understanding guides the development of our crop recommendation model, which integrates multiple factors to provide nuanced guidance to agricultural practitioners seeking optimal crop selections.

Feature Engineering. It involved selecting pertinent features from the initial dataset columns such as province, territory, and cultivation, among others. In the process of model formulation, the features Province, Territory, Crop, Periodicity, Region, and Harvest were elected as the focal points. The exclusion of the Year column was driven by the project's non-requirement for a time series model, given the presence of a dedicated dashboard to address temporal aspects. Additionally, 'Area Units' and 'Production Units' were omitted due to their consistent values (e.g., Hectare and Tonnes), which minimally impacted the model. Feature selection utilized a correlation matrix and data distribution chart to identify influential variables. These analyses guided the selection of features that maximized the model's predictive capability, resulting in a refined feature set contributing significantly to model performance.

Model Selection. It encompassed multiple algorithms: LR, SVM, DT, and AdaBoost. While these algorithms initially displayed promise, overfitting emerged as a concern, particularly with SVM, DT, and AdaBoost. To counter overfitting, a rigorous cross-validation technique, grid search, was employed. This systematic approach systematically explored hyperparameter combinations, fine-tuning SVM, DT, and AdaBoost. This process effectively curbed overfitting tendencies and bolstered model generalization.

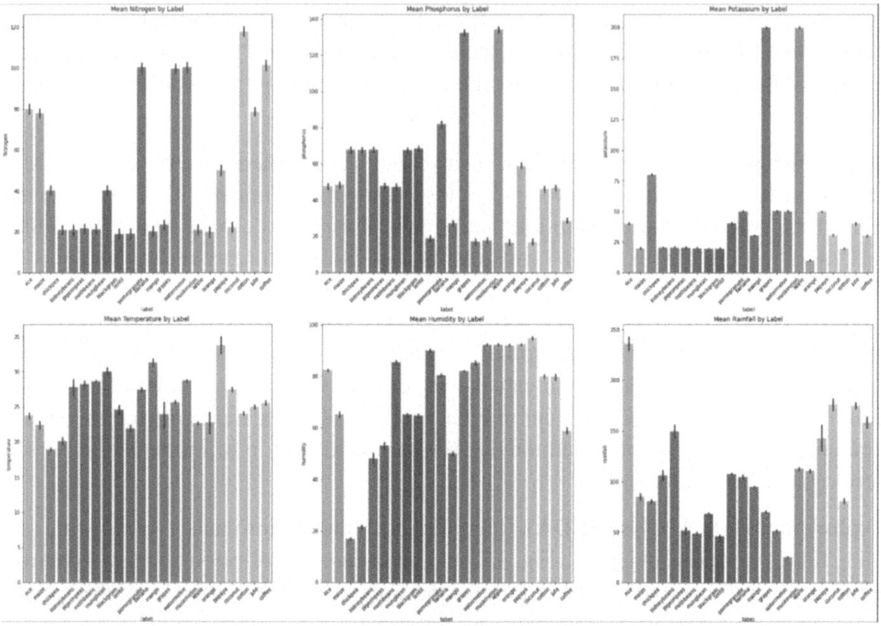

Fig. 3. Histogram plot for attributes of crop recommendation

Model Architecture and Training Procedures. In this subsection, we will provide a detailed overview of the structural framework and step-by-step guidance for each of the models utilized in our research.

Logistic Regression. It is a simple, yet interpretable algorithm used for binary and multi-class classification tasks. It's suitable for cases where the relationship between the features and the target variable is primarily linear.

Training and Testing Algorithm:

1. LR estimates the probabilities of each class and assigns the class with the highest probability as the predicted crop type.
2. During each iteration, LR updates the coefficients in the direction that reduces the prediction error.
3. The process continues until the algorithm converges to a set of coefficients that provide the best fit to the data.

Support Vector Machine. It is a powerful classification algorithm that aims to find the optimal hyperplane that maximizes the margin between different classes.

Training Algorithm:

1. It finds the hyperplane that best separates different crop types in the feature space.
2. Its training aims to find the hyperplane that best separates different classes while maximizing the margin between them.

Testing Algorithm:

3. For testing, the input features are transformed into the same feature space, and the position of the input vector relative to the hyperplane is used to predict the crop type.
4. The distance between the data point and the hyperplane is used to make the classification decision.
5. SVM assigns the class label corresponding to the side of the hyperplane where the data point lies.

Decision Tree. These models recursively split the dataset based on feature conditions to form a tree-like structure. They are suitable for capturing non-linear relationships.

Training Algorithm:

1. It learns to split the feature space based on the most discriminative features to create a decision tree.
2. The process continues until a stopping criterion is met, such as a maximum tree depth or a minimum number of data points per leaf.
3. The result is a tree-like structure that captures complex relationships within the data.

Testing Algorithm:

During testing, the input features traverse the decision tree, and at each node, a decision is made based on the feature conditions until a leaf node (crop type) is reached.

AdaBoost. It is an ensemble method that combines multiple weak learners (typically decision trees) to create a strong classifier. It excels at improving classification accuracy.

Training Algorithm:

1. It sequentially adds weak learners to the ensemble, giving more weight to instances that were misclassified by the previous learners. It focuses on difficult-to-classify samples.
2. The weak learners are combined into an ensemble, with each having a weight proportional to its accuracy.
3. The final model is created by summing the weighted predictions of the weak learners.

Testing Algorithm:

During testing, each weak learner in the ensemble provides a prediction, and their predictions are combined to make the final crop-type prediction.

Hyperparameter Tuning. It was executed to enhance the efficacy of the utilized models in the agricultural suggestion undertaking. A grid search was employed with the following parameters:

Support Vector Machine [26]. It included 'C' value from the range of 1 to 100 to balance margin and error trade-off, 'gamma' values of 1, 0.1, 0.01, 0.001, 0.0001 to influence decision boundary flexibility, and 'kernel' including radial basis function, linear, sigmoidal, polynomial to determine data transformation type.

Decision Tree Classifier [27]. It included 'criterion' of 'gini' and 'entropy' for impurity measure, and 'max_depth' values of 4, 5, 6, 7, 8, 9, 10, 11, 12, 15, 20, 30, 40, 50, 70, 90, 120, 150 for controlling tree depth.

Adaptive Boosting [28]. It included 'n_estimators' of 50, 100, 150 for the number of weak learners, and 'learning_rate' of 0.01, 0.1,0.2 for controlling the contribution of each weak learner.

Classifier Implementation. It involved modelling a multi-class classifier and aimed at predicting 'crop_type' based on a given set of features. Our algorithm selection encompassed LR, SVM, DT, and AdaBoost, each chosen to harmonize with our dataset's characteristics. Logistic Regression, known for its straightforward interpretability, complemented SVM, DT, and Ada- Boost, which excelled in navigating intricate relationships, effectively counteracting overfitting tendencies through built-in regularization mechanisms. To preempt and address potential overfitting issues, we implemented a robust cross-validation strategy, coupling seamlessly with a grid search mechanism for hyperparameter tuning. This process effectively fine- tuned the algorithms by exploring diverse hyperparameter combinations, yielding optimal configurations that not only mitigated overfitting but also enhanced model generalization and predictive accuracy.

Training and Validation. It underwent a thorough process. During training, a subset of the dataset was utilized to teach the models intricate relationships between features and crop types. Hyperparameter tuning was achieved through an exhaustive grid search, optimizing model configurations. Validation assessed the models' predictive capabilities on unseen data, affirming their robustness and real-world utility.

Performance Evaluation. The model involved critical metrics: accuracy, precision, recall, and F1 score. By employing these metrics, a comprehensive understanding of model performance was achieved, aiding model selection and effectiveness assessment.

3.3 Market Analysis

Market analysis for the dashboard involved collecting data from government agency. After cleaning and formatting, data analysis utilized Power BI to create visualizations which are explained thoroughly as below.

Data Collection. For the dashboard, the data was collected from government agencies [29]. The data was amassed over a quinquennium, spanning from 2018 to 2022. The collected data encompassed the valuation of goods, the produce output, and the preliminary date of appearance.

Data Cleaning and Formatting. It involved removing any blank or duplicate rows and converting the data types to the correct format. The data analysis includes creating line graphs to show the trend of the price of commodities, bar charts to compare the yield of crops in different districts, clustered bar charts to show the price of market, and display cards to show the earliest arrival date, min price, max price, and price.

Data Analysis. It was performed using Power BI's KPI visualization and cards. The visualizations were designed for easy interpretation and to offer insights into the crop price data across the period. The visualizations for market analysis have been discussed in sect. 4.3.

Software Requirement Table 3 presents a comprehensive overview of all the essential prerequisites that must be fulfilled in order to successfully execute this study within the designated environment.

Table 3. Software Requirements

Tools	Programming Language	Machine Learning Libraries	Web Framework	Database	Frontend	Deployment
Crop Production Prediction & Crop Recommendation System	Python 3.x	scikit-learn, XGBoost, LightGBM	Flask or Django	PostgreSQL or MySQL	HTML, CSS, JavaScript	Docker, AWS/Azure/Google Cloud
Market Analysis Dashboard	Not applicable	Not applicable	Power BI	CSV, SQL databases	Not applicable	Power BI Service

4 Results and Discussion

In this section, our investigation unveils the effectiveness of advanced machine learning models in both yield prediction and crop recommendation for the agricultural sector as discussed in sect. 3. The examination thoroughly explores the detailed performance metrics of each model, highlighting their distinctive strengths and contributions to improving agricultural decision-making processes. This scrutiny demonstrates the valuable impact these models have in enhancing decision-making within the realm of agriculture.

4.1 Yield Prediction

Extra Tree Regressor. The performance in predicting crop yield was carried out by gauging key metrics. The calculated Mean Absolute Deviation (MAD) for the Extra Tree Regressor was determined to be 0.0511, the Mean Squared Deviation (MSD) was ascertained to be 0.0127, and the co-efficient of determination (R2) value was observed to have achieved 0.9906. These obtained outcomes serve as an indication of the robust predictive ability of the model in appraising crop yield, among features and improved model performance. These steps optimized the dataset for effective crop recommendation model development and evaluation.

Extreme Gradient Boosting (XGBoost). The model demonstrated exceptional performance in predicting crop yield. The calculated Mean Absolute Deviation (MAD) was 0.0369, the Mean Squared Deviation (MSD) reached 0.0034, and the R2 value was an impressive 0.9975. These results indicate that the XGBoost algorithm excels in capturing intricate patterns and non-linear interactions within the dataset, thus outperforming other models in predictive accuracy and goodness of fit.

StackNet Regressor. The strengths of GradientBoostingRegressor, LGBMRegressor, and CatBoostRegressor at level 0 was leveraged, each with carefully tuned hyperparameters. At level 1, RandomForestRegressor was employed, incorporating 250 estimators, and optimizing for features. Demonstrated exceptional performance in predicting crop yield. The calculated Mean Absolute Error (MAE) was 0.0572, the Mean Squared Error (MSE) reached 0.0095, and the R-squared (R2) value was an impressive 0.9929.

In conclusion, our XGBoost-based yield prediction model showcases remarkable accuracy and effectiveness, outperforming the findings of [22, 23] having an accuracy of 94.2% and 94% respectively. This advancement underscores the potential of our approach to provide more precise and reliable predictions, thereby empowering agricultural stakeholders with enhanced decision-making capabilities. The comparisons of models are shown in Table 4.

Table 4. Comparative Analysis of Yield Predications Models.

Model	MAE/MAD	MSE/MSD	R2 Squared
Extra Tree Regressor	0.051	0.013	0.991
LightGBM	0.037	0.003	0.997
StackNet Regressor	0.057	0.009	0.993

4.2 Crop Recommendation

Logistic Regression. This has delivered encouraging outcomes. Such results were attained without the requirement of grid search as a consequence of the dearth of overfitting indications. The Logistic Regression model has been able to obtain a precision rate of 97.62%, which bears testament to its ability to accurately categorize varied crop types, based on the input features that it receives. Moreover, the model has demonstrated a recall rate of 97.5% and an F1-score of 97.50%, thereby affirming its efficacy in identifying pertinent crop recommendations.

Support Vector Machine. The model was optimized through the selection of the most suitable hyperparameters, namely C = 2 and gamma = 1, and the application of a radial basis function kernel, has been demonstrated to manifest remarkably impressive performance. Specifically, the precision attained by the Support Vector Machine model has reached a highly commendable percentage of 98.72%, clearly revealing its exceptional accuracy in effectively classifying various types of crops. Furthermore, the equally noteworthy recall rate of 98.64% and F1-score of 98.64% substantiate the model's outstanding capability in providing robust crop recommendations that rely heavily on input features.

Decision Tree. It was primed to avert surpassing a depth of 10 and implemented the 'gini' criterion, demonstrating exceptional levels of accuracy, retrieval, and the harmonic mean of precision and recall. It is noteworthy to mention that the model's precision achieved an outstanding percentage of 99.15%, thereby highlighting its ability to recommend crop types with utmost precision. In addition, the model's recall rate of 99.09% and F1-score of 99.09% further underscore its robust capacity to classify crop types accurately and provide valuable recommendations.

AdaBoost. This algorithm, employing a learning rate of 0.2 and utilizing 150 estimators, demonstrated remarkable proficiency. The model achieved a F1- score of 83.32%, precision of 90.23%, and recall rate of 85.0%, signifying its aptitude to offer relevant agricultural recommendations. Despite being marginally less than alternative models, AdaBoost nevertheless demonstrates dependable predictive precision.

Model Performance Analysis: The varying performances of Decision Tree (DT), Support Vector Machine (SVM), and Logistic Regression (LR) models can be attributed to their inherent characteristics and the nature of the dataset. Decision Trees, while achieving exceptional precision and recall, may have a tendency to overfit the data if not carefully pruned, which could lead to its remarkable accuracy. SVM, on the other hand, excels in classification tasks due to its ability to find complex decision boundaries but is highly sensitive to the choice of hyperparameters like C and gamma, which were expertly tuned in this case. Logistic Regression, being a simpler linear model, can perform remarkably well when the data has linear separability, as it does not require complex decision boundaries.

AdaBoost's superiority may be explained by its ensemble nature, which combines multiple weaker learners (typically Decision Trees) to create a stronger, more robust model. While its individual learners may not be as accurate as SVM, DT, or LR, AdaBoost focuses on improving the areas where these models may falter, achieving a balanced trade-off between precision and recall. Its ability to adapt and prioritize previously misclassified samples during training allows it to excel in capturing complex patterns and delivering reliable recommendations, ultimately surpassing other models in this particular context.

In conclusion, our present findings evince the efficiency of the ensemble algorithms shown in Table 5., specifically the Decision Tree model, in providing accurate and robust crop recommendations. A comparative analysis with the literature survey [6–9, 14–17] evinces that our research outperforms the extant work. For instance, [8] referring to the literature survey, similar algorithms had lower F1-score values, recall, and precision with an accuracy of 80.97% which suggests that our investigation has better crop recommendation accuracy.

Table 5. Comparative Analysis of Crop Recommendation Models.

Model	Best Hypermeters	Accuracy	Precision	Recall	F1 Score
LR	-	96.81	97.61	97.5	97.5
SVM	'C': 2, 'gamma': 1, 'kernel': 'rbf'	98.86	98.72	98.64	98.64
DT	'criterion': 'gini', 'max_depth': 10	99.03	99.15	99.09	99.09
AdaBoost	'learning_rate': 0.2, 'n_estimators': 150	91.42	90.23	85.0	83.32

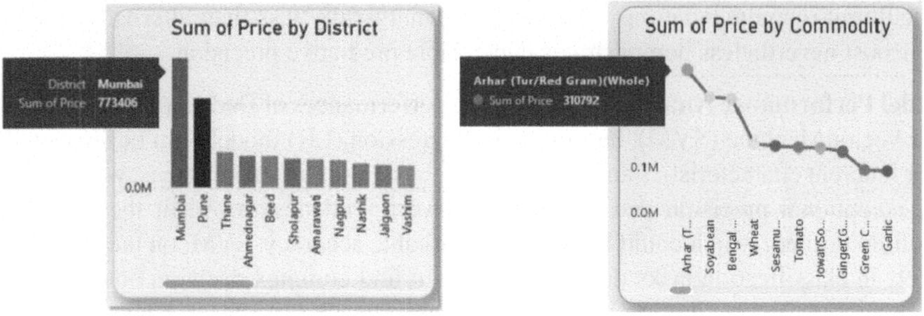

Fig. 4. Sum of Price by District (left). Sum of Price by Commodity (right)

4.3 Market Analysis

In Fig. 4 (left), it was perceived that Mumbai, with a Total Value of 773406, demonstrated the utmost worth and was 64,350.50% greater than Kolhapur, which recorded the minimum Total Value at 1200. Remarkably, Mumbai contributed 20.05% of the collective Total Value. The Array of Total Value across all 31 Districts ranged from 1200 to 773406.

For Fig. 4 (right), it was observed that Arhar (Tur/ Red Gram) (Whole) manifested the highest Total of Value at 310792, with Soyabean and Bengal Gram (Gram)(Whole) trailing closely behind. Arhar (Tur/ Red Gram) (Whole) was accountable for comprising 8.06% of the Total of Value. Among the 101 Commodity, the Total of Value fluctuated between 306 to 310792.

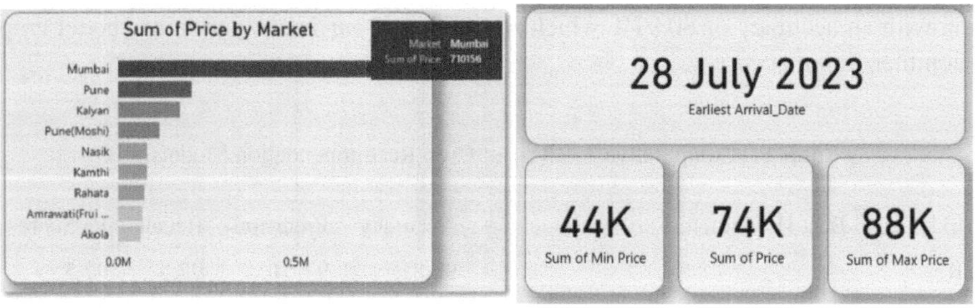

Fig. 5. Sum of Price by Market (left). Display of Price(right)

Regarding Fig. 5 (left), it is noteworthy that at the numerical value of 710156, the bustling metropolis of Mumbai boasted the most exorbitant Sum of Price. Mumbai represented a staggering 18.41% of the total Sum of Price. It is of utmost significance to underscore that the aggregate of Price fluctuated within the range of 500 to 710156 throughout all markets.

The earliest date of arrival shown in Fig. 5 (right) is the latest showcasing in the Dashboard. We can also visualize the min price, actual price, and max price.

5 Conclusion

In conclusion, our research project has the potential to make significant positive contributions to the agriculture domain in India by addressing the challenges faced by farmers in crop planning and management through data-driven decision-making. The tools we have developed empower farmers with accurate crop production predictions, personalized crop recommendations, and valuable insights into crop price trends. This enables them to optimize their crop selection, maximize yields, and make informed decisions about their produce.

Throughout our implementation, we discovered that multiple linear regression was insufficient for handling complex data in crop production prediction, while the XGBoost model from ensemble learning performed the best since it achieved the lowest values for both Mean Squared Deviation (MSD), Mean Absolute Deviation (MAD), to indicating superior precision and accuracy in its predictions. Additionally, its high coefficient of Determination (R2) value of 0.9975 suggests an excellent fit to the underlying data distribution. As for crop recommendation, the Decision Tree model achieved the highest F1-score, precision, and recall among the evaluated algorithms and proved to be the most effective, as it displayed exceptional F1-score, precision, and recall values of 99.09%, 99.09%, and 99.15%, respectively, suggesting robust classification performance. The Agriculture Dashboard is a vital tool for visualizing and analyzing crop price data. The dashboard enables users to track commodity prices, crop yields, cultivated land, arrival dates, and price trends. Looking ahead, potential future research directions may involve refining and integrating more advanced machine learning algorithms and incorporating real-time data streams for enhanced accuracy and efficiency. Moreover, improving farming means using special tools like soil sensors and apps on phones to check and track the nutrients in the soil for better crops and working together with agricultural experts and policymakers can make these tools even more effective for the agriculture sector, promoting sustainable and resilient farming practices.

References

1. Importance of India's Agriculture Economy. https://www.ibef.org/research/case-study/importance-of-india-s-agriculture-economy
2. India GDP Sector Wise GDP Composition 2021–22. https://statisticstimes.com/economy/country/india-gdp-sectorwise.php
3. Cereal crop yield by hectare by country, around the world. https://www.theglobaleconomy.com/rankings/cereal_yield/
4. Kunchakuri, S., Pallerla, S., Kande, S., Sirisala, N.R.: An efficient crop yield prediction system using machine learning algorithm. In: 4th Smart Cities Symposium (SCS 2021), pp. 120–125. Online Conference, Bahrain (2021)
5. Nagini, S., Kanth, T.V.R., Kiranmayee, B.V.: Agriculture yield prediction using predictive analytic techniques. In: 2016 2nd International Conference on Contemporary Computing and Informatics (IC3I), pp. 783–788. Greater Noida, India (2016)
6. Sreemathy, J., Prasath, N.: Crop recommendation with BiLSTM-MERNN algorithm for precision agriculture. In: 2023 International Conference on Networking and Communications (ICNWC), pp. 1–5. Chennai, India (2023)

7. Shedthi, B.S., Shetty, V., Anusha, Shetty, R.R., Shetty, A., Alva, B.A.D.: Crop and nutrient recommendation system using machine learning for precision agriculture. In: 2022 International Conference on Artificial Intelligence and Data Engineering (AIDE), pp. 101–106. Karkala, India (2022)
8. Sagana, C., et al.: Machine learning-based crop recommendations for precision farming to maximize crop yields. In: 2023 International Conference on Computer Communication and Informatics (ICCCI), pp. 1–5. Coimbatore, India (2023)
9. Chakraborty, S., Kumar, A., Pooniwala, O.R.: Intelligent crop recommendation system using machine learning. In: 2021 5th International Conference on Computing Methodologies and Communication (ICCMC), pp. 843–848. Erode, India (2021)
10. Aravatagimath, A., Sutagundar, A.V., Yalavigi, V.: Agriculture product marketing data analysis using machine learning. In: 2021 International Conference on Forensics, Analytics, Big Data, Security (FABS), pp. 1–6. Bengaluru, India (2021)
11. Sunil, G.L., Nagaveni, V., Shruthi, U.: A review on prediction of crop yield using machine learning techniques. In: 2022 IEEE Region 10 Symposium (TENSYMP), pp. 1–5. Mumbai, India (2022)
12. Deshmukh, A., Srivatsa, A., Monteiro, A., Gajakosh, C.: Crop yield prediction to achieve precision agriculture using machine learning. In: 2022 IEEE 2nd International Conference on Mobile Networks and Wireless Communications (ICMNWC), pp. 1–6. Tumkur, Karnataka, India (2022)
13. Karn, R.K., Suresh, A.: Prediction of crops based on a machine learning algorithm. In: 2023 International Conference on Computer Communication and Informatics (ICCCI), pp. 1–8. Coimbatore, India (2023)
14. Sharma, A., Tamrakar, A., Dewasi, S., Naik, N.S.: Early prediction of crop yield in india using machine learning. In: 2022 IEEE Region 10 Symposium (TENSYMP), pp. 1–6. Mumbai, India (2022)
15. Sivanandam, K., Prasanth, M., Naveen, B., Saran, S.: An efficient machine learning approaches for crop recommendation based on soil characteristics. In: 2023 Second International Conference on Electronics and Renewable Systems (ICEARS), pp. 71–76. Tuticorin, India (2023)
16. Katarya, R., Raturi, A., Mehndiratta, A., Thapper, A.: Impact of machine learning techniques in precision agriculture. In: 2020 3rd International Conference on Emerging Technologies in Computer Engineering: Machine Learning and Internet of Things (ICETCE), pp. 1–6. Jaipur, India (2020)
17. Kavitha, R., Kavitha, M., Srinivasan, R.: Crop recommendation in precision agriculture using supervised learning algorithms. In: 2022 3rd International Conference for Emerging Technology (INCET), pp. 1–4. Belgaum, India (2022)
18. Sai, N., Venkat, P.S., Avinash, B.L., Jabber, B.: Crop yield prediction based on indian agriculture using machine learning. In: 2020 International Conference for Emerging Technology (INCET), pp. 1–4. Belgaum, India (2020)
19. Aruna Devi, M., Suresh, D., Jeyakumar, D., Swamydoss, D., Lilly Florence, M.: Agriculture crop selection and yield prediction using machine learning algorithms. In: 2022 Second International Conference on Artificial Intelligence and Smart Energy (ICAIS), pp. 510–517. Coimbatore, India (2022)
20. Muthoni, F., Thierfelder, C., Mudereri, B., Manda, J., Bekunda, M., Hoeschle-Zeledon, I.: Machine learning model accurately predicts maize grain yields in conservation agriculture systems in Southern Africa. In: 2021 9th International Conference on Agro-Geoinformatics (Agro-Geoinformatics), pp. 1–5. Shenzhen, China (2021)

21. Venkatesh, A., Saravaan, M.S.: Prediction of crop yield production with novel lasso regression against polynomial regression algorithm to achieve higher accuracy.In: 2022 International Conference on Edge Computing and Applications (ICECAA), pp. 1478–1482. Tamilnadu, India (2022)
22. Rai, S., Nandre, J., Kanawade, B.R.: A comparative analysis of crop yield prediction using regression. In: 2022 2nd International Conference on Intelligent Technologies (CONIT), pp. 1–4. Hubli, India (2022)
23. Rani, S., Kumar, S., Jain, A., Swathi, A.: Commodities price prediction using various ML techniques. In: 2022 2nd International Conference on Technological Advancements in Computational Sciences (ICTACS), pp. 277–282. Tashkent, Uzbekistan (2022)
24. India Agriculture Crop Production. https://www.kaggle.com/datasets/pyatakov/india-agriculture-crop-production
25. Crop Recommendation Dataset. https://www.kaggle.com/datasets/aksahaha/crop-recommendation
26. 3.2. Tuning the Hyper-parameters of an Estimator. https://scikitlearn/stable/modules/grid_search.html
27. Sklearn.tree.decisiontreeclassifier. https://scikit-learn/stable/modules/generated/sklearn.tree.DecisionTreeClassifier.html
28. Sklearn.ensemble.adaboostclassifier. https://scikitlearn/stable/modules/generated/sklearn.ensemble.AdaBoostClassifier.html
29. Current Daily Price of Various Commodities in Various Markets (mandi). https://data.gov.in/resource/current-daily-price-various-commodities-various-markets-mandi#api
30. H2oai. GitHub - H2oai/Pystacknet. GitHub. https://github.com/h2oai/pystacknet

Artificial Intelligence with MRI-Guided Radiation Therapy for Cancer Treatment

K. Priyadharshini[1(✉)], Divya Mohan[1], and Amulya S. Bhat[2]

[1] Department of Computing Technologies, SRM Institute of Science and Technology, Kattankulathur Campus, Chennai, India
priyakannan.it@gmail.com

[2] Department of Computer Science and Engineering, Sahyadri College of Engineering & Management, Mangaluru, India

Abstract. Radiation oncology is a complex and highly technological area of medicine that relies heavily on digital data processing and computer software; New developments in offline and online MRI-guided RT are described in this paper, along with the exciting prospects they present for advancing research and clinical care, the challenges they present, and the necessity for interdisciplinary collaboration to address those challenges. One of the cornerstones of cancer care is radiation therapy (RT), which is still widely used today. Recently, a lot of work has gone into making MRI a regular part of clinical RT preparation and evaluation. MRI-guided RT is a hybrid technique that combines MRI's capabilities with those of RT because to MRI's enhanced contrast imaging of soft tissues, imaging of moving organs, and monitoring of physiological changes in tumors and other tissues. Clinical contexts can be evaluated quantitatively rather than qualitatively, allowing for the discovery of complicated patterns in medical data. Artificial Intelligence (AI) systems are ideally suited. As molecular imaging and personalized medicine have progressed it has interest in and importance placed on nuclear theranostic approaches, which can aid in providing individualized care for a wide range of illnesses by facilitating more accurate diagnosis, better prognosis forecasting, and more precise patient selection. In the case of prostate cancer (PC), imaging is utilized both to detect local recurrence for salvage therapy and to rule out metastases, which require systemic treatment. Hence this paper proposes AI based MRI-RT (AI-MRI-RT) to mitigate the challenges prescribed above. The purpose of this research is to investigate how prostate cancer radiation relates to diagnostic imaging and it provides a narrative and analytical report of the impact of cutting-edge approaches in the management of prostate cancer, particularly in pretreatment prognostication, radiation planning, and the incorporation of systemic therapy.

Keywords: Artificial Intelligence · Magnetic resonance imaging · Cancer Treatment · Nuclear Imaging

1 Introduction

Radiation therapy (RT) is still widely used as a standard method for treating cancer [1]. At least once for either curative or palliative purposes, radiotherapy (RT) is used on more than 40% of cancer patients worldwide [2]. For almost half of cancer patients, radiation

T. Shreekumar et al. (Eds.): ISCComm 2023, CCIS 2231, pp. 72–85, 2025.
https://doi.org/10.1007/978-3-031-75605-4_5

therapy is an essential part of treatment. Millions of patients, according to estimates, don't have access to this crucial therapeutic method because of obstacles such a lack of infrastructure, technology, and human resources [3]. In addition, technical developments over the past few decades have led to radiation therapy's rising complexity, calling for heavy emphasis on software and hardware designed to facilitate interaction between humans and machines [4]. Another major development is the incorporation of imaging technologies like orthogonal radiography, CT, and cone-beam CT into RT rooms [5]. As a result, the tumor's position in relation to surrounding important normal tissues and the intended dose distribution can be checked often. High-dose clinical RT with image guidance for intensity modulation has become the standard of therapy [6]. Recent years have seen extensive research on the feasibility of using MRI into the RT process [7]. MRI-guided RT has become increasingly common because MRI is more effective than CT for imaging tissue and tumor physiology, as well as at imaging the motion of organs. Advances in MRI-Guided Clinical Practice Offline MRI (done outside of a treatment session) are used to plan the RT that has already begun [8].

Recent years have seen an uptick in the use of phrases like "personalized medicine," "targeted medicine," and "precision medicine" to describe this trend [9]. Experts in radiation therapy, radiation oncology, medical physics, and medical dosimetry all play important roles in the radiation therapy process, and although technology has advanced, the vast majority of processes still necessitate, manual input [10]. Radiation oncology professionals are in short supply around the world due to a combination of factors, including the human-machine interactions are becoming increasingly complicated and the rising prevalence of cancer [11]. In final analysis, imaging is being used more than ever before in the treatment process. The field of theranostics has seen extensive application in this setting [12].

Depending on clinical priorities and main therapy, imaging should give a step-by-step multimodal strategy that supports local and systemic staging. Clinical guidelines recommend MRI and PET/CT scans for determining the presence and spread of disease [13]. After radical prostatectomy (RP), biochemical recurrence (BCR) alone, without imaging, is often sufficient to warrant salvage radiation therapy (RT). This is due to the low threshold at which imaging techniques become useful for metastasis prediction, as well as the well-documented low sensitivity of imaging at such low levels [14]. Prostate Imaging for Recurrence Reporting (PI-RR) is developed to standardize the collection, analysis, and reporting of MRI findings in cases of local PC recurrence following RT and RP. The primary focus of radiotherapy over the past decade has been on increasing the therapeutic dose to the tumor while decreasing the radiation to nearby healthy tissue. The benefit-to-risk ratio is significantly enhanced, with the greatest improvement seen in prostate cancer due to its unique radiobiology. Modern imaging techniques have allowed for these advancements because of their diagnostic usefulness and the capability of merging their knowledge with conventional radiation planning. Our goal in writing this research is to provide a high-level summary of the ways in which diagnostic imaging has contributed to advances in radiation for prostate cancer.

The following are the paper's most significant contributions:

1. The research presents AI-based MRI-RT (AI-MRI-RT), which can aid in providing tailored therapy for a variety of illnesses by permitting more precise diagnosis, better prognosis predicting, and more exact patient selection.
2. This is accomplished by presenting ainterpretive and quantitative report of the impact of novel methods for treating prostate cancer, with particular attention paid to pre-treatment prognostication, radiation planning, and the incorporation of systemic therapy.
3. MRI-guided RTis prompted by MRI's improved soft-tissue contrary, organ movement imaging, and observation of tumor and tissues physiologic changes.

The remaining portion of the paper is built upon this outline. Radiation therapy and magnetic resonance imaging are discussed in detail in Sect. 1, while the disease's historical development is discussed in Sect. 2. Section 3 explains what AI-MRI-RT is and how it works. In Sect. 4, we show the data that supports our hypothesis. Section 5 concludes the research by discussing potential next steps.

2 Literature Review

Here, we'll reflect on the most recent work that scientists all over the world have accomplished.

Gomes Marin, J. F et al. [15]. Incorporated and provide an introduction to nuclear medicine theranostics [NMT] by discussing the history of theranostic agents and their development from their first applications to the present day, with examples highlighting their imaging properties. With the rise of molecular imaging and personalized medicine, the nuclear theranostic approach has gained popularity and traction as a means to better tailor the diagnosis, treatment, and prognosis of a wide range of diseases to the specific needs of individual patients, thereby reducing the need for unnecessary tests and procedures and saving substantial resources. When a diagnostic biomarker and a therapeutic medication have the same target in sick cells or tissues, we call this theranostics. In both clinical and scientific settings, nuclear medicine is today one of the most important components of the theranostic paradigm.

Otazo, R et al. [16]. Developed Radiotherapy guided by MRI the ability of magnetic resonance imaging (MRI) to provide improved this consolidation is driven by applications like soft-tissue contrast, organ motion imaging, and the monitoring of tumor and tissue physiologic changes., known as MRI-guided radiation therapy (RT). Radiation therapy (RT) is still widely used as a standard method for treating cancer. Recently, a lot of work has gone into making MRI a regular part of clinical RT planning and monitoring. Many medical centers have already begun using offline MRI for patient care planning. In addition, when compared with CT guidance, MRI-guided linear accelerator systems permit better response to anatomic changes between RT fractions. Researchers are working on ways to use MRI to guide intrafraction adaptive radiation therapy for cancers that are mobile and to use MRI-derived biomarkers to track patient response and perhaps adjust therapy based on physiological fluctuations. These advances in MRI guiding lay the groundwork for a radical transformation in the ways in which treatment is conceived of, tracked, and adjusted. Improving MRI-guided RT requires several processes, including the acquisition of real-time volumetric anatomic images, the correction of image distortion induced by magnetic field inhomogeneities, the acquisition of repeatable quantitative images across multiple MRI systems, and the biological validation of quantitative images.

Huynh, E et al. [17]. Elaborated Artificial Intelligence (AI) where the use of AI in medicine has the potential to significantly change the field. Two areas where AI systems really shine are the detection of intricate trends in medical information and the provision of a quantitative rather than qualitative evaluation of clinical situations. Because of its complexity and high level of technology, radiation oncology is increasingly dependent on computing, software, and the processing of digital data as a result, AI could have particularly game-changing applications in this area of medicine. Cancer patients who undergo radiation therapy may benefit from AI's increased precision, efficiency, and overall treatment quality. In this Perspective, we first give a broad review of AI methodologies, and then we analyze the potential effects AI will have on each stage of the radiation therapy workflow.

Panebianco, V et al. [18]. Detailed Radiation Therapy and Radical Prostatectomy followed by magnetic resonance imaging (RT-RP). Prostate cancer (PCa) recurrence can be detected locally for salvage therapy and metastases, which require systemic treatment, can be ruled out with imaging techniques. Reduced variability in MRI capture, interpretation, and reporting is needed to detect local PCa recurrence in males who have biochemical relapse following curative-intent local therapy. Based on the available research, the Prostate Imaging for Recurrence Reporting (PI-RR) criteria are developed. A group of professionals from all across the world surveyed the literature without following any particular methodology. The PI-RR method was developed through consensus reached through in-person and online conversations. This article offers a standardized method of reporting (PI-RR) for MRI assessment of PCa local recurrence following RT and RP.

Francolini, G et al. [19]. Introduced in this research is to look into how prostate cancer screening and treatment with radiation relate to one another [DIRA-PC]. We did a non-systematic search of studies examining the predictive utility of imaging before treatment to offer a review of the integration between modern scanning and radiotherapy approaches. The use of radionics in individualized treatment plans is discussed. The genuine clinical value of complicated workflows in clinical practice should be investigated in future clinical trials.

Modifications could enhance the usefulness of some widely used models, such as NMT, MRI-RT, AI, RT-RP, and DIRA-PC. The following have been identified as knowledge gaps in the field: Few studies have looked into how using multiple imaging modalities together can improve cancer diagnosis. Despite its importance for medical and decision-making applications, uncertainty estimates and confidence intervals are often overlooked in cancer detection design. Impact of erroneous positive and negative results on patient outcomes and clinical workflow in cancer diagnosis is under-analyzed. Their acceptance as well as dependability in clinical settings have been restricted by the produced models' weak accessibility and explain the ability.

3 AI Based MRI-RT

There is a range of therapeutic usefulness for these rule-based AI systems. However, there has been a significant algorithmic paradigm shift underlying the computerization of tasks involving images over the past decade. There is a huge range of potential uses and we haven't even scratched the surface here. Numerous complicated jobs within the radiation therapy process have been digitized and consequently automated over the years.

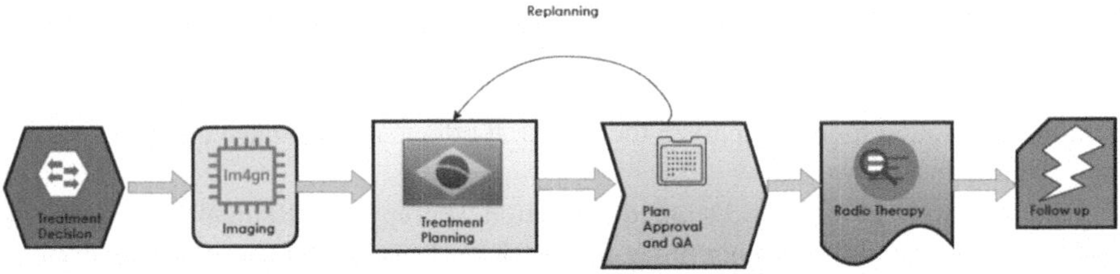

Fig. 1. AI used in the process of Radiation Therapy

The Fig. 1 presents an overarching picture of the procedures used in radiotherapy, with quickly resuming anticipated AI uses at each phase. Precision oncology has great potential to benefit from decision-making aid technologies that combine genetic, imaging, and clinical information. After deciding to use radiation therapy, the next step in the process is to schedule a simulation visit, where diagnostic images will be taken for use in the treatment planning process. Radiation therapy for a patient begins with the development of an individualized treatment plan, where in the planning procedure may be simplified with the use of automated tumor and organ segmentation and optimal dose prediction which is then subject to approval, review, and quality assurance (QA) processes before it is administered. Artificial intelligence systems have the potential to speed up the QA process and uncover uncommon errors, particularly for complicated therapies. The patient then goes on to undergo subsequent care. Clinical productivity, patient outcomes, and satisfaction could all benefit from enhanced image guiding, motion control, and scheduling.

$$\mathrm{STF} = \alpha - (\partial\sqrt{2\mathrm{logl}}) * \alpha) \tag{1}$$

where STF is the signal threshold factor, α is the wavelength of the specific input, ∂ is the deviation angle, and l is the length of the sample, principal component analysis is performed to eliminate irrelevant information in Eq. (1) as shown above.

$$\mathrm{JP(x, y)} = \frac{1}{\mathrm{R}}\exp\left((\mathrm{x, y}) + \log(\mathrm{R})\right) \tag{2}$$

From the function above Eq. (2) and the MRI characteristics of a signal be x and y in that particular coordinates, we may calculate the joint probability of delivery as JP(x, y) where R be the radiation constant.

$$\mathrm{CT_P} = \frac{\sum(\mathrm{P_{xy}} + \beta_{\mathrm{xy}})^2}{\mathrm{z}} - \frac{\mathrm{dy}}{\mathrm{dx}}(\mathrm{x} + \mathrm{y}), (\mathrm{z} = 1,2,3\ldots\ldots\mathrm{n}) \tag{3}$$

Mathematical representations of the data are saved and transmitted to subsequent radiation levels in AI logic. The signals at the cancer treatment patient is denoted by (CT_P) are described by the aforementioned Eq. (3). Each input d in the specified layer uses the angle β to determine how heavily the data input xy should be weighted.

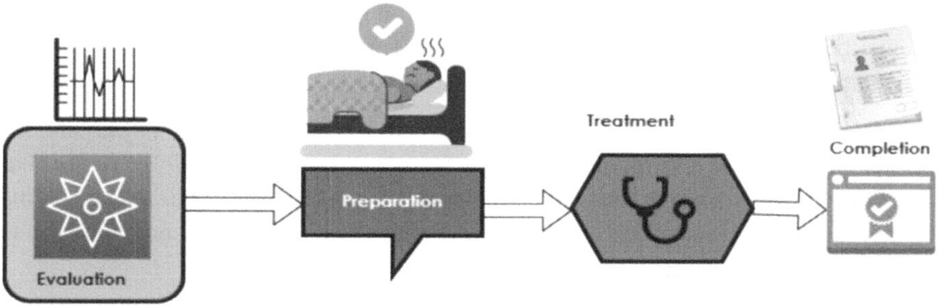

Fig. 2. Patient participation and Radiation Therapy workflow considerations

Patient involvement in radiation therapy planning is depicted in Fig. 2. Clinical evaluation, treatment strategy development, treatment method implementation, and finalization; these are the four basic phases that make up the radiation therapy workflow. Front-of-house staff employees that interact directly with patients are highlighted.

$$C(P) = \frac{1}{N}\sum_{x=0}^{N-1}(E + P + (T(xy) - x))^2 \tag{4}$$

From the above Eq. (4) it is clear that C(P) be the completion of the patient information in detection and diagnosis of the cancer treatment. N be the number of patients. E be the evaluation scenario, P be the preparation, T(xy) is the treatment of the patient taken for consideration.

$$x(y) = C(P)(Y/t)) * \theta(t) + C(P) \tag{5}$$

Equation (5) represents the logic used in the hospital industry using sensors and IoT with the help of Artificial Intelligence. x(y) be the signals taken for consideration, Y/t be the radiation constant, θ(t) be the angle of deviation and C(P) be the complete patient history and information.

Substituting Eq. (4) in Eq. (5) we get

$$x(y) = [\frac{1}{N}\sum_{x=0}^{N-1}(E + P + (T(xy) - x))^2](Y/t)) * \theta(t) + [\frac{1}{N}\sum_{x=0}^{N-1}(E + P + (T(xy) - x))^2 \tag{6}$$

This is the condition of patient information after finishing the detection of cancer using MRI.

Diagnosis of the Patient: Patient intake and evaluation is the first step in the clinical radiation treatment workflow. The radiation oncologist will consult with the patient and go over their symptoms before making a treatment recommendation.

Preparation: During simulation consultations, usually, medical photographs are taken of the patient after they have been immobilized to avoid major movement. Radiation oncologists and medical physicists are often involved in this procedure since it can be highly difficult depending on the illness site and because the ideal patient immobilization is subjective.

Treatment: Wait periods have a negative impact on the clinic's productivity as well as patients' peace of mind and sense of satisfaction. Tumor growth and shrinking, as well as anatomical variables can all cause radiation doses to the tumor and organs to change.

Completion: Among the several methods used to assess whether or not a patient's treatment has been successful, the Criteria for Response Evaluation in Solid Tumors has gained the greatest traction. The use of AI algorithms could improve our understanding of tumor responses during radiation therapy. Patients with bladder and pancreatic cancers have had their treatment outcomes predicted by AI.

Artificial intelligence algorithms may be able to pick up on subtle lung alterations associated with local recurrence that doctors might otherwise dismiss as radiation-induced fibrosis. Improved outcomes might be achieved by earlier, individualized therapy treatments if this additional data are available.

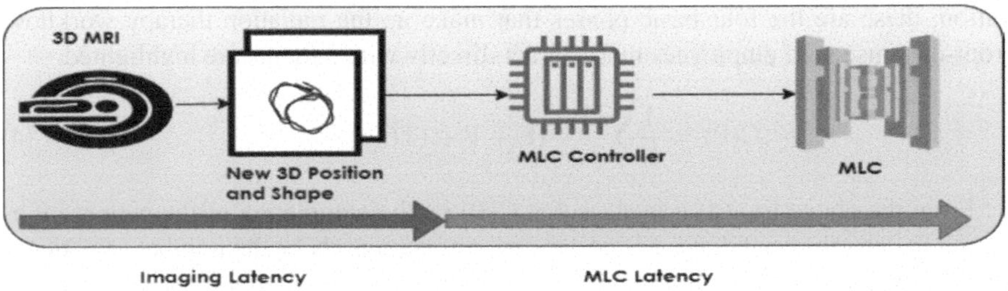

Fig. 3. Conceptual Framework for Intra Fraction shape-Adaptive therapy by 3D MRI-RT

The concept of using real-time 3D MRI techniques for multi leaf collimator tracking to accommodate changes in volumetric location and form is depicted in Fig. 3. If the requirements for tumor coverage or organs at danger are not met, inter fractional or even intra fractional therapy modification would be possible with accurate volumetric mapping of dose buildup. There are a number of variables, including total length of therapy and quality of radiation beam, which can alter the amount of damage caused by radiation to a specific area of tissue.

Circular electron motion in a magnetic field results in an increased dosage at air-tissue interfaces and other inhomogeneous anatomic regions. Secondary electrons, which have received energy from the primary beam, are primarily responsible for delivering the radiation dose. Downstream, straight line segments are where secondary electrons deposit their energy. The target's volumetric position and form will be provided via a quick 3D MRI technique and sent to the MLC controller. The goal of rapid 3D MRI research is to reduce imaging delay to enable real-time 3D adaption.

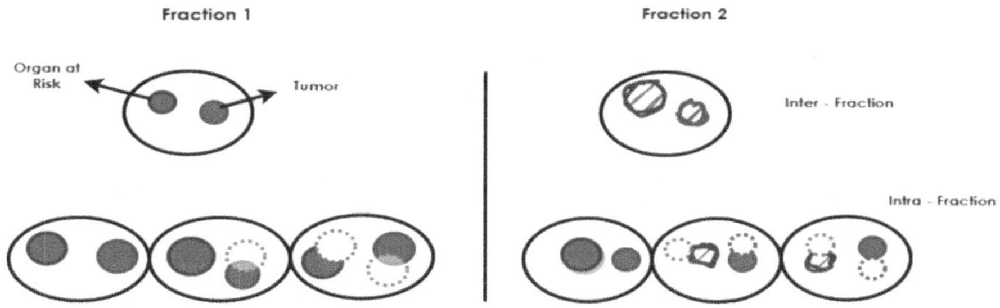

Fig. 4. Online MRI-guided radiation treatment with Inter- and Intra-Fractional adaptation to organ motion

Figure 4 illustrates the concept of inter- and intra-fractional response to organ movement with continuous MRI-guided RT. Cancers impacted by constant movement and by random movement have the potential for actual time responsive therapy within each segment with it. The idea of modifying RT in light of imaging results is not innovative or exclusive to MRI. However, MRI is expected to considerably improve patients' adaptability to changes throughout treatment, especially during individual sessions, because of its advantages over CT. Adaptation to anatomical changes is a primary focus of modern online MRI-guided RT. Our review's focus is on recent advances in clinical MRI-guided RT and the potential applications of this technique in cancer treatment and research. It discusses the need of multidisciplinary collaboration in bringing MRI-guided RT's potential to fruition and some of the major hurdles in the field's path.

$$C(I_F, IR_F) = \frac{1}{N}\sum\nolimits_{x=0}^{M-1}\left[F_x \ln \int F_x((I_F, IR_F))\right] \tag{7}$$

I_F and IR_F are signatures of the relevant fractions in the networks. In this context, I_F, IR_F stand for the Inter fraction and Intra fraction, respectively. F_x is the fraction of values taken for consideration. When comparing the actual labels to the predicted results, the detection error can be quantified using Eq. (7).

$$S_x\left(e^{S+1}\right) = \sum\nolimits_x^M F\left[s_x(BR) * S_x + HS_x R_x(y)\right] \tag{8}$$

Distributed sensors in the right places gather S_m of information about a certain cancer related area. Hence to facilitate this procedure, sensors S are dispersed uniformly across the area. Bed rest BR is a common occurrence because of ongoing health problems. According to Eqs. (8) bed rest causes patients discomfort, stress, and restlessness. Sensors in patients' bodies and on their hospital beds allow for constant monitoring. Healthcare sensors HS_m and medical equipment are susceptible to hostile assaults such eavesdropping, interference, malware, and worm threats due to their restricted resources R_x.

Fig. 5 illustrates While AI will have a greater impact on "back-of-house" staff, such as those involved in the RT-specific technicalities (such as tumor/organ segmentation, treatment plan formulation, and quality assurance) "front-of-house" staff, such as physicians, radiation therapists, and nurses, will be less affected. As a profession that deals

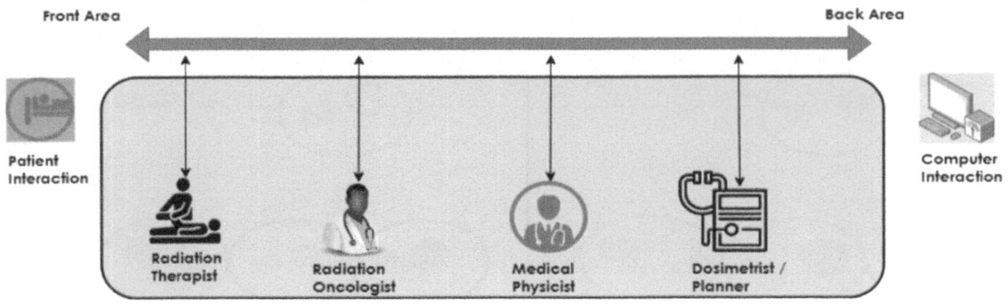

Fig. 5. Implications for the Current Radiation Therapy Workforce of AI Adoption in Radiation Oncology

directly with patients, nursing is unlikely to undergo significant changes as a result of the introduction of AI into the clinic.

$$Precision = F * \sqrt{\frac{y-1}{x-1}\sum(NIM - (MRI^2) - 1} \tag{9}$$

Equation (9) determines the precision factor denoted as *Precision* taken for consideration in radiation therapy using MRI. F be the Fraction of tumor area, NIM be the nuclear imaging sample, MRI be the Magnetic radiation imaging component.

Additionally, the use of AI solutions is likely to lead to enhanced standardization of tumor segmentation and less unnecessary variance, especially in low-resource healthcare settings; this may lead to better clinical outcomes and quality of treatment.

Artificial intelligence-based magnetic resonance imaging radiotherapy (AI-MRI-RT) is a promising method for the early detection of cancer. As a result, this article assesses the best identification frameworks and highlights the many recent platform experiments that have advanced the area of AI. The proposed model, AI based MRI-RT (AI-MRI-RT), used AI to improve upon previous attempts.

4 AI Based MRI-RT

Treatment planning with MRI is the primary focus of early clinical studies of MRI-guided RT, and prostate cancer is a primary target. This clinical investigation found that compared to using both CT and MRI, using MRI alone resulted in more reliably pinpointed target localization. 270 prostate cancer patients from different hospitals participated in a bigger multicenter and multivendor prospective clinical research. Some popular models, including NMT, MRI-RT, AI, RT-RP, and DIRA-PC could benefit from refinements that would make them more effective compared to the proposed model AI-MRI-RT.

4.1 Cost Analysis

The last considerations to make are related to money: cost, cost-benefit analysis, and cost-effectiveness. There is hope that MRI guidance can increase the therapeutic benefit of radiation therapy which is shown above in the Fig. 6a) and 6b). In some cases, distinct and dedicated systems may allow for increased patient throughput despite the higher expense. Hence AI-MRI-RT gains less cost and effective than the existing models available.

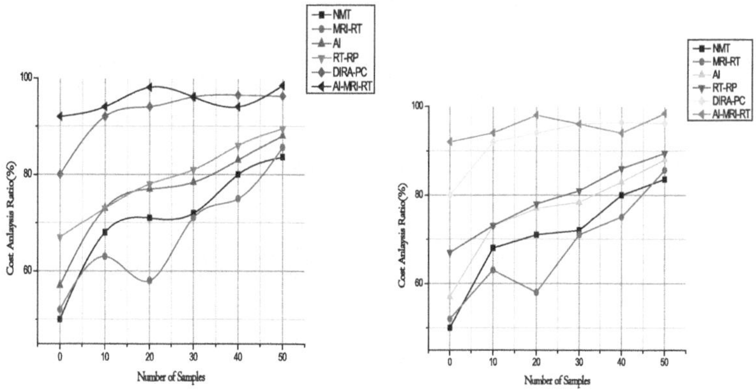

a) b)

Fig. 6. Cost Analysis

4.2 Efficiency Analysis

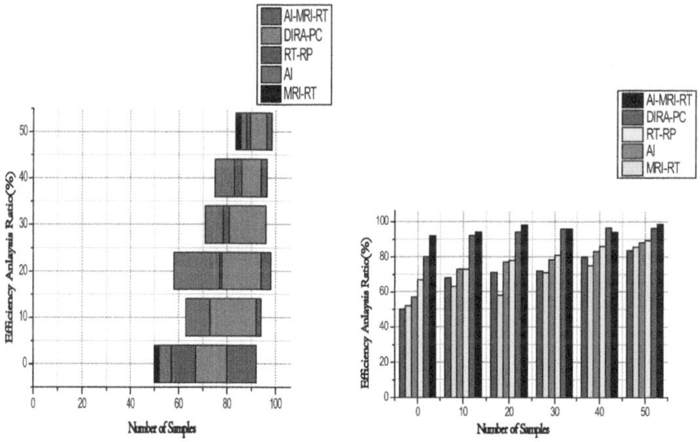

a) **b)**

Fig. 7. Efficiency Analysis

An effectiveness evaluation of the proposed framework is performed with the provided data, and the results are shown in the Fig. 7a) and 7b). Above is a graphical representation of the suggested model's parameters. AI-MRI-RT's greater efficiency sets it apart from other technologies currently in use. This is accomplished by thinking about a number of different things.

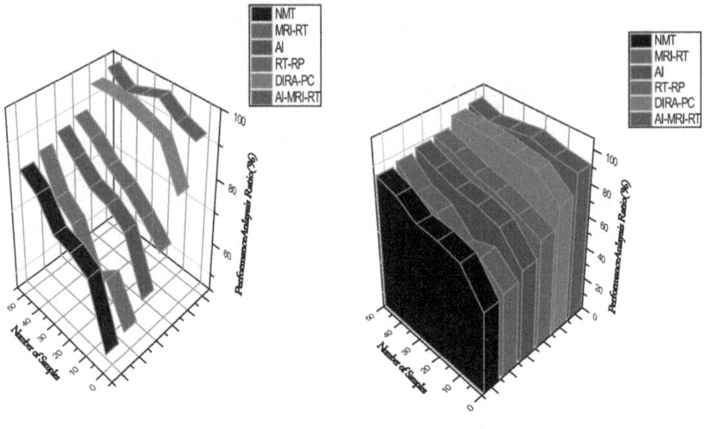

a) b)

Fig. 8. Performance Analysis

4.3 Performance Analysis

Due to the high standards and complexity of modern machinery, mechanical failure diagnostics is essential in industrial settings, and both industry and academia are making strides to improve defect detection systems (see Fig. 8a) and 8b) for a visual representation of the input-to-output performance discrepancies that arise from incorrect diagnosis.

4.4 Detection Analysis

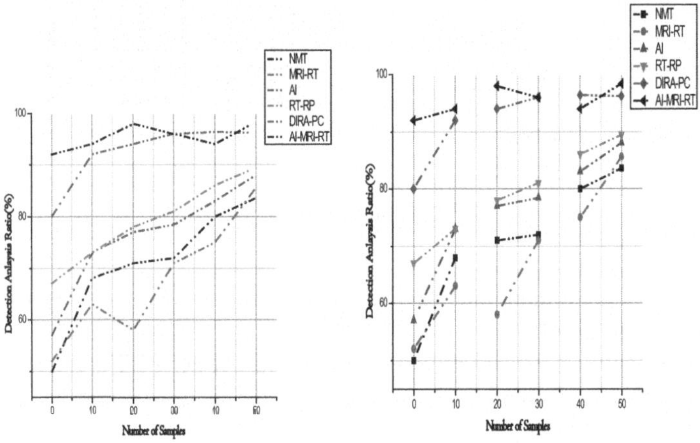

a) b)

Fig. 9. Detection Analysis

Figure 9a) and 9b) demonstrates that this approach is the most useful tool for troubleshooting input-based MRI. Identification and detection, the first stage in solving any issue, is essential. During MRI testing, it is essential to perform detection tests to guarantee that the system can manage and recover from a variety of faults.

4.5 Diagnosis Analysis

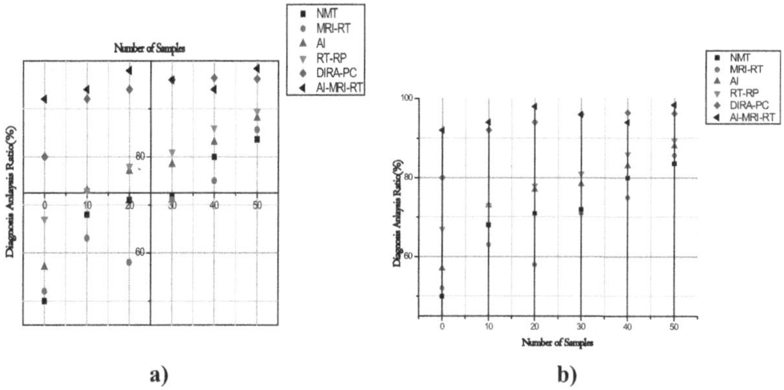

Fig. 10. Diagnosis Analysis

Important technical infrastructure can continue to function well with a process in place to periodically inspect all equipment, as shown in Fig. 10 a) and 10b) illustration of the diagnostic flexibility. This not serves as oversight; it verifies that all conditions for a valid evaluation are met. Since using vibration sensors has been shown to improve test results, they see widespread application. The projected success rate of diagnosis is maximized for a given input level when compared to the precision with which one can convey knowledge.

The diagnostic accuracy and technical/operational efficacy of various models are compared and contrasted. NMT, MRI-RT, AI, RT-RP, and DIRA-PC are all examples of popular models of this type. Hence to attain high precision, enhanced performance, and efficient use of MRI-RT, the results indicate that simulations be utilized in tandem with the diagnostic strategy for data characterization and forecasting.

5 Conclusion

Overcoming the fundamental constraints of CT guidance and providing chances for enhanced inter fraction and intra fraction adaptation. Hence to achieve this goal, it may be necessary to create interdisciplinary and multi centric educational and research platforms. Potential advantages of MRI-guided RT could be verified through the MRI-guided method. Hence AI-MRI-RT is used as a clear evident in tracking, detection, diagnosis of cancer patients and gives a complete solution. In addition to improving accuracy, repeatability, and consistency, radiation therapy could benefit greatly from combining human intuition with AI's ability to exploit different information from vast databases. Interventions based on AI have the potential to improve the health of populations everywhere. The majority of the world's cancer patients, 55%, are in countries with low or moderate per capita income. Because to a lack of skilled staff and expensive machinery, up to 90% of patients in low-income countries do not have access to radio therapy despite the fact that they would likely benefit from it.

References

1. Chamunyonga, C., Edwards, C., Caldwell, P., Rutledge, P., Burbery, J.: The impact of artificial intelligence and machine learning in radiation therapy: considerations for future curriculum enhancement. J. Medical Imaging and Radiation Sci. **51**(2), 214–220 (2020)
2. Beaton, L., Bandula, S., Gaze, M.N., Sharma, R.A.: How rapid advances in imaging are defining the future of precision radiation oncology. Br. J. Cancer **120**(8), 779–790 (2019)
3. Fiorino, C., Guckenberger, M., Schwarz, M., van der Heide, U.A., Heijmen, B.: Technology-driven research for radiotherapy innovation. Mol. Oncol. **14**(7), 1500–1513 (2020)
4. Bucknell, N.W., et al.: Functional lung imaging in radiation therapy for lung cancer: a systematic review and meta-analysis. Radiother. Oncol. **129**(2), 196–208 (2018)
5. Zaidi, H., Vees, H., Wissmeyer, M.: Molecular PET/CT imaging-guided radiation therapy treatment planning. Acad. Radiol. **16**(9), 1108–1133 (2009)
6. Seifert, R., Weber, M., Kocakavuk, E., Rischpler, C., Kersting, D.: Artificial intelligence and machine learning in nuclear medicine: future perspectives. In: Seminars in Nuclear Medicine **51**(2), 170–177 (2021). WB Saunders
7. Law, M.Y., Liu, B.: DICOM-RT and its utilization in radiation therapy. Radiographics **29**(3), 655–667 (2009)
8. Holzinger, A., Haibe-Kains, B., Jurisica, I.: Why imaging data alone is not enough: AI-based integration of imaging, omics, and clinical data. Eur. J. Nucl. Med. Mol. Imaging **46**, 2722–2730 (2019)
9. Weber, W.A., et al.: The future of nuclear medicine, molecular imaging, and theranostics. J. Nucl. Med. **61**(Supplement 2), 263S-272S (2020)
10. Hendrickson, K., Phillips, M., Smith, W., Peterson, L., Krohn, K., Rajendran, J.: Hypoxia imaging with [F-18] FMISO-PET in head and neck cancer: potential for guiding intensity modulated radiation therapy in overcoming hypoxia-induced treatment resistance. Radiother. Oncol. **101**(3), 369–375 (2011)
11. Buvat, I., Weber, W.: Nuclear medicine from a novel perspective: buvat and weber talk with OpenAI's ChatGPT. J. Nucl. Med. **64**(4), 505–507 (2023)
12. Glatstein, E., Lichter, A.S., Fraass, B.A., Kelly, B.A., van de Geijn, J.: The imaging revolution and radiation oncology: use of CT, ultrasound, and NMR for localization, treatment planning and treatment delivery. International Journal of Radiation Oncology* Biology* Physics **11**(2), 299–314 (1985)
13. Leech, M., Osman, S., Jain, S., Marignol, L.: Mini review: personalization of the radiation therapy management of prostate cancer using MRI-based radiomics. Cancer Lett. **498**, 210–216 (2021)
14. Sorace, A.G., et al.: Imaging for response assessment in cancer clinical trials. In: Seminars in Nuclear Medicine **50**(6), 488–504 (2020). WB Saunders
15. Gomes Marin, J.F., et al.: Theranostics in nuclear medicine: emerging and re-emerging integrated imaging and therapies in the era of precision oncology. Radiographics **40**(6), 1715–1740 (2020)

16. Otazo, R., et al.: MRI-guided radiation therapy: an emerging paradigm in adaptive radiation oncology. Radiology **298**(2), 248–260 (2021)
17. Huynh, E., et al.: Artificial intelligence in radiation oncology. Nat. Rev. Clin. Oncol. **17**(12), 771–781 (2020)
18. Panebianco, V., et al.: Prostate magnetic resonance imaging for local recurrence reporting (PI-RR): International consensus-based guidelines on multiparametric magnetic resonance imaging for prostate cancer recurrence after radiation therapy and radical prostatectomy. European Urology Oncology **4**(6), 868–876 (2021)
19. Francolini, G., et al.: Integration between novel imaging technologies and modern radiotherapy techniques: how the eye drove the chisel. Cancers **14**(21), 5277 (2022)

Software Requirements to UML Class Diagrams Using Machine Learning and Rule-Based Approach

Feisal Alaswad[1(✉)], E. Poovammal[1], Batoul Aljaddouh[1], and B. Supriya[2]

[1] Department of Computing Technologies, SRM Institute of Science and Technology, Kattankulathur, Chennai 603203, India
{fa5788,poovamme,ba5329}@srmist.edu.in

[2] Sahyadri College of Engineering and Management, Mangalore University, Mangaluru, Karnataka 574199, India

Abstract. The conversion of textual software requirements into UML class diagrams is a critical step in the development and design process in the field of software engineering. This study provides a novel technique for automatically transforming textual software requirements into UML class diagrams using a model-driven engineering methodology. The main goal is to automate converting textual requirements into a visual representation that can be easily understood and implemented, which will speed up the software development process. In our methodology, we employ a combination of rule-based techniques and Naïve Bayes for named entity recognition. This process assists in resolving ambiguity that arises when predefined rules are unable to determine the type of UML element. This makes it possible for us to precisely distinguish the different parts of a class diagram. Using software requirements and the related UML class diagrams, a wide range of experiments were carried out to assess the efficiency of the suggested methodology. The results demonstrate the superiority of the proposed method, achieving an impressive recall metric of 89.44% and a precision metric of 83.72%. These metrics indicate the high accuracy and reliability of the method in capturing the essential elements of the textual requirements and translating them into UML class diagrams.

Keywords: UML Class Diagram · Model-driven engineering · Software engineering · Natural language processing · Machine Learning

1 Introduction

Model-driven engineering (MDE) is an approach to software development that emphasizes using models as central artifacts throughout the development process. [1] It involves creating abstract representations of various aspects of a system, such as its structure, behavior, and functionality, and then using these models to derive the actual implementation code, documentation, and other

T. Shreekumar et al. (Eds.): ISCComm 2023, CCIS 2231, pp. 86–101, 2025.
https://doi.org/10.1007/978-3-031-75605-4_6

artifacts. The manual translation of textual software specifications into UML class diagrams can be time-consuming and error-prone. Capturing and visualizing a software's requirements is essential for its successful execution in software development. Diagrams created using the Unified Modelling Language (UML) effectively express and illustrate software systems' linkages, behavior, and organizational structure. Understanding is improved, and planning, design, and implementation are made easier with the help of creating UML diagrams from software requirements. Use cases, classes, objects, interactions, and other components of software systems can all be represented using UML diagrams, which offer a standardized syntax for doing so. They provide a visual representation that breaks down complex concepts and encourages interaction between stakeholders, such as developers, designers, testers, and clients. Teams may generate a shared understanding and a plan for developing reliable and scalable software solutions by converting software requirements into UML diagrams. To create UML diagrams from software requirements, it is necessary to thoroughly examine and extract the pertinent information from the requirements documents [2]. Once located, these components are converted into the proper UML diagram representations using the chosen software program. Natural language explanations of requirements frequently contain unclear, imprecise, incomplete, inconsistent, and incoherent terminology. Furthermore, fixing errors that occurred in the earlier stages of the software development process can be highly expensive. Therefore, addressing these issues sooner and for less money is preferable. As a result, creating UML diagrams and conducting requirements analysis are challenging tasks that require automated or semi-automatic support [3].

This work proposes an automated system to convert the software requirements in a non-structured form into class diagrams. Requirements are written in a non-structured form, which makes it challenging to extract information from them. This difficulty is represented by the need for a dynamic system capable of extracting information from different requirements patterns. The implementation of the proposed system will be based on a set of rules that help in the conversion. There is a general tendency to use machine learning techniques in the software development process [4]. Also, modern deep learning techniques such as transformers will increase the system's efficiency in understanding the roles and actions within the software requirements. Transformers have become an effective natural language processing (NLP) tool [5], showcasing excellent ability in a variety of tasks like part-of-speech (POS) tagging. The contributions made by this work:

1. Introduce an automatic system for converting software requirements written in a non-structural language into UML Class Diagrams.
2. Provide a method that accepts free-flowing text while fusing machine learning and rules.
3. Discuss the possibility of generalizing this approach to other UML diagrams.

The work is organized as follows: Sect. 2 presents a thorough literature review. The system description can be found in Sect. 3. Our suggested method is evaluated, and the results and discussions are reported in Sect. 4. The study is concluded in Sect. 5.

2 Literature Review

Growing interest in AI-based software engineering technology has been observed during the previous two decades. According to our examination of this field, there haven't been many methods that offer automated tools to support the early stages of software development, primarily analysis and design.

A conceptual model are used to describe the class diagram. It would be ideal to have a tool that builds class diagrams and processes requirements automatically. The requirements, diagram, and system being created are thus understood by a variety of technical and non-technical persons [6] There are two main approaches: the semi-automatic approach and the automatic approach. Both approaches aim to facilitate the translation of textual requirements into visual UML representations, but they differ in the level of human involvement and the degree of automation.

2.1 Semi-automatic Approach

The semi-automatic approach involves a combination of manual effort and automated tools to convert textual requirements to UML diagrams. A method for converting unstructured software requirements into UML class diagrams is pointed out in [7]. The approach is applied through a tool named RAPID, which utilizes diverse NLP technologies. This includes OpenNLP for lexical and syntactic analysis, the Stemming Algorithm for identifying word roots, and WordNet for conducting semantic analysis and checking semantic correctness. The Class Extraction Engine module then uses the output of the preceding module and a set of heuristic criteria to produce class diagrams, which are then improved using domain ontology. Nevertheless, since every sentence in the requirements document is required to follow a predetermined structure, the tool can only handle straightforward statements. In [8] To create a class diagram from the natural language NL problem statement, Relative Extraction Methodology REM is suggested. This information will be utilized to create a dependency graph, a visual representation that serves as a knowledge base. The user is given the option to add a new attribute or remove it. However, this method has a shortcoming in requiring human interaction and the developer's participation in the refinement process. Additionally, the created classes' and their constituent parts' accuracy for complex issue statements declines. Sharma et al. [9] suggest a technique for creating a high-level class diagram from a structured software requirements document. The FDCT tool employs heuristic rules and a lexicon specific to the domain to carry out this approach. To perform lexical and semantic analysis and impose constraints on requirement sentences, a tool called the Requirements Analysis Tool (RAT) has been developed. Three stages make up the process. First, using user-defined glossaries, the requirements statement is translated into a collection of tokens. Second, the state machines examine the grammar of the requirement statements. Depending on a domain-specific ontology, the third phase comprises semantic analysis. The tool simply distinguishes between association and generalization as types of relationships. Professional assistance is needed to convert the generated classes into implementation-level fine-grained classes. The three parts of a sentence-the subject, object, and predicate can be extracted using NLP using the REM approach. The authors of [10] introduced a three stages method. First, the Natural Language SRS is parsed, followed by the identification of object-oriented (OO) elements to create an OO analysis model. Lastly, a diagram is generated, which undergoes review by a human reviewer to identify any ambiguities or inconsistencies. However, the current approach employs a static parser that is limited to handling static relationships and cannot handle the dynamic model. As a result, the obtained models are significantly incomplete, with numerous unconnected components, isolated classes, and unnecessary classes and relationships. The diagrams lack several pertinent classes and relationships. Furthermore, some of the identified domain classes pose difficulties in being semantically classified as domain classes, as they appear to violate the encapsulation principle. MOVA is a software tool developed to create, measure, and validate class diagrams according to requirement [11]. It enables users to examine invariants and evaluate OCL metrics. However, the tool cannot automatically identify object-oriented (OO) concepts, necessitating user involvement in these processes. Also, the class and object diagrams generated by MOVA are saved in an XML format, which restricts their compatibility with other tools. MOVA heavily relies on human intervention due to its inability to automatically identify OO concepts. The MOVA Meta model is a subset of the UML metamodel and has limited support for OCL and UML, lacking full support for the complete OCL syntax. Additionally, MOVA does not include advanced relationships like aggregation, composition, and dependency. LIDA is a tool that offers a semi-automatic approach for extracting class and object diagrams from unrestricted natural language requirements [12]. The software description is analyzed to identify imported information and parts of speech (POS) from the NL text. Subsequently, the analyst manually identifies potential classes and eliminates inadequate classes or candidates that should be attributed from the noun list. Then, the analyst identifies candidate methods and roles from the verb list, utilizing the LIDA Modeler to visually associate the extracted components with the appropriate classes. However, LIDA requires significant user involvement and only serves as an aid in generating class diagrams rather than automatically generating complete models. The approach lacks support for traceability and does not identify aggregation and generalization.

2.2 Automatic Approach

The automatic approach aims to automate the conversion process as much as possible, minimizing the need for manual effort. In [13], the author proposed the UMLG system. This system is divided into six modules: acquiring text input, comprehending text, knowledge extraction, producing UML diagrams, and multilingual code generation. The requirements text is first scanned by UMLG, which then tokenizes it. Part-of-Speech POS tags, such as nouns and adjectives, are then detected. The primary components of a sentence, such as the subjects, objects, etc., are noted. The natural language text is then semantically analyzed to get the diagram. The system then uses the information gathered to create a class diagram and offers the corresponding source code blocks. The system has to be improved more to extract more classes and diagrams. One notable limitation is Attributes, multiplicity, and relationships are not identified by the system. UML Model Generator from Analysis of Requirements (UMGAR) can be viewed as an enhanced iteration of SUGAR, crafted through the utilization of Stanford Parser, WordNet, and JavaRap. This tool processes natural language requirements, transforming them into diverse UML diagrams like class and collaboration diagrams. UMGAR allows the visualization of these diagrams exclusively in modeling tools compatible with XMI import [14]. Zhou et al. have developed a system aimed at generating a class diagram from unstructured requirements. Their objective was to establish the relationship between the two classes. To achieve this, they proposed an automatic approach for acquiring linguistic patterns [15]. SUGAR [16] is a tool designed to aid developers in creating static UML diagrams. However, it has a limitation in that it imposes a constraint on requirement specifications, requiring each sentence to be expressed in the active voice. The accuracy of identifying candidate classes and relationships is limited, and it relies on human interaction, particularly when eliminating irrelevant classes. This reliance on human decision-making can sometimes lead to misunderstandings or incorrect outcomes if domain knowledge is lacking. In the study conducted by Thakur et al. [17], the researchers propose an approach to generate a class diagram using use case specifications, POS tags, and typed dependencies (TD). They employ TD rules to extract design components like classes, attributes, relationships, and operations. However, their tool primarily analyzes simple sentences, and the rules used for attribute extraction are not applicable to most sentence structures due to issues with consecutive name processing. In the work of [18], they develop a Python-based tool called Visual Narrator, which generates a conceptual model in the form of a Prolog program or OWL ontology. However, their tool only generates entities and relationships and does not create a class diagram. Additionally, the extraction of attributes for each entity is not addressed. The authors of [19] employ the Tree-Tagger analyzer, an NLP tool, and develop a JAVA plugin to generate a use case diagram from user stories. In the study by Javed et al. [20], a conceptual model is generated from unrestricted formats such as user stories, or use cases. They define new rules for extracting attributes from the requirements text, but these rules are based on a predefined set of verbs, which proves to be ineffective for attribute extraction.

3 Material and Methods

This research aims to systematically map software specifications to UML class diagrams by leveraging rule-based and machine learning approaches. Specifically, we employ Transformers for part-of-speech (POS) tagging, which facilitates extracting relevant information from the input text. Additionally, we utilize

Naive Bayes, a popular probabilistic algorithm, to leverage prior knowledge and improve the accuracy of the mapping process. The proposed system consists of three key engines or modules. Firstly, we employ NLP techniques to pre-process the textual specifications, including tasks such as sentence segmentation, tokenization, and POS tagging. These steps ensure the input text is structured and prepared for further analysis.

Next, we apply rule-based techniques to extract the relevant classes, methods, relationships, and attributes from the pre-processed specifications. These rules are designed based on domain-specific knowledge and guidelines provided by the UML specification. By employing rule-based methods, we can capture explicit patterns and rules governing the mapping process, thus ensuring the correctness and consistency of the generated UML class diagrams. Python's versatility and ease of use, coupled with its rich ecosystem of NLP and machine learning tools, make it an ideal choice for implementing the proposed methodology. Figure 1 shows our Software requirements to Class diagram system (SRTC) architecture.

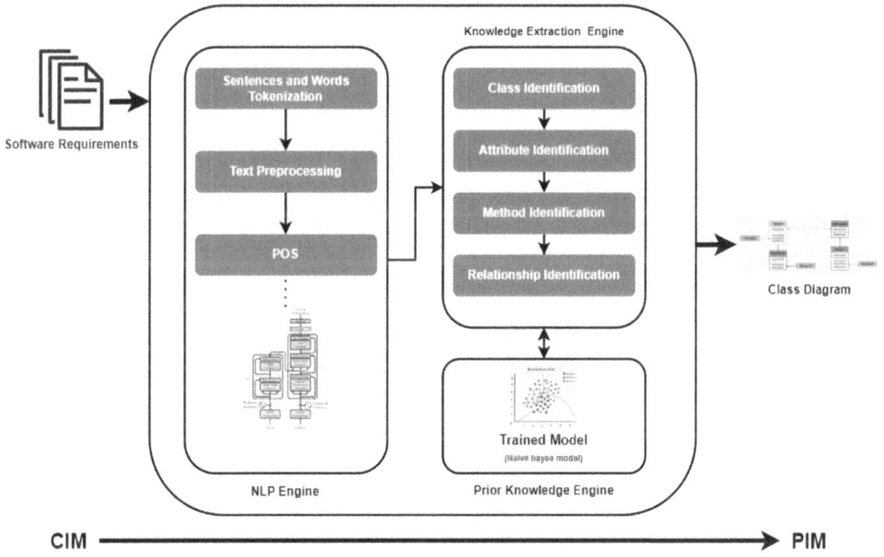

Fig. 1. SRTC system architecture.

3.1 NLP Engine

The NLP technologies are a powerful tools that leverages NLP techniques to analyze and understand human language. With its advanced algorithms and models, the NLP engine can perform various language-related tasks.

Sentences and Words Tokenization. Creating reliable systems requires accurately capturing software requirements and translating them into formal representations, such as UML class diagrams. Tokenizing the textual requirements into sentences and words as the first step in achieving this goal paves the way for subsequent processing and analysis. NLP tokenization technique divides the text into smaller, more comprehensible tokens. Tokenization or word segmentation enables a more in-depth text analysis in the context of software requirements. Sentences are broken down into words, so that the subsequent conversion procedures can work at a more streamlined and controllable level.

Text Preprocessing. Software requirements are expressed in natural language, which can be ambiguous, inconsistent, and noisy. Text preprocessing uses several strategies to enhance the textual data's quality and usability to address these issues. By preprocessing the requirements, we can improve the precision, effectiveness, and dependability of the subsequent stages of converting requirements to class diagrams. Typical Text Preprocessing Methods:

- **Stop Word Removal:** Removing words that are frequently used but have little significance or don't help the reader understand the criteria, like "a," "the," or "and."
- **Removal of Special Characters and Punctuation:** It removes any symbols, special characters, or punctuation that aren't necessary for converting a class diagram.
- **Domain-Specific Preprocessing:** Adding domain-specific dictionaries helped to resolve domain-specific terminology.
- **Coffree:** It is accomplished by utilising [21], a Python utility that performs coreference resolution, including pronoun substitution.

Part-of-Speech Tagging (POS). Part-of-speech (POS) tagging is a fundamental task in NLP that involves assigning a grammatical label (such as noun, verb, adjective, etc.) to each word in a given sentence. While traditional approaches to POS tagging rely on handcrafted rules or statistical models, recent advancements in deep learning have introduced powerful models like Transformers that have shown remarkable performance in various NLP tasks, including POS tagging. The BERT pre-trained Transformer model used in this study which have been trained on large corpora of text. The model used in this work is trained on Penn Treebank - WSJ dataset [22]. The Penn Treebank dataset is popular because of its size, diversity of topics, and high-quality annotations. It has been widely used as a benchmark for evaluating the performance of different POS tagging algorithms.

3.2 Knowledge Engine

In some cases, the possibility of identifying the entity and categorizing it into predefined categories is not possible using pre-defined rules, so we resorted in this research to take advantage of the capabilities of machine learning by training the Named Entity Recognition (NER) model based on Naive Bayes (NB) and prior knowledge represented by text and labeled entities. Using the Naive Bayes theory, NER applies the Bayes rule to identify and categorize named entities in text. In NER, specific entities, such as names of classes, attributes, methods, relationships, and other preset categories, are identified and categorized within specific textual contexts. The Naive Bayes theory is a probabilistic model that works well for NER tasks since it assumes that features are independent. This method bases the algorithm's calculations of the probabilities of various entity labels on the features found in the input text. Text processing, information extraction, and text analysis are just a few of the applications that NER systems can make possible by utilizing the Naive Bayes theory. This approach employs the Bayes rule to express $P(label|features)$ in terms of P (label) and $P(features|label)$ in order to determine the probability for a label. Naive Bayes is computationally efficient and can work well with relatively small datasets. The NB classifier requires training data samples in the following format: (xi, yi), where xi denotes the class of the word or document and yi denotes its context. Eqs. (1) and (2).

$$P(y_j \mid x_1, x_2, \ldots, x_n) = \frac{P(y_j) \times \prod_{i=1}^{n} P(x_i \mid y_j)}{P(x_1, x_2, \ldots, x_n)} \tag{1}$$

Since $P(x_1, x_2, x_3, ..., x_n)$ is constant for given input, we can use the following classification rule:

$$\begin{gathered} P(y_j \mid x_1, x_2, \ldots, x_n) \propto P(y_j) \times \prod_{i=1}^{n} P(x_i \mid y_j) \\ \Downarrow \\ y = \arg\max_{y_j \in Y} \prod_{i=1}^{n} P(x_i \mid y_j) \end{gathered} \tag{2}$$

The relative frequency of class yj in the training set is P (xi — yj). y is the greatest likelihood that instance xi will be generated for class yj under consideration. The probability values for each class and label are calculated depending on TF-IDF rule shown in Eqs. (3), (4) and (5)

$$TF - IDF(t, d) = TF(t, d) * IDF(t) \tag{3}$$

$$IDF(t) = \log \frac{N}{1 + df} \tag{4}$$

$$TF(t, d) = \frac{\text{number of times } t \text{ appears in } d}{\text{total number of terms in } d} \tag{5}$$

where df is the number of documents that contain word t and N is the total number of documents. Word frequency scores called TF-IDF attempt to draw attention to words that are more intriguing. Words that stand out in a given paper are highlighted due to the scoring.

Table 1. Predefined rules to identify the classes.

Rule Ref	Rule Form
RC1	If t_i frequency < 3% (statistically) $\Rightarrow t_i \notin$ C
RC2	If $t_i \in$ {keywords set} $\Rightarrow t_i \notin$ C
RC3	If $t_i \in$ {proper nouns set} $\Rightarrow t_i \notin$ C
RC4	If $t_i \in$ {general concepts set} $\Rightarrow t_i \notin$ C
RC5	If $t_i \in$ A $\Rightarrow t_i \notin$ C
RC6	If $t_i \in$ M $\Rightarrow t_i \notin$ C
RC7	If $t_i \in$ R $\Rightarrow t_i \notin$ C
RC8	If $t_i \in$ N & $t_{i+1} \in \{has, have, consists\} \Rightarrow t_i \in$ C
RC9	If $t_i \in$ N & $t_{i+1} \in$ N $\Rightarrow t_i \in$ C , $t_{i+1} \in$ A
RC10	If $t_i \in$ N & $t_{i-1} \in$ V $\Rightarrow t_{i-1} \in$ M & $t_i \in$ C
RC11	If $t_i \in$ N & All [RC1..RC7] are not True $\Rightarrow t_i \in$ C
RC12	If $t_i \in$ G & All [RC1..RC7] are not True $\Rightarrow t_i \in$ C

3.3 Rule-Based Engine

This engine uses predefined rules or patterns to analyze textual requirements and automatically generate corresponding class diagrams, which depict the structure and relationships between classes in an object-oriented system. The rule set comprises a collection of predefined rules and patterns that are designed to capture common linguistic patterns found in software requirements. These rules can be created based on domain-specific knowledge, best practices, or linguistic analysis.

First, some variables and sets should be defined. Let t_i represents the currently parsed token within the text sequence of one of the software requirements sentence. The token can be a single word or a series of related connected words meaning a single concept. t_{i+1} represent the next token. Let the sets C, A, M, and R also represent the sets of words reserved for Classes, Attributes, Methods, and Relationships, respectively. the sets N, G and V represent the sets of POS for Verbs, Gerunds and Nouns, respectively.

Class-Identification Rules. Class-identification rules are a set of predefined rules that help in identifying classes or entities within software requirements. These rules are made to identify patterns, words, or linguistic clues in the requirements language that points to the existence of a class or other entity. Table 1 Shows the predefined rules to identify the classes within the requirements.

The first rule means if the token is repeated within the software requirements by less than 3% (statistically) compared to the length of the full text of the requirements, it cannot be considered a class, since the threshold at which the token can be viewed as a class has been determined based on its frequency. Rules 2, 3, and 4 are similar and state that a token cannot be considered a class

Table 2. Predefined rules to identify the attributes and methods.

Rule Ref	Rule Form
RA1	If t_i frequency > 5% (statistically) $\Rightarrow t_i \notin$ A
RA2	If $t_i \in$ N & t_i has "Snake/Camel" style $\Rightarrow t_i \in$ A
RA3	If $t_{i-1} \in N \cap C$ & $t_i \in \{has, have, consists\} \Rightarrow t_{i+1} \in$ A
RA4	If $t_{i-1} \in A$ & $t_i \in$ N $\Rightarrow t_i \in$ A
RA5	If $t_{i-1} \in N \cap C$ & $t_i \in$ N $\Rightarrow t_i \in$ A
RM1	If $t_i \in$ V & $t_{i+1} \in N \cap C \Rightarrow t_i \in$ M
RM2	If $t_{i-1} \in N \cap C$ & $t_i \in \{Modal\ auxiliary\ set\}$ & $t_{i+1} \in V \Rightarrow t_{i+1} \in$ M

Table 3. Predefined rules to identify the relations between classes.

Rule Ref	Rule Form
RR1	If $t_{i-1} \in N \cap C$ & $t_i \in V$ & $t_{i+1} \in N \cap C \Rightarrow t_i \in R_{association}$
RR2	If $t_{i-1} \in N \cap C$ & $t_i \in \{depends\ on, based\ on, require,$ $rely\ on, uses, follows\}$ & $t_{i+1} \in N \cap C \Rightarrow t_i \in R_{composition}$

if it is in one of the predefined sets. These sets are the set of keywords (such as button, window, database, system, etc.) , the set of pronouns (such as she, he, etc.) and the set of general concept (such as education, ministry, etc.). These sets are unlimited and grow in size with the software requirements data used to train the model. Rules 5, 6, and 7 state that if the parsed token is found in one of the sets, it cannot be considered again in another set. That means $M \cap A \cap C \cap R = \Phi$. Rules 8, 9, and 10 depend mainly on frequent patterns in the studied data set. These patterns are derived from the linguistic rules that depend mainly on the outputs of the POS process. Rules 11 and 12 check if the parsed token is Ground or Noun, and if all previous rules have not been satisfied, then the token is classified as Class.

Attribute-Method-Identification Rules. Attribute-Method-identification rules are a set of predefined rules that help in identifying attributes and methods within software requirements. Table 2 Shows the predefined rules found to identify the attributes and methods within the requirements.

Relations-Identification Rules. Relations-identification rules are a set of predefined rules that help in identifying Relations between classes within software requirements. Table 3 Shows the predefined rules found to identify the attributes and methods within the requirements.

Depending on Tables 1, 2, and 3, the elements of the diagram can be defined. The rules for defining relationships, attributes and methods within the diagram depend mainly on the rules for defining classes, so the classes are extracted first.

Algorithm 1. Class diagram elements extraction algorithm

procedure CREATECLASSDIAGRAM(Software Requirements SRs)
Require: Textual Software Requirements
  Initialize C, A, M, and R empty sets
  SRs = StopWordRemoval(SRs)
  SRs = RemoveSpecialCharsPuncs(SRs)
  SRs = Coffree(SRs)
  N, V, G = POS(SRs)
  for each SR in SRs **do**
    for each t in SR **do**
      if RC_1, RC_2,.. or RC_7 is True **then**
        t not belongs to C
      else if RC_8, RC_9,.. or RC_{12} is True **then**
        Add t to C
      end if
      if RA_1, is True **then**
        t not belongs to A
      else if RA_2, RA_4 or RA_5 is True **then**
        Add t to A
      else if RA_3 is True **then**
        Add t_{i+1} to A
      else if RM_1 is True **then**
        Add t to M
      else if RM_2 is True **then**
        Add t_{i+1} to M
      end if
      if RR_1 is True **then**
        Add t to R as Association relation
      else if RR_2 is True **then**
        Add t to R as Composition relation
      end if
      if t not classified **then**
        CP, AP, MP, RP = ProbabilitiesUsingKnowledgeEngine(SR, t)
        C, A, M, R = ClassifyUsingProbabilities(t, CP, AP, MP, RP)
      end if
    end for
  end for
  return C, A, M, R
end procedure

Then they are relied upon to extract the remaining elements. In the event that the rules fail to identify the category of the term, the probabilistic classification of pre-defined categories based on prior knowledge is relied upon. Algorithm 1 shows Class diagram elements extraction procedure.

4 Experiments and Results

The experimental results section presents the outcomes of our study, which aims to evaluate the effectiveness of a novel approach in a specific domain. This section provides an analysis of the evaluation metrics used, describes the dataset employed, and presents the results obtained. Additionally, we discuss the implications of these findings and their significance with the research objectives.

4.1 Software Requirements Datasets

This dataset originates from a new startup that develops specialized web business apps. Nine team members work together to create applications during biweekly Scrum sprints iteratively. In 2014, WebCompany provided 98 user stories to create a complete web application centered on interactive storytelling. Overall, 79 of these 98 user stories were syntactically sound, employable, and pertinent for creating conceptual models [23]. Archivists developed an open-source software called ArchivesSpace from the British Royal Archive. Online resources include this project's user stories [24]. UML class diagrams from the AtlanMod Zoo can be found in the research's data collection [25]. A collection of 305 top-notch UML class diagrams that model different domains may be found in the AtlanMod Zoo. The diagrams range in size from a handful to hundreds of classes. There are 8706 UML fragments in total in this collection. Each fragment is an English sentence that can be used to infer details about the diagram. In this dataset, only 62 diagrams have labels.

4.2 Evaluation Metrics, Results and Discussions

In this section, we present the outcomes of a practical assessment of our system's performance in converting textual software requirements into class diagrams. The purpose of this evaluation is to compare the effectiveness and efficiency of our approach with other methods currently used in this field. To gauge the performance of our approach, we employed a comprehensive set of evaluation metrics. These metrics were carefully chosen to encompass different aspects of the system's effectiveness and provide a comprehensive understanding of its performance. Recall measures the system's ability to construct all classes, as indicated by Eq. (6). Precision assesses the accuracy and relevance of the identified categories in the suggested conceptual model, as demonstrated in Eq. (7). The over-specification rate (OverSR) quantifies the number of classes that the computerization process adds to the developed conceptual model, as indicated by Eq. (8).

$$Recall = \frac{N_{correct}}{N_{correct} + N_{missing} + N_{implicit}} \tag{6}$$

$$Precision = \frac{N_{correct}}{N_{correct} + N_{incorrect}} \tag{7}$$

Table 4. Performance evaluation of our system compared to state-of-the-art

Method	Recall	Precision	OverSR
LIDA	64.16%	58.3%	52.55%
GOOAL	74.2%	67.87%	62.72%
CM-Builder	68.7%	61.4%	54.4%
NL-OOML	72.42%	65.7%	59.21%
READ	86.77%	78.30%	66.7%
DC-Builder	78.51%	70.05%	65.9%
Proposed System (SRTC)	**89.44%**	**83.72%**	**68.4%**

$$OverSR = \frac{extra(valid)}{N_{correct} + N_{incorrect} + N_{implicit}} \quad (8)$$

where $Ncorrect$ stands for the number of real classes that are acknowledged, and $N_{missing}$ for the number of classes that the proposed conceptual method ignores because they were chosen by a human expert. $Nincorrect$ is the number of correct levels that were determined to be inaccurate. The number of accurate additional classes regained is represented by $extra(valid)$. Implicit refers to the count of classes that are computed accurately and appropriately but are not explicitly mentioned in the requirements text. hence, $N_{implicit}$ refers to the number of calculated classes that are accurate and correct but are no longer reported in the requirements text.

We selected a diverse set of software requirements as problem statements from different domains to conduct the evaluation. These requirements encompassed various complexity levels, ensuring a comprehensive assessment of our system's capabilities. We then applied our system and the alternative approaches to convert these textual requirements into class diagrams.

Our Named Entity Recognition model can identify 2116 named entities in the training dataset.achieving an 87% accuracy during validation. Our system consistently outperformed the alternative approaches in terms of effectiveness. The class diagrams produced by our system demonstrated a higher degree of adherence to design principles and captured the system requirements more comprehensively. Table 4 shows the performance evaluation of our system compared to the state-of-the-art. Our suggested system achieved a Recall rate of 89.44%, a Precision rate of 83.72%, and an OverSR rate of 68.4%. The superior performance of our model can be attributed to its utilization of a probabilistic knowledge engine, which effectively identifies elements even when the software requirements are ambiguous.

When considering the potential application of our approach in converting software requirements to other UML diagrams, success can be achieved by encompassing various patterns that represent the desired elements of the target diagram. The more comprehensive the range of these pattern rules, the greater the transferability of the approach. Through theoretical analysis, we discovered that converting requirements into structural diagrams is comparatively easier than converting them into behavioral diagrams. This is primarily due to the presence of relationships and temporal sequences among events in behavioral diagrams. Incorporating the time or sequence dimension becomes necessary to reveal the order in which events occur. While these events are not strictly sequential, each behavior group has a starting point, making it challenging to convert the sequential text as software requirements into overlapped groups of sequential behaviors.

5 Conclusions

This research paper introduces a model-driven engineering approach that presents a novel method for automatically converting textual software requirements into UML class diagrams. Our methodology combines rule-based techniques and Naive Bayes for named entity recognition, allowing us to accurately identify the various components of a class diagram. Through comprehensive experiments using real-world software requirements and corresponding UML class diagrams, we evaluated the effectiveness of the proposed method. The results obtained from the experiments demonstrate the superiority of the proposed method, achieving an impressive recall metric of 89.44% and a precision metric of 83.72%. These metrics indicate the high accuracy and reliability of the method in capturing the essential elements of the textual requirements and translating them into UML class diagrams. However, it is important to acknowledge the limitations of our system. In certain rare cases, the system encountered challenges in accurately capturing complex relationships or interpreting ambiguous requirements, resulting in minor discrepancies in the generated class diagrams. These limitations will be addressed in future iterations to further enhance the system's performance.

References

1. Alberto, S.: Modeldriven engineering: a survey supported by the unified conceptual model. Comput. Lang. Syst. Struct. **43**, 139–155 (2015). https://doi.org/10.1016/j.cl.2015.06.001
2. Ciraci, S., Nanthaamornphong, A., Carver, J., Morris, K., Filippone, S.: Extracting UML class diagrams from objectoriented fortran: ForUML. Sci. Programm. **2015**, 421816 (2015). https://doi.org/10.1155/2015/421816
3. Salih, O., Sahraoui, A.: From requirements engineering to uml using natural language processing - survey study. Europ. J. Eng. Res. Sci. **2**, 44 (2017). https://doi.org/10.24018/ejers.2017.2.1.236
4. Alaswad, F., Poovammal, E.: Software quality prediction using machine learning. In: International Conference on Innovative Technology for Sustainable Development. vol. 62, 4714–4720 (2022). https://doi.org/10.1016/j.matpr.2022.03.165

5. Patwardhan, N., Marrone, S., Sansone, C.: Transformers in the real world: a survey on NLP applications. Information. **14**, 242 (2023). https://doi.org/10.3390/info14040242
6. N. Pattanotai: Automatic class description generation. In: 2016 2nd IEEE International Conference on Computer and Communications (ICCC), pp. 105–109 . https://doi.org/10.1109/CompComm.2016.7924674
7. More, P., Phalnikar, R.: Generating UML diagrams from natural language specifications. Int. J. Appl. Inform. Syst. **1**, 19–23 (2012). https://doi.org/10.5120/ijais12-450222
8. Krishnan, H.A., Samuel, P.: Relative extraction methodology for class diagram generation using dependency graph. In: International Conference on Communication Control and Computing Technologies (2010). https://doi.org/10.1109/icccct.2010.5670730
9. Sharma, V.S., Sarkar, S., Verma, K., Arun Panayappan, Kass, A.: Extracting high-level functional design from software requirements. In: Asia-Pacific Conference on Software Engineering (2009). https://doi.org/10.1109/apsec.2009.63
10. Popescu, D., Rugaber, S., Nenad Medvidovic, Berry, D.J.: Reducing ambiguities in requirements specifications via automatically created object-oriented models. In: Lecture Notes in Computer Science, pp. 103–124. Springer Science+Business Media (2008). https://doi.org/10.1007/978-3-540-89778-1_10
11. Clavel, M., Marina Cecilia González, Silva: The MOVA tool: a rewriting-based uml modeling, measuring, and validation tool. In: XII Jornadas de Ingeniería del Software y Bases de Datos, pp. 393–394 (2007)
12. Overmyer, S.P., Benoit, L., Owen, R.: Conceptual modeling through linguistic analysis using LIDA. In: Proceedings of the 23rd International Conference on Software Engineering. ICSE 2001, pp. 401–410 (2001). https://doi.org/10.1109/ICSE.2001.919113
13. Sarwar, I., Samad, A., Mumtaz, S.: Object oriented software modeling using nlp based knowledge extraction. Eur. J. Sci. Res. **35**, 22–33 (2009)
14. Ontology, D.: UDRA: Reflecting natural language text in to UML diagrams
15. Zhou, N.: Automatic acquisition of linguistic patterns for conceptual modeling. Presented at the (2005)
16. D. Deva Kumar, Sanyal, R.M.: Static uml model generator from analysis of requirements (SUGAR). In: Advanced Software Engineering and its Applications, ASEA (2008). https://doi.org/10.1109/asea.2008.25
17. Thakur, J.K., Gupta, A.: Automatic generation of analysis class diagrams from use case specifications. In: International Conference on Knowledge and Systems Engineering (KSE) (2017)
18. Lucassen, G., Robeer, M., Dalpiaz, F., van der Werf, J.M.E.M., Brinkkemper, S.: Extracting conceptual models from user stories with Visual Narrator. Require. Eng. **22**, 339–358 (2017). https://doi.org/10.1007/s00766-017-0270-1
19. Elallaoui, M., Nafil, K., Touahni, R.: Automatic transformation of user stories into UML use case diagrams using NLP Techniques. Proc. Comput. Sci. **130**, 42–49 (2018). https://doi.org/10.1016/j.procs.2018.04.010
20. Javed, M., Lin, Y.: Iterative process for generating ER diagram from unrestricted requirements. Presented at the (2018). https://doi.org/10.5220/0006778701920204
21. Hudson, R.: Coreferee · spaCy Universe. https://spacy.io/universe/project/coreferee
22. Marcus, M., Marcinkiewicz, M., Santorini, B.: Building a large annotated corpus of English: the penn treebank. Comput. Linguist. **19**, 313–330 (2002)

23. Lucassen, G., Dalpiaz, F., van der Werf, J.M.E.M., Brinkkemper, S.: Improving agile requirements: the quality user story framework and tool. Require. Eng. **21**, 383–403 (2016). https://doi.org/10.1007/s00766-016-0250-x
24. archivesspace: Confluence. https://archivesspace.atlassian.net/wiki/spaces. Accessed 15 July 2023
25. AtlanMod: Zoos - AtlanMod. https://web.imt-atlantique.fr/x-info/atlanmod/index.php?title=Zoos. Accessed 15 July 2023

CFALEA_LSTM: Adaptive Lotus Effect Algorithm Enabled Long Short-Term Memory for Rainfall Prediction Using Time Series Data

P. Vijaya[1(✉)], Satish Chander[2], Roshan Fernandes[3], Anisha P. Rodrigues[3], and B. Supriya[4]

[1] Modern College of Business and Science, Bowshar, Oman
pvvijaya@gmail.com
[2] Biral Institute of Technology, Mesra, Ranchi, India
[3] NITTE (Deemed to be University), NMAMIT, Karkala, Karnataka, India
[4] Shayadri College of Engineering and Management, Mangaluru, Karnataka, India

Abstract. Rainfall prediction (RP) utilizing machine learning (ML) is an imperative tool for agricultural and economic practices. RP is also considered as a challenging chore in weather forecasting. An accurate RP can assist farmers to make knowledgeable decisions regarding crop planting, harvesting and fertilization. It also helps in the prevention of floods, disaster management and water resource management. Therefore, the prediction of precise rainfall is significant for diverse applications of a country. Here, Adaptive Lotus Effect algorithm-based Long Short-Term Memory (ALEA_LSTM) is presented for RP utilizing time series data (TSD). Initially, TSD is obtained from a specific database and thereafter technical indicators for RP are extracted from the considered input TSD. Afterwards, feature selection (FS) is accomplished to choose suitable features employing mutual information (MI). Then, data augmentation (DA) is conducted by the oversampling method to augment the dimensionality of data. At last, RP is executed by Long Short-Term Memory (LSTM) that is trained by the Adaptive Lotus Effect Algorithm (ALEA). Moreover, ALEA is devised by integrating the adaptive concept with the Lotus Effect Algorithm (LEA). In addition, ALEA_LSTM achieved minimal values of Mean absolute percentage error (MAPE) of about 0.359, Mean squared error (MSE) of about 0.133 and Root mean squared error (RMSE) of about 0.364.

Keywords: Rainfall prediction (RP) · Lotus Effect Algorithm (LEA) · oversampling method · mutual information (MI) · Long Short-Term Memory (LSTM)

1 Introduction

Data gathered with the time stamps in a certain pattern is termed as TSD [1–4]. These kinds of time-based data are collected with definite time periods, like hourly, weekly and daily basis. TSD can be employed efficiently for making predictions in several domains

T. Shreekumar et al. (Eds.): ISCComm 2023, CCIS 2231, pp. 102–118, 2025.
https://doi.org/10.1007/978-3-031-75605-4_7

and fields that include stock market trends, climate changes, foreign currency rates and estimation of energy consumption [4]. The time series assessment is recognized to be an excellent alternate solution while long-term RP is required. Rainfall is to imperative process in the hydrological cycle and it is one of the most studied elements of hydrological as well as climate science, as it affects the community in a direct or indirect manner [5, 6]. RP is mostly relevant to enhancing the resilience and comfort of societies as it directly influences agriculture, drinking water provisions, electricity production, sanitation and the incidence of landslides and flooding that obstruct the sustainable growth of environmental systems [7, 8]. Nevertheless, reliable and effective RP is indefinable as it is the result of complicated non-linear processing within the atmosphere, which is mostly vulnerable to beginning states [8]. Nowadays, rainfall is considered as one of the prime causes of several critical incidents across the globe [9].

RP must be performed in advance for installing rainwater harvesters in a few regions, wherein there is a scarcity of water [9]. Seasonal rainfall and its amount make it a challenging chore for RP. Normally, dynamical and empirical techniques are two techniques for RP. An empirical method utilized an association of earlier historical information for forecasting. The common empirical schemes for predicting climate are Regression as well as artificial neural network (ANN). The dynamical technique employed manual systems and statistical schemes for predicting seasonal rainfalls. Recently, researchers have successfully developed classification and prediction models in various areas utilizing ML namely k-nearest neighbours (KNN) and support vector machine (SVM) [10]. Hence, several ML approaches have been utilized for hydrometeorological time series forecasting namely ANN, Prophet, random forest (RF) and support vector regression (SVR) methods [6]. Deep learning (DL) approaches like LSTM, Convolutional Neural Network (CNN), Recurrent Neural Network (RNN) and Generative Adversarial Network (GAN) play a significant part in assessing and processing larger data volumes to offer important information [10]. DL schemes namely Restricted Boltzmann Machine (RBM), Autoencoder, Conditional RBM, Hidden Markov system, Convolution as well as pooling and RNN have already revealed promising outcomes in TSD predictions [9]. Currently, time series decomposition approaches are utilized for extracting features and prediction processes concerning hydrometeorological time series [6].

The crucial intention is to present ALEA_LSTM for RP utilizing TSD. Here, input TSD is taken from certain databases and technical indicators are extracted from input TSD. The extracted technical indicators are the Relative Strength Index (RSI), WILLIAMS % R (WILLR), Time Series Forecast (TSF), Trade Volume Index (TVI) and Triple Exponential Average (TRIX). Then, FS is done utilizing MI. Afterwards FS process, DA is executed to enhance the dimensionality of selected data employing the oversampling technique. Finally, RP is accomplished by LSTM and its tuning process is conducted employing ALEA. However, ALEA is an incorporation of adaptive concept with LEA.

The important contribution is revealed below.

- **Proposed ALEA_LSTM for RP using TSD,** Rainfall has highly significant meaning for the life, thus accurate prediction of rainfall is required. Here, LSTM is employed to predict rainfall, which is tuned by ALEA. Moreover, ALEA is joining adaptive concept and LEA.

The organizing of the below structure is, Reviewed approaches and its shortcomings are interpreted in Sect. 2, ALEA_LSTM methodology is specified in Sect. 3, ALEA_LSTM results are elucidated in Sect. 4 whereas Sect. 5 describes the conclusion of ALEA_LSTM.

2 Motivation

Precise prediction of rainfall is significant in everyday life, prevention of natural disasters, water resources management planning and risk assessment. However, RP is a complicated chore owing to dynamic complexities and non-stationary conditions of assessed hydrological data. Thus, a new approach is introduced to predict rainfall using TSD and reviewed methods and its limitations are expounded in this segment.

2.1 Literature Survey

Venkatesh, R., *et al.* [10] developed a GAN for RP that performed superior predictions on both annual and monthly averages with 99% accuracy. However, it failed to decrease computational time for tuning phases and this method did not meet the requirements of real-world processing. Rahman, A.U., *et al.* [4] designed fused ML for RP in smart cities and this model was capable of extracting real-world feature-enabled weather data from highly sensible and modern weather sensors for real-time RP. The accuracy obtained by this scheme was 0.94. Still, it did not fuse ensemble ML approaches on many different databases. Li, D., *et al.* [6] presented a discrete wavelet transform (DWT)-SVR-Prophet Model for RP using TSD. It was readily adaptable for other TSD predictions with RMSE of 6.1704, but it failed to enhance the stability and accuracy of RP. Li, H., *et al.* [8] introduced an optimally pruned extreme learning machine (OP-ELM) for the prediction of rainfall. This approach showed promising outcomes with MAE 13.76 mm, even though it did not reveal the time complexity for RP. Shahab S. Band, *et al.* [11] applied the differentiation operations for the effective prediction of rainfall. It obtained a minimum MAE of 0.510. The complexity of the model was high. Albenis Pérez-Alarcón, *et al.* [18] implemented a hybrid model by combining Artificial Neural Network and Autoregressive Integrated Moving Average (ANN + ARIMA). This approach had high reliability. It was complex for a large size of data.

2.2 Challenges

The experienced limitations by reviewed conventional schemes for RP are described below.

- Fused ML developed in [4] failed to integrate a monitoring scheme with an information protective system to validate the functioning of weather sensors for ensuring data integrity until it is employed for prediction. In the proposed method, the oversampling technique is used for balancing the database, in which the newer samples include significant information about minority classes.

- In [6], the method did not optimize the selection of DWT wavelet fundamental operations and decomposition levels for rainfall time series with diverse features for adequate extraction. In the implemented scheme, the MI is used for FS, which selects highly consistent, distinctive, and appropriate features.
- Precise RP can be employed to predict threats and lessen the sufferers. However, there are several challenges to constructing a prediction system owing to complications and diverse factors that affect rainfall. In the proposed model, the LEA is combined with the adaptive concept to predict rainfall effectively.

3 Proposed ALEA_LSTM for RP Using TSD

The prediction of rainfall is more challenging owing to the unsure environment as well as dynamic climate alteration. Here, ALEA_LSTM is designed for RP using TSD. Initially, TSD is acquired from the database and technical indicators like RSI, WILLR, TSF, TVI and TRIX are extracted. Then, MI is employed to select features and the oversampling method is utilized for DA. Lastly, RP is accomplished by LSTM which is tuned by ALEA. In addition, ALEA is the combination of adaptive concept with LEA. The diagrammatical presentation of ALEA_LSTM for RP using TSD is exposed in Fig. 1.

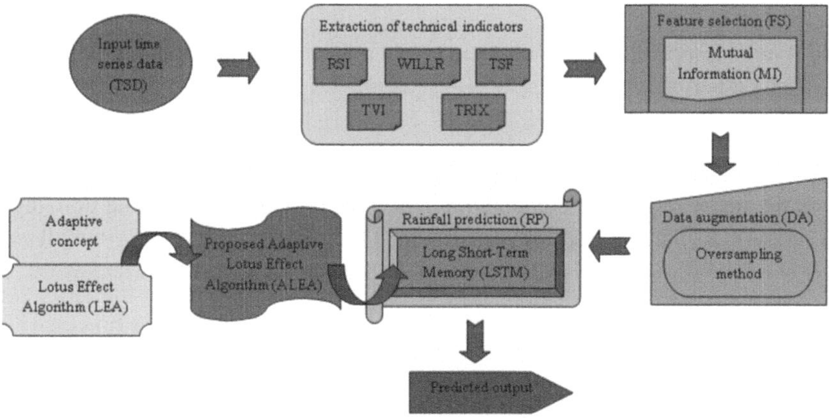

Fig. 1. Diagrammatic representation of ALEA_LSTM for RP using TSD

3.1 Acquisition of TSD

For conducting the RP process, TSD is acquired from the database [13] that can be modelled as follows.

$$F = \{F_1, F_2, ...F_r...F_p\} \tag{1}$$

where, total TSD in the database F is implied as F_p and F_r denotes r^{th} input TSD.

3.2 Extraction of Technical Indicators

The technical indicators refer to mathematical model derived from past data utilized to predict prospective rainfall. Here, technical indicators [14] such as RSI, WILLR, TSF, TVI and TRIX are extracted.

3.2.1 RSI

RSI defines as the ratio of a smoothed average η-period rain to the absolute value of the smoothed average of η-period non-rain and it is illustrated by,

$$I_1 = 100 - (100/(1 + B)) \tag{2}$$

Here, B indicates an average of up closes over rainfall at η-period to the average of down closes over rainfall at η-period.

3.2.2 WILLR

WILLR normalizes rainfall as percentage amongst 0 and 100, which is manifested by,

$$I_2 = -100 \times \left(\frac{\mathrm{H} - \mathrm{E}}{\mathrm{H} - \mathrm{T}}\right) \tag{3}$$

Here, E represents current rainfall, H and T mentions the highest high and lowest low rain in past η-periods.

3.2.3 TSF

TSF is the linear regression computation that plots present rainfall value utilizing the least square fit technique. It specifies to excellent fit line associated with η-rainfall points $(v1, w1)$, $(v2, w2)$....(vi, wi) is given as,

$$I_3 = w = \varphi v + \ell \tag{4}$$

3.2.4 TVI

TVI leverages rainfall data within a day and TVI can be calculated utilizing the below expression.

$$I_4 = \begin{cases} X1 = X1_{m-1} + \left(\left(\frac{2}{(m+1)}\right) * (\mathrm{E} - X1_{m-1})\right) \\ X2 = X2_{m-1} + \left(\left(\frac{2}{(m+1)}\right) * (X1_{m} - X2_{m-1})\right) \\ X3 = X3_{m-1} + \left(\left(\frac{2}{(m+1)}\right) * (X2_{m} - X3_{m-1})\right) \end{cases} \tag{5}$$

where, $X1_{m-1}$, $X2_{m-1}$ and $X3_{m-1}$ denote the exponential moving average (EMA) value of η rainfall periods back and E implies current rainfall.

3.2.5 TRIX

TRIX reveals the percentage rate of rainfall change among two triple smoothed EMAs that is evaluated as,

$$I_5 = \left[(X3_{mt} - X3_{mt-1})/(X3_{mt-1}) \right] \tag{6}$$

The extracted technical indicators are signified as I_r with dimension $a \times b$, such that,

$$I_r = \{I_1, I_2, I_3, I_4, I_5\} \tag{7}$$

3.3 FS Utilizing MI

FS process is carried out to select highly consistent, distinctive and appropriate features. Here, I_r with dimension $a \times b$ is given as input and MI is employed for FS. MI [15] is a common measure of dependence among two arbitrary variables and it can be estimated by,

$$M_r = \mathrm{I}(K; Z) = \sum_{k \in K} \sum_{z \in Z} \mathrm{P}[K = k, Z = z] \,.\, \log \left(\frac{\mathrm{P}[K = k,\ Z = z]}{\mathrm{P}[K = k] \,.\, \mathrm{P}[Z = z]} \right) \tag{8}$$

Here, K and Z defines candidate feature and target, $\mathrm{P}[K = k, Z = z]$ specifies conditional probability, $\mathrm{P}[K = k,\ Z = z]$ denotes joint distribution, $\mathrm{P}[K = k] \,.\, \mathrm{P}[Z = z]$ implies product distribution and $\mathrm{P}[Z = z]$ refers to marginal probability. Afterwards, calculation of MI for individual features, choose top ξ features having higher values. M_r with dimension $a \times x$ symbolizes selected features, where $b > x$.

3.4 DA Utilizing Oversampling Method

DA specifies to augmentation of the group of investigational data to complete data. Here, M_r with dimension $a \times x$ as input to execute DA employing the oversampling method. In the oversampling technique [16], newer samples are included to the minority class for balancing the database. These approaches are classified as Synthetic Minority Oversampling Technique (SMOTE) and random oversampling. In SMOTE, artificial samples are produced for minority-class samples. The newer samples include significant information to minority classes and avoid instances from misclassification. The augmented data utilizing oversampling method is symbolized as V_r with dimension $j \times x$, where $a < j$.

3.5 RP Utilizing ALEA_LSTM

Nowadays, rainfall is extensively considered as a critical influence of several incidents across the globe. In this view, RP is essential as it is a variable that is closely associated with natural disasters namely floods, landslides and so on. Here, rainfall is predicted utilizing LSTM, and its training process is accomplished by ALEA, which is formed by assimilating adaptive concept and LEA. The input considered for performing RP is V_r with dimension $j \times x$.

3.5.1 LSTM Structure

LSTM [17] is a significant and recent approach in DL that comprises of three basic segments namely input, output and forget gates. These gates are utilized with info flow for identifying prediction outcomes. The sigmoid activation operation refers to LSTM gates. An input gate represents newer information that is going to be stored in the cell state that is illustrated in Eq. (8). The hidden gate assists in keeping suitable information, which is computed by Eq. (9). An output gate is utilized for providing activation to last output of LSTM block at the timestamp s is evaluated by Eq. (10).

$$m_s = \sigma\left(N_{\chi m}\chi_s + N_{\hbar m}\hbar_{s-1} + N_{\zeta m}\zeta_{s-1} + \beta_m\right) \tag{9}$$

$$f_s = \sigma\left(N_{\chi f}\chi_s + N_{\hbar f}\hbar_{s-1} + N_{\zeta f}\zeta_{s-1} + \beta_f\right) \tag{10}$$

$$t_s = \sigma\left(N_{\chi t}\chi_s + N_{\hbar t}\hbar_{s-1} + N_{\zeta t}\zeta_{s-1} + \beta_t\right) \tag{11}$$

To acquire a memory vector for the present timestamp ($\zeta\ \{s\}$), the temperature is computed. The rainfall predicted output using LSTM is implied by S_r and LSTM architecture is shown in Fig. 2.

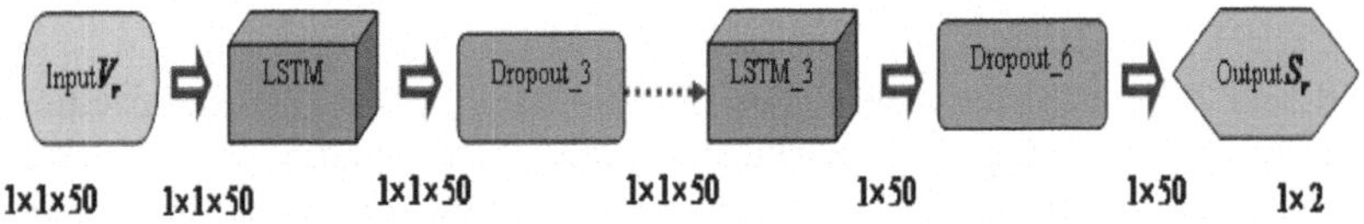

Fig. 2. LSTM architecture

3.5.2 Training of LSTM Utilizing ALEA

The LSTM remember the data for a long duration. So, this model efficiently handles the long-term dependencies. Also, it is substantially less sensitive to the vanishing gradient issue. LEA [18] is enthused by the Lotus flower, where propagation of this plant together with the self-cleaning attribute of its leaves is a major stimulation for this approach. LEA is the best approach for tracking or monitoring applications that need very precise data. Here, LEA is combined with the adaptive concept to design ALEA, which is more effectual in training LSTM to predict rainfall.

Lotus Effect Encoding

The superior solution is attained in $\Im$ search space, where τ implies LSTM learning parameter, such a way that $\Im = [1 \times \tau]$.

Fitness Measure

It is specified by computing variation amid LSTM and target outputs that can be illustrated as,

$$\phi = \frac{1}{p}\sum_{r=1}^{p}[\varUpsilon_r - S_r]^2 \tag{12}$$

Here, p reveals total samples whereas $/T_r$ and S_r mentions target output and LSTM output.

ALEA obtained excellent solutions by performing the following steps.

Step 1, Initializing of solution

At first, the initializing of the solution is generated in $\Im$ search space, which is manifested by,

$$R = \{R_1, R_2, ..., R_l, ..., R_o\} \tag{13}$$

Here,R_l signifies l^{th} solution,o indicates overall variables and R denotes population.

Step 2, Computing objective function

The estimation of the objective function is conducted by determining the difference amid the target and LSTM output utilizing Eq. (12).

Step 3, Exploration stage

Global pollination executed by the dragonflies is considered as an exploration stage. Each should be attracted towards food and diverted from enemies. With concern to these attributes, five factors are involved in updating individual locations in the swarm. Separation is computed by,

$$P_l^q = \sum_{n=1}^{Y} R_l^q - R_n^q \tag{14}$$

Here, R_l specifies the individual's present location having index l in iteration q, R_n reveals the location of nth individual in neighbour at iteration q whereas Y signifies total individual in the neighborhood. Alignment can be calculated as follows.

$$G_l^q = \frac{\sum_{n=1}^{Y} R_n^q}{Y} \tag{15}$$

Here, R_n illustrates the velocity of nth individual in the neighbourhood at iteration q. Cohesion is given as,

$$H_l^q = \frac{\sum_{n=1}^{Y} R_n^q}{Y} - R_l^q \tag{16}$$

where, Y signifies total neighbours. The attraction towards food sources can be evaluated by,

$$U_l^q = R_+^q - R_l^q \tag{17}$$

Here, R_+^q indicates the position of the food source. The distraction of the enemy is modelled as shown below.

$$D_l^q = R_-^q + R_l^q \tag{18}$$

where, R_{-}^{q} implies enemy location. A step or the velocity vector reveals the dragonfly's direction of movement, which is manifested by,

$$\Delta R_{l}^{q+1} = \left(\rho P_{l}^{q} + gG_{l}^{q} + hH_{l}^{q} + uU_{l}^{q} + dD_{l}^{q}\right) + \omega \Delta R_{l}^{q} \tag{19}$$

Here, ρ, g, h denote coefficients of separation, alignment and cohesion coefficients whereas u and d symbolizes food and enemy factors, ω specifies inertia weight. After computing the step vector, the location vector is illustrated by,

$$R_{l}^{q+1} = R_{l}^{q} + \omega \Delta R_{l}^{q+1} \tag{20}$$

where, q signifies iteration counter. The dragonfly's location is updated as mentioned below.

$$R_{l}^{q+1} = R_{l}^{q} + L(\mu) \times R_{l}^{q} \tag{21}$$

An updated formulation of ALEA is mentioned as,

$$R_{l}^{q+1} = R_{l}^{q} + \left(0.01 \times \frac{\gamma_1 - \sigma}{|\gamma_2|^{\frac{1}{\varsigma}}}\right) \times R_{l}^{q} \tag{22}$$

Here, γ_1 is made adaptive and can be specified as,

$$\gamma_1 = c - \left(0.01 \times \left(1 - \frac{q}{A}\right)\right) \tag{23}$$

where, c denotes the speed increment coefficient and A reveals the ultimate iteration number.

Step 4, Exploitation stage

Local pollination refers to the extraction stage of the approach and it is modelled by,

$$R_{l}^{q+1} = R_{l}^{q} + W\left(R_{l}^{q} - \vartheta^{*}\right) \tag{24}$$

Here, R_{l}^{q+1} implies the position of pollen in $(q+1)^{th}$ iteration, ϑ^{*} mentions the best-identified pollen position amongst every evolution iteration and W specifies the growth region that is estimated by,

$$W = 2e^{-\left(\frac{\Delta q}{A}\right)^2} \tag{25}$$

Step 5, Exploitation stage reinforcement

The local search is concerned as water movement across the leaves of lotus flowers utilizing water drops. The capacity of every pit can be calculated by,

$$\alpha_l^q = \frac{\left(\left|\phi_l^q - \phi_{\max}\right|\right) \times \aleph}{\left(\left|\phi_{\min} - \phi_{\max}\right|\right)} \tag{26}$$

Here, α_l^q denotes the capacity of pit l in iteration q, ϕ_l^q indicates the size of lth pit in iteration q, $\phi_{\min}$ and $\phi_{\max}$ signifies maximal and minimal fitness amid pits whereas $\aleph$ specifies a constant number representing the maximal capacity of pit. Equation (27) captures the probability of choosing a pit amongst present pits.

$$\mathcal{S}_l^q = \frac{\alpha_l^q}{\sum_{n=0}^{k-} \alpha_l^q} \tag{27}$$

where, k- reveals the count of pits whose capacity is higher than overflowed pit. The velocity and location of the drop's movement at local search are evaluated using Eq. (28). Equation (29) is utilized for calculating the velocity and location of the drop's movement overflowing from the pit on a surface.

$$\left.\begin{aligned} \mathcal{N}_l^{q+1} &= c \times \mathcal{N}_l^q \\ R_l^{q+1} &= R_l^q + \mathcal{N}_l^{q+1} \end{aligned}\right\} \tag{28}$$

$$\left.\begin{aligned} \mathcal{N}_l^{q+1} &= \mathcal{N}_l^q + \Re\left(R_E^q - R_l^q\right) \\ R_l^{q+1} &= R_l^q + \mathcal{N}_l^{q+1} \end{aligned}\right\} \tag{29}$$

In above equations, R_E^q represents the present location of the deepest pit in iteration q and $\mathcal{N}_l^q$ indicates the present velocity of drop l in iteration q.

Step 6, Termination

ALEA follows the aforementioned steps to acquire superior outcomes. Algorithm 1 exhibits the pseudo-code of ALEA.

Algorithm 1. Pseudo code of ALEA

SL. No	Pseudo code of ALEA
1	Input, $R_i^q, c, g, h, u, \rho, \omega, d$
2	Output, R_i^{q+1}
3	Begin
4	if $q \leq Q$, compute fitness measure utilizing Eq. (12)
5	//Global pollination, Update g, h, u, ρ, ω, d and P, G, H, U, D values utilizing Eq. (14) to Eq. (18)
6	If intentional dragonfly is presented in neighborhood radius
7	Update velocity and location vectors utilizing Eq. (19) and Eq. (20)
8	else
9	Update location vector utilizing Eq. (22)
10	else
11	//Local pollination, Update growth region of flowers and location of new flower using Eq. (24)
12	for $l : 1$ to y
13	Compute local search using Eq. (28) and individual pit capacity utilizing Eq. (26)
14	then, choose pit utilizing Eq. (27) and calculate water movement employing Eq. (29)
15	end
16	Return ϑ^*

4 Results and Discussion

The acquired outcomes by ALEA_LSTM for RP along with the dataset are interpreted in this part.

4.1 Experimentation Setup

ALEA_LSTM designed for RP is experimentally implemented in the PYTHON tool. The parameter details are provided in Table 1.

4.2 Description of Dataset

Rain in Australia dataset [13] comprises about ten years of everyday weather investigations from various areas all over Australia. The features considered for making this dataset are date, location, MinTemp, MaxTemp, Rainfall, Evaporation, Sunshine, WindGustDir, WindGustSpeed, and WindDir9am.

Table 1. Experimental parameters

Parameter	Value
GAN	
Epoch	20
Batch size	32
Fused_ML	
Epoch	25
Batch size	64
DWT_SVR	
Epoch	25
Batch size	32
Learning rate	0.0001
OP_ELM	
Epoch	25
Batch size	64
Devised scheme	
Epoch	25
Batch size	50
Learning rate	0.0001

4.3 Performance Measures

MAPE, MSE and RMSE are the evaluation measures considered to estimate ALEA_LSTM for RP.

4.3.1 MAPE

MAPE is computed by taking variation amid actual and predicted values of ALEA_LSTM and dividing it by actual value, which is formulated as,

$$\partial_1 = \frac{1}{p} \sum_{r=1}^{p} \left| \frac{C_r - J_r}{C_r} \right| \tag{30}$$

Here, J denotes the predicted value and C indicates the actual value.

4.3.2 MSE

MSE evaluates an error by squaring a variation amid predicted and actual values of ALEA_LSTM that is given as,

$$\partial_2 = \frac{1}{p} \sum_{r=1}^{p} (J_r - C_r) \tag{31}$$

4.3.3 RMSE

RMSE quantifies the variations among predicted and actual values of ALEA_LSTM by squaring errors as shown below.

$$\partial_3 = \sqrt{\frac{1}{p}\sum_{r=1}^{p}(J_r - C_r)} \quad (32)$$

4.4 Comparative Techniques

GAN [10], Fused ML [4], DWT-SVR-Prophet Model [6] and OP-ELM [8] are concerned with existing methods for evaluation.

4.5 Comparative Analysis

The estimation of ALEA_LSTM is accomplished by varying delay and training data.

4.5.1 Estimation Based on Delay

Figure 3 shows the evaluation of ALEA_LSTM in relation to metrics by altering delay. For delay = 5000, attained values by comparative and designed schemes are elaborated. Figure 3 a) describes the analysis in terms of MAPE. ALEA_LSTM achieved MAPE of 0.359 whereas GAN, Fused ML, DWT-SVR-Prophet Model and OP-ELM acquired 0.411, 0.399, 0.388 and 0.384. Estimation of ALEA_LSTM with respective to MSE is elucidated in Fig. 3 b). MSE acquired by ALEA_LSTM is 0.133 whereas MSE attained by GAN is 0.284, Fused ML is 0.235, DWT-SVR-Prophet Model is 0.207 and OP-ELM is 0.152. Figure 3 c) demonstrates an assessment of ALEA_LSTM considering RMSE. GAN, Fused ML, DWT-SVR-Prophet Model and OP-ELM attained RMSE of 0.533, 0.485, 0.455 and 0.390 while ALEA_LSTM obtained RMSE of 0.364.

4.5.2 Estimation Based on Training Data

Analysis of ALEA_LSTM in respective of metrics by altering training data is described in Fig. 4. The obtained values by designed and comparative techniques while training data is 90% are discussed. Evaluation of ALEA_LSTM concerning MAPE is indicated in Fig. 4 a). MAPE acquired by ALEA_LSTM is 0.376 while MAPE achieved by GAN is 0.455, Fused ML is 0.413, DWT-SVR-Prophet Model is 0.405 and OP-ELM is 0.398. Figure 4 b) displays an assessment of ALEA_LSTM by means of MSE. GAN, Fused ML, DWT-SVR-Prophet Model and OP-ELM attained MSE of 0.287, 0.278, 0.218 and 0.193 whereas ALEA_LSTM obtained MSE of 0.143. The estimation of ALEA_LSTM according to RMSE is explained in Fig. 4 c). ALEA_LSTM attained RMSE of 0.378 whereas GAN, Fused ML, DWT-SVR-Prophet Model and OP-ELM acquired MSE 0.536, 0.527, 0.467 and 0.439.

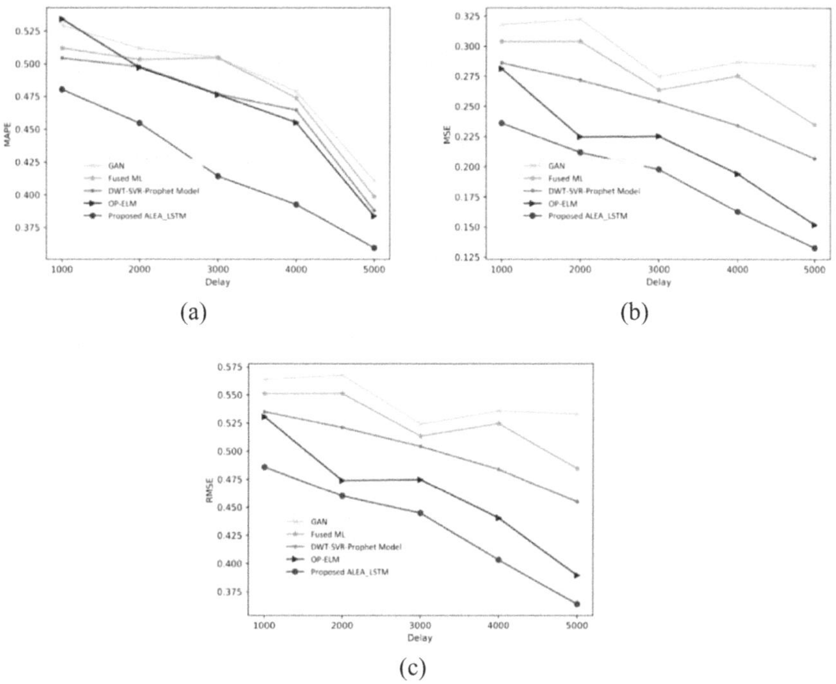

Fig. 3. Comparative estimation regarding delay, a) MAPE, b) MSE, c) RMSE

4.6 Comparative Discussion

ALEA_LSTM showed excellent outcomes when compared with GAN, Fused ML, DWT-SVR-Prophet Model and OP-ELM in terms of MAPE, MSE and RMSE. Table 2 indicates a discussion of acquired values by existing and designed approaches. It is recognized that ALEA_LSTM achieved minimal MAPE, MSE and RMSE of 0.359, 0.133 and 0.364 for delay = 5000.

The reasons behind the superior performance in comparison to existing models are listed below:

In the proposed method, the newer samples include significant information to minority classes and avoid instances of misclassification. Also, the prediction is done by LSTM, which efficiently handles the long-term dependencies. Moreover, LEA is combined with the adaptive concept for effectual RP by training the LSTM. Therefore, ALEA_LSTM is revealed as effectual technique for predicting rainfall.

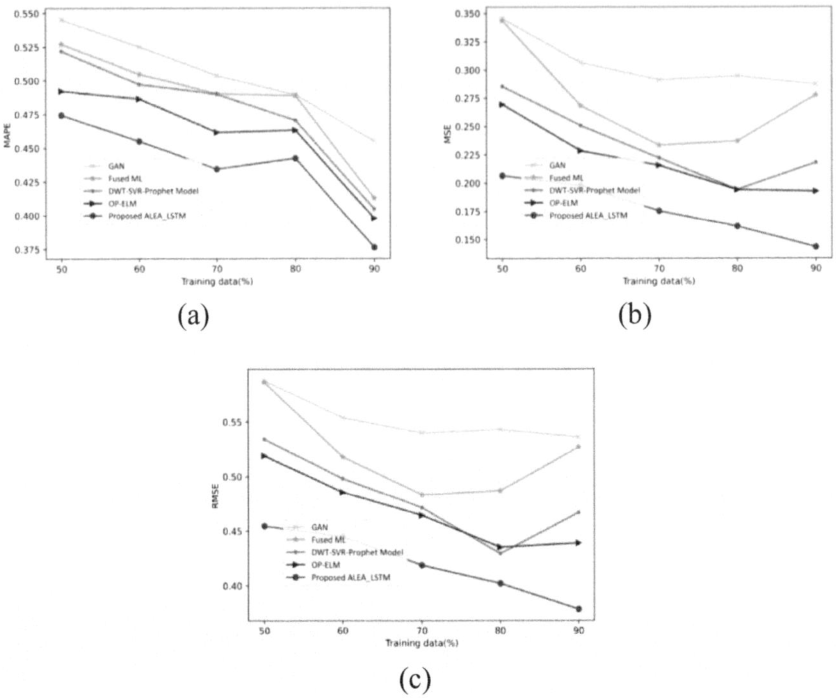

Fig. 4. Comparative evaluation regarding training data, a) MAPE, b) MSE, c) RMSE

Table 2. Comparative discussion of ALEA_LSTM

Analysis based on	Metrics/ Methods	GAN	Fused ML	DWT-SVR-Prophet Model	OP-ELM	Proposed ALEA_LSTM
Delay	MAPE	0.411	0.399	0.388	0.384	0.359
	MSE	0.284	0.235	0.207	0.152	0.133
	RMSE	0.533	0.485	0.455	0.390	0.364
Training data	MAPE	0.455	0.413	0.405	0.398	0.376
	MSE	0.287	0.278	0.218	0.193	0.143
	RMSE	0.536	0.527	0.467	0.439	0.378

5 Conclusion

Rainfall impacts the local water quality and quantity. The precise and timely RP is more desirable in hydrogeologic hazard mitigation and water management which are significant for environmentally sustainable progression. Earlier RP processes are complicated and computationally expensive due to its inherent higher variability and uncertainty.

In this research, ALEA_LSTM is presented for RP utilizing TSD. The processes conducted to execute RP are technical indicators extraction, FS, DA and RP. Firstly, input TSD is acquired from a specific database. After that, technical indicators are extracted that include RSI, WILLR, TSF, TVI and TRIX. Then, appropriate features are selected in the FS stage utilizing MI. The top selected features are given to the DA stage. In the DA stage, the oversampling method is employed to enhance dimensionality. Lastly, augmented data is fed to the RP stage which is accomplished by LSTM. The tuning of LSTM is done by ALEA, which is devised by merging the adaptive concept with LEA. Furthermore, ALEA_LSTM obtained minimum MAPE, MSE and RMSE about 0.359, 0.133 and 0.364 while considering delay as 5000. In future, some additional features will be considered for extraction to predict rainfall. Also, a hybrid network structure will be designed for differentiating light-rain and heavy-rain scenarios.

References

1. Aftab, S., Ahmad, M., Hameed, N., Bashir, M.S., Ali, I., Nawaz, Z.: Rainfall prediction using data mining techniques, a systematic literature review. International journal of advanced computer science and applications **9**(5) (2018)
2. Yue, T., Zhang, S., Zhang, J., Zhang, B., Li, R.: Variation of representative rainfall time series length for rainwater harvesting modelling in different climatic zones. J. Environ. Manage. **269**, 110731 (2020)
3. Mishra, N., Soni, H.K., Sharma, S., Upadhyay, A.K., A comprehensive survey of data mining techniques on time series data for rainfall prediction. J. Ict Research & Applications **11**(2) (2017)
4. Rahman, A.U., et al.: Rainfall prediction system using machine learning fusion for smart cities. Sensors **9**, 3504 (2022)
5. Poornima, S., Pushpalatha, M., Jana, R.B., Patti, L.A: Rainfall Forecast and Drought Analysis for Recent and Forthcoming Years in India **15**(3), 592 (2023)
6. Li, D., et al.: Prediction of rainfall time series using the hybrid DWT-SVR-Prophet model. Water **15**(10), 1935 (2023)
7. Tian, X., Schleiss, M., Bouwens, C., van de Giesen, N.: Critical rainfall thresholds for urban pluvial flooding inferred from citizen observations. Sci. Total. Environ. **689**, 258–268 (2019)
8. Li, H., He, Y., Yang, H., Wei, Y., Li, S., Xu, J.: Rainfall prediction using optimally pruned extreme learning machines. Nat. Hazards **108**, 799–817 (2021)
9. Soundararajan, S.: A novel deep learning framework for rainfall prediction in weather forecasting. Turkish J. Computer and Mathematics Education (TURCOMAT) **12**(11), 2685–2692 (2021)
10. Venkatesh, R., Balasubramanian, C., Kaliappan, M.: Rainfall prediction using generative adversarial networks with convolution neural network. Soft. Comput. **25**, 4725–4738 (2021)
11. Rain in Australia dataset is taken. https.//www.kaggle.com/jsphyg/weather-dataset-rattle-package (2023)
12. Technical indicators is taken. https.//www.tradingtechnologies.com/xtrader-help/x-study/technical-indicator-definitions/list-of-technical-indicators/ (2023)
13. Batina, L., Gierlichs, B., Prouff, E., Rivain, M., Standaert, F.X., Veyrat-Charvillon, N.: Mutual information analysis, a comprehensive study. J. Cryptol. **24**(2), 269–291 (2011)
14. Shelke, M.S., Deshmukh, P.R., Shandilya, V.K.: A review on imbalanced data handling using undersampling and oversampling technique. International Journal of Recent Trends in Engineering & Research **3**(4), 444–449 (2017)

15. Salehin, I., Talha, I.M., Hasan, M.M., Dip, S.T., Saifuzzaman, M., Moon, N.N.: An artificial intelligence-based rainfall prediction using LSTM and neural network. In: proceedings of 2020 IEEE International Women in Engineering (WIE) Conference on Electrical and Computer Engineering (WIECON-ECE) IEEE, pp. 5–8 (2020)
16. Dalirinia, E., Jalali, M., Yaghoobi, M., Tabatabaee, H.: Lotus effect optimization algorithm (LEA), a lotus nature-inspired algorithm for engineering design optimization. The Journal of Supercomputing, pp. 1–39 (2023)
17. Band, S.S., et al.: Evaluation of time series models in simulating different monthly scales of drought index for improving their forecast accuracy. Frontiers in Earth Science **10** (2022)
18. Pérez-Alarcón, A., Garcia-Cortes, D., Fernández-Alvarez, J.C., Martínez-González, Y.: Improving monthly rainfall forecast in a watershed by combining neural networks and autoregressive models. Environmental Processes **9** (2022)

A Deep Learning Survey on Diseases Prediction and Detection in Health Care

R. Priyanka and M. R. Pooja(✉)

Vidyavardhaka College of Engineering, Mysuru, Karnataka 570002, India
{4vv22scs,pooja.mr}@vvce.ac.in

Abstract. In the field of medical diagnosis and detection, computer-aided diagnosis and detection is a vibrant research topic that is gaining popularity. Computer scientists have proposed and created an extensive variety of AI procedures for the identification and diagnosis of illnesses like cancer, lung conditions, rheumatoid arthritis, diabetic retinopathy, heart conditions, Alzheimer's disease, hepatitis, dengue, liver disease, and Parkinson's disease. Modern machine learning scholars have recognized two problems that need to be addressed: increased perception accuracy and disease diagnosis. In terms of diagnosing diseases, deep learning procedures are better than conventional computation methods. Having the capacity to alter their hyper-parameters as new data becomes available, deep learning makes use of massive neural networks with the connected neurons.

Keywords: Deep learning · Diseases detection · Diseases prediction · Machine Learning · Artificial Neural Network · Image Processing

1 Introduction

Deep learning and deep neural networks are terms for Artificial Neural Networks (ANN) with numerous layers. It has been grown to be among the most influential tools in recent decades and has become very well-liked in literature because of its capacity to organize a lot of material [1]. Now, profound hidden layers are beginning to beat conventional methods in several submissions, including pattern recognition. Among the most frequently employed Deep Neural Networks are the Convolutional Neural Network (CNN). The convolution of matrices in mathematics is whence it derives its name. CNN is made up of numerous layers, some of which include convolutional, non-linear, pooling then fully connected layers [16, 17].

The difference between convolutional and fully connected layers and those with pooling and nonlinearity is that the latter contain parameters. CNN does quite well when it comes to machine learning challenges. The results from submissions that deal with picture information, like the largest image classification data set (Image Net), computer vision and Natural Language Processing (NLP), were excellent [2]. This essay will go into great length into each of CNN's components, along with some of the major issues they encounter. Additionally, it will describe how they all operate. Also included is a summary of the variables affecting CNN's effectiveness [6]. This post will assume the fact that readers are aware of machine learning and artificial neural networks (Fig. 1).

T. Shreekumar et al. (Eds.): ISCComm 2023, CCIS 2231, pp. 119–127, 2025.
https://doi.org/10.1007/978-3-031-75605-4_8

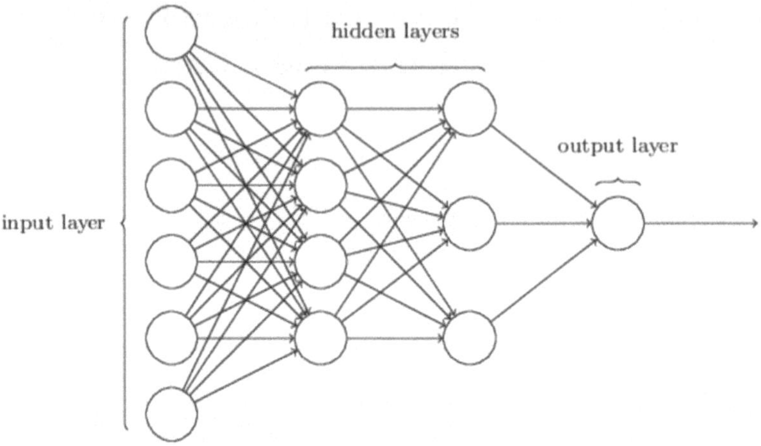

Fig. 1. Deep Learning Layers with many inputs passed to many hidden layers to get the accurate result in the output layer.

Over the past ten years, in several sectors connected to pattern recognition, such as voice and image processing, CNN has achieved groundbreaking successes. The greatest benefit of CNNs is their ability to reduce the complexity of ANN parameters. This accomplishment has encouraged both scholars and creators to use larger models to tackle interesting problems that could not be solved with traditional ANN [3]. The most crucial premise regarding problems that CNN solves is that there shouldn't be any spatially dependent features [18, 19].

Figure 1 shows how deep learning deals with layers where it contains input layer followed by all the hidden layers then output layer.

2 Literature Survey

A good heart disease guess system can be demonstrated using a data analysis strategy. Software on the internet is utilized to guess cardiac illness automatically. The security and effectiveness of user tasks are also considerably enhanced. The difficulty of the encryption raised by using the advised methodology with CNN and LSTM heart disease prediction prior to automatic diagnosis. It also used to retrieve results more quickly, offers high-quality services, and reduces the price of preserving a life [1, 20, 22].

This presents a strong framework that may recognize the perspectives for patient risk factor prediction considering their accessible health reasons. The persistence of the learning to clarify the best steps that specialized doctors can take to reduce their chance of developing heart disease. SVM, CMAR, Bayesian Classifiers, and C4.5 are examples of methods employed [2].

A model for heart illness guess using CNN on the UCI dataset and got an accuracy of 94%. They afterwards compared the model accuracy of various traditional machine learning algorithms like SVM, KNN, and Naive Bayes with the proposed model [3].

Made predictions about cardiac illness using a machine learning algorithm. In this study, categorization techniques are used to project cardiac disease in male patients. This document provides a thorough discussion of the facts, common types, and risk factors associated with coronary heart disease. Heart disease is forecasted using the three WEKA interfaces available in this system: Artificial Neural Networks, Decision Trees, and Naive Bayes [21]. Decision trees like CART, C4.5, CHAID, J48, ID3 Algorithms, and Naive Bayes Techniques are frequently utilized in prediction techniques. [4].

A method has been developed the NB, SVM, and OPF algorithms are used by the system [5]. The precision of this suggested model is 94.30%. And in, constructed a deep residual network model to categorize the heartbeat, and it managed to be accurate score of 99.06% [7].

The study examined how deep learning networks and acoustic signal processing can be used to diagnose lung disorders. It's an interesting development to learn that obstructive lung illnesses can be diagnosed using DLCNN. To predict and diagnose obstructive phenotypes or clinical effects, DLCNN algorithms look for similarities in the results of diagnostic tests [8].

The current study presented the use of neural networks in treating various lung conditions. This work lays a significant research foundation for future researchers by describing the state of DNN research in medical pictures, summarizing and analyzing the framework's development, and thoroughly examining the kinds of research that produced fruitful results in these domains [9].

In order to learn about complex tumors utilizing the Deep CNN method, which achieved 95.5% accuracy, this paper predictions CNN has been used in multiple optimization and varied activation functions [10]. This research reports that CNN has been used in several feature extractions using DCNN, and that the results of utilizing various optimizers have varied the outcome for accurate detection of both normal and anomalous [11].

An extensive investigation into brain tumors employing MRI, the anatomy of brain tumors, and publicly accessible datasets utilizing image-enhancing approaches, segmentation, feature extraction, classification, DL, and TL with quantum ML are demonstrated in this work [12]. Utilizing six different datasets with over fitting issues that arise during cross-model identification, this method employs CNN and CAD deep learning-based algorithms to identify CNN-based early tumors [13].

The computerized system uses k-means for clustering and feature extraction along with major module study and discrete wavelength transform, and it achieved 99% accuracy [14]. To identify malignancies using brain MR images is non-invasive and flexible. Using 98.90% accuracy and datasets from MRI and CT scan brain pictures, entropy analyzes two crucial textural and morphological features from the segmented image [15] (Table 1).

Table 1. Summarizes the overall literature survey where paper illustrates about the various algorithms and methodologies used to predict and detect the various diseases with much accurate results.

Type of Disease detected and predicted	Algorithm and method used	Result and Accuracy
Cardiovascular disease	CNN and LSTM	Offers high-quality services and quick results
Cardiovascular disease	CNN, SVM, KNN and NB	94% accurate
Cardiovascular disease	NB, SVM and OPF	99.6% accurate
Brain disease	DCNN	95.5% accurate
Brain disease	DL, TL with quantum ML	Early tumor detection successful
Brain disease	K-means	99% accurate
Lung disease	DLCNN	Quick results and more accurate

3 Review of Methodologies

1. Methodology Used for Cardiovascular Disease

Artificial Neural Network (ANN)
The term "Artificial Neural Network" comes from organic neural networks that help to form how the human brain is organized. Comparable to the interconnected neurons in the human brain, a human brain's neurons, ANN too display network with linked neurons at various layers. Nodes are the term used to describe these neurons [23, 24].

ANN primarily consists of **3 layers:**

Various formats of programmer's input can be received by the input layer. A hidden layer exists in between the input layer and the output layer. To reveal hidden characteristics and patterns, the input is altered utilizing hidden layers through a series of calculations. Then, using this layer, the result is sent to the output layer. When receiving input, an artificial neural network computes a weighted sum of the inputs while accounting for biases. To signify the outcomes of this calculation, a transfer function is utilized [16] (Fig. 2).

i. **Feature Selection:** By selecting pertinent data and removing noise from the data, feature selection is a technique for minimizing the total number of input variables in a model. In this scenario, the machine learning model's attributes that are crucial to solving the problem at hand are automatically chosen [20]. Filtering methods, packing methods, and embedding methods are three categories that encompass a variety of features selection strategies.
ii. **Wrapper Method:** Each attribute's value is assessed by the wrapper method using a machine learning approach. The algorithm is trained using a selection of features,

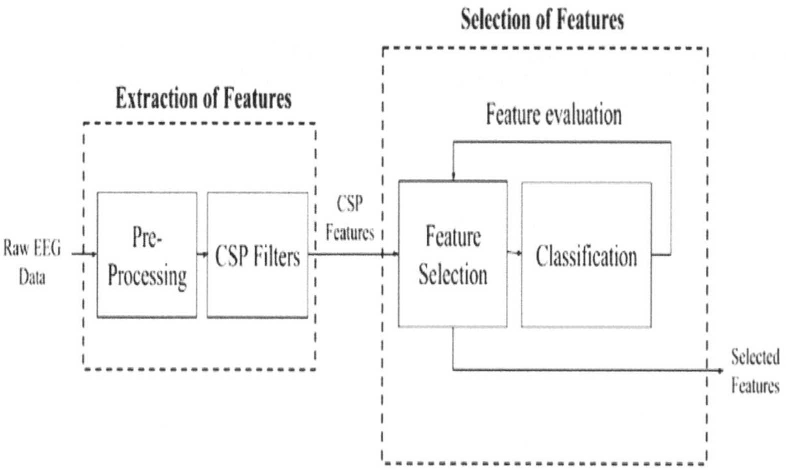

Fig. 2. Feature extraction and selection in ANN where the features will be preprocessed and selecting the feature based on classification.

and its efficacy is assessed using a collection of parameters including accuracy and F1 score. Selected are the components that perform better [20].

iii. **Embedding Method:** During the training stage of the machine learning algorithm, the embedding method chooses the features. The algorithm chooses the attribute that is most important for the task at hand.

2. Methodology Used for Lung Disease

Clinical techniques and computer-based procedures are the ways utilized to treat the respiratory system. There are three categories of clinical assessment methods: traditional general examination methods, history-based methods, and methods depending on the histopathology of images. Conversely, computer-based diagnosis methods can be categorized into four primary groups: wavelet, image analysis, image processing, and CNN research. We emphasize CNN-based audio processing as a fascinating field as significant components may now be automatically recognized by this technology without the need for human input.

In this work, we address the previous research in relation to the following issues:

i. **Dataset Selection:** A noise-free database must be acquired and kept up to date because it is the foundation of the entire model. It's significant to adequately preprocess the training data.

ii. **Deep learning algorithms choice:** It's crucial to comprehend the goal of the study. It is possible to assess the top algorithms to determine which ones produce outcomes that are most like what is needed.

iii. **Feature extraction strategies:** To be able to create effective models, it is a vital duty. It is successful when high model accuracy and ideal feature selection are essential since these two elements help in creation of duplicated data during each cycle of data analysis.

3. Methodology Used for Brain Disease

The primary phases of this approach are the gathering of data, pre-processing, and neural network detection. Most of the time, detection requires precise automatic and semi-automatic techniques. CNN-based fully automatic segmentation is employed due to these reasons (Fig. 3).

- **Data Collection**
  The system has acquired information as MRI pictures because it is primarily focused on the detection of brain tumors [26].
- **Pre-processing**
  The main objective is to enhance image highlights that are required for extra processing. For all subsequent preprocessing needs, the input image is here turned into a gray scale image. After threshold the image, more erosion and dilation are performed to the threshold image. To extract the contours and extreme points from this image.
- **Convolutional Neural Network**
  CNN is used to produce better results. To obtain the included map, the signal was convolved with kernels. With the kernel's weights, previous layers are connected. to improve the characteristics of the information image through back propagation analysis. Since all unit feature maps are shared by kernels [26]. The goal is to lessen over fitting. Kernels are used to collect all neighborhood data. Kernel is a key resource for context data. The neural network's output is subjected to an activation function.

A. Convolutional Layers

The Convolution layer's goal is to extract features from the input image with just the relevant portion of the image being linked to the subsequent convolution layer.

B. Padding

By adding a zero layer outside the input volume, padding allows us to obtain an output with dimensions that are comparable to the input volume while preserving border data. Nothing is being padded here.

C. Activation Function

A nonlinear activation function for more accurate results than traditional sigmoid functions, the Rectifier Activation function, or ReLU, is employed.

$$F(x) = \max(x, 0)1.$$

D. Pooling Layer

Combining spatially close features is what it is used for. Typically, features are joined via max-pooling.

It manages over-fitting and reduces the supplied image's dimension.

E. Full Connected Layer

Ultimately, the fully connected layer provides the categorized image using the training set.

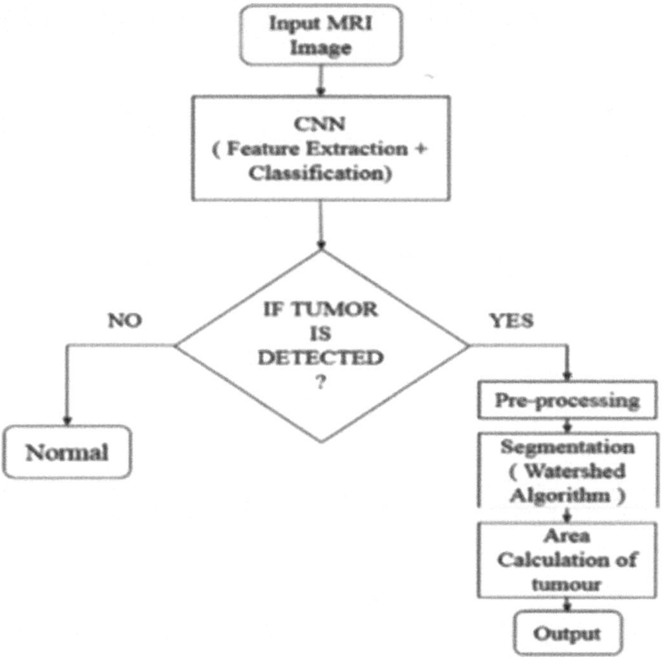

Fig. 3. Flow chart for image processing in Brain disease where we use CNN to detect that tumor is detected or not if yes then preprocessing and segmentation will be done to calculate the area of tumor and produces the output.

4 Conclusion

This paper is a review of different methodology and approach to detect or predict the diseases in heart, lung and brain using Deep learning as the main research area. Different techniques such as feature selection, deep belief networks, CNN, or some reinforcement and ensemble techniques can also help improve the level of precision of prediction and detection of disease.

References

1. Franklin, R.G., Muthukumar, B.: Survey of heart disease prediction and identification using machine learning approaches. In: Proceedings of the Third International Conference on Intelligent Sustainable Systems [ICISS 2020] (2020)
2. Purushottama. C., Saxena, K., Sharma, R.: Efficient Heart Disease Prediction System. Elsevier, Procedia Comput. Sci. **85**, 962–969 (2020)
3. Kalluri, H.K., Sajja, T.K.: A deep learning method for prediction of cardiovascular disease using convolutional neural network. Revue d intel. Artificielle **34** 601–606 (2020). https://doi.org/10.18280/ria.340510
4. Krishnan, S., Geetha. S.: Prediction of heart disease using machine learning algorithms. In: 1st International Conference on Innovations in Information and Communication Technology (ICIICT) (2019)

5. Marinho, L.B., de MM Nascimento, N., Souza, J.W.M., Gurgel, M.V., Rebouças Filho, P.P., de Albuquerque, V.H.C.: A novel electrocardiogram feature extraction approach for cardiac arrhythmia classification. Future Gener. Comput. Syst. **97**, 564–577 (2019)
6. Li, Z., Zhou, D., Wan, L., Li, J., Mou, W.: Heartbeat classification using deep residual convolutional neural network from 2- lead electrocardiogram. J. Electro cardiol. **58**, 105–112 (2020)
7. Pandey, S.K., Janghel, R.R.: Automatic arrhythmia recognition from electrocardiogram signals using different feature methods with long short-term memory network model. Signal Image Video Process. **14**(6), 1–9 (2020)
8. Sfayyih, A.H., Sulaiman, N., Sabry, A.H.: A review on lung disease recognition by acoustic signal analysis with deep learning networks (2023)
9. Ma, J., Song, Y., Tian, X., Hua, Y., Zhang, R., Wu, J.: Survey on deep learning for pulmonary medical imaging (2019)
10. Huang, Z., et al.: Convolutional neural network based on complex networks for brain tumor image classification with a modified activation function. IEEE Access **8**, 89281–89290 (2020)
11. Banerjee, S., Mitra, S., Masulli, F., Rovetta, S.: 'Deep radiomics for brain tumor detection and classification from multi-sequence MRI.' Social Netw. Comput. Sci. **1**(4), 1–15 (2019)
12. Amin, J., Sharif, M., Haldorai, A., Yasmin, M., Nayak, R.S.: Brain tumor detection and classification using machine learning: a comprehensive survey. Complex Intell. Syst. **8**, 3163–3183 (2022)
13. Naseer, A., Yasir, T., Azhar, A., Shakeel, T., Zafar, K.: 'Computer-aided brain tumor diagnosis: performance evaluation of deep learner CNN using augmented brain MRI,' Int. J. Biomed. Imag. **2021**, 1–11 (2021)
14. Polly, P., Shil, S.K., Hossain, M.A., Ayman, A., Jang, Y.M.: 'Detection and classification of HGG and LGG brain tumor using machine learning. In: (ICOIN), Jan (2018)
15. Gupta., Khanna, P.: A non-invasive and adaptive CAD system to detect brain tumor from T2-weighted MRIs using customized Otsu's thresholding with prominent features and supervised learning. Signal Process. Image Commun. **59**, 18–26 (2017)
16. Johnson, K.W., BS, Torres Soto, J., MS, Glicksberg, B.S.: Artificial intelligence in cardiology., Elsevier. J. A. Coll. Cardiol. **71**(23), 2668−2679 (2018)
17. Kodati, S., Vivekanandam, R.:Analysis of heart disease using in data mining tools orange and WEKA. Glob. J. Comput. Sci. Technol. **18** (2018)
18. ChalaBeyene, P.K.: Survey on prediction and analysis the occurrence of heart disease using data mining techniques. Int. J. Pure Appl. Math. **118**(8), 165–174 (2018)
19. Bagavathy, S., Gomathy, V., Sheeba Rani, S., Sujatha, K., Bhuvana, M.K., Murugesan, M.: Early heart disease detection using data mining techniques with hadoop map reduce. Int. J. Pure Appl. Math. **119**(12), 1915–1920 (2018)
20. Ukey, P., More, V., Patil, P., Sankhe, A.: Heart disease prediction using data mining. IOSR J. Eng. **4**, 64–67 (2018)
21. Arun, R., Deepa, N.: Heart disease prediction system using naive bayes. Int. J. Pure Appl. Math. **119**(16), 3053–3065 (2018)
22. Yildirim, Ö.: A novel wavelet sequence based on a deep bidirectional LSTM network model for ECG signal classification. Comput. Biol. Med. **96**, 189–202 (2018)
23. Marinho, L.B., de MM Nascimento, N., Souza, J.W.M., Gurgel, M.V., Rebouças Filho, P.P., de Albuquerque, V.H.C.: A novel electrocardiogram feature extraction approach for cardiac arrhythmia classification. Future Gener. Comput. Syst. **97**, 564-577 (2019)

24. Kumar, K.H., Sajja, T.K.: A deep learning method for prediction of cardiovascular disease using convolutional neural network. Revue d intel. Artif. **34** 601–606 (2020). https://doi.org/10.18280/ria.340510
25. Dinesh, K.G., Arumugaraj, K., Santhosh, K.D., Mareeswari, V.: Prediction of cardiovascular disease using machine learning algorithms. In: (2018) International Conference on Current Trends towards Converging Technologies (ICCTCT), Coimbatore, India, (2018)
26. Brain Tumor Detection using Convolutional Neural Networks by Aditya Jambhale, Abhay Pawar, Tanuj Patil, Snehal Andhare e-ISSN: 2395–0056 Volume: 07 Issue: 05 May (2020)

Fault Diagnosis in Belts Using Signal Processing Techniques and Machine Learning

M. V. Kirankumar[1](✉), Rajeshwari[1], Nishmitha[1], Pavitra Kumari[1], B. K. Raghavendra[2], and Kulkarni Sourabh Vyankatesh[3]

[1] Department of Mechatronics Engineering, Mangalore Institute of Technology and Engineering, (Visvesvaraya Technological University, Belagavi), Moodbidri, Karnataka 574225, India
kirankumarmv2008@gmail.com

[2] Electrical and Electronics Engineering, Govternment Polytechnic, Koppal, Karnataka 583231, India

[3] KPIT Technologies, Pune, India

Abstract. In the era of modern industries, use of advanced machines has increased the productivity of plants. Defect of such complex machines can increase the downtime and hence reduce productivity and ultimately the profit. Proper maintenance of systems is very crucial to ensure trouble free operation of machines. Predictive Maintenance techniques are those techniques which are used for identification of fault in its early stage. This avoids the sudden unavailability of machines and reduces the downtime. Predictive Maintenance techniques using vibration analysis is one of the very popular methods of Condition Monitoring. These techniques are evolving rapidly due to their efficient detection of fault in its incipient stage. However, non-stationary of the vibration signal produced particularly in belts in belt drive systems using conventional techniques of signal processing poses challenges. In this project work, the application of advanced signal processing techniques is investigated for the automatic fault detection in belts. This project work is focused on developing a novel concept of vibration signal processing for the automatic fault detection in belts using wavelet analysis and the belt condition classification and prediction using Machine Learning algorithms.

Keywords: Condition Monitoring · Machine Learning · Fault Diagnosis

1 Introduction

As belt transmission is being used in almost every industry, its maintenance has become topic of utmost importance. Sudden failure of a belt can cause unpredictable downtime which can cause financial loss to the firm. Here, continuous condition monitoring becomes crucial for predicting the machine's failure. Techniques for vibration analysis can be utilized to find the fault at an early stage. Important information like the source of failure and its severity can be extracted from vibration signatures of the machine.

T. Shreekumar et al. (Eds.): ISCComm 2023, CCIS 2231, pp. 128–144, 2025.
https://doi.org/10.1007/978-3-031-75605-4_9

Typical belt failure is a result of some common faults such as rapid sidewall wear, worn cover on back, underside cracked, slip, cracked bushings and belt fatigue. The causes for these faults can be improper tension, worn out pulleys, misalignment, etc. The corrective action can be just replacing the pulleys, maintaining optimum tension in the belt and even replacing belts.

The problem here in conventional techniques of maintenance is it can only be done after the belt is observed as damaged or after the indication of loss due to belt failure. So, the identification of faults in belts at its incipient stage can avoid the loss and the sudden downtime of machine.

Also, the current vibration analysis techniques for fault diagnosis are majorly FFT (Fast Fourier Transform) based. One of the major disadvantages of using the FFT is it can't efficiently analyze transient features in time.

Condition Based Maintenance(CBM)

Sensors are used in CBM a sort of diagnosis method, to track an asset's status across time as it is being used. This maintenance method uses sensors to periodically or continuously (also known as "real-time condition monitoring") check mechanical and operational conditions. The collected data is then processed. For each monitored variable, thresholds are set in order to distinguish between various machine states. An alert is sent when a parameter value exceeds its threshold, the component is determined to be defective, and it is removed for maintenance. This strategy can also be strengthened by an algorithm that sets off warnings in response to aberrant behaviors and trends in historical data. This can predict an impending failure, allowing us to plan an immediate repair intervention. It is challenging to distinguish between condition-based maintenance and predictive maintenance because artificial intelligence approaches are also being used for pattern recognition and to predict failures; however, some authors classify condition-based maintenance as a predictive maintenance approach.

This maintenance technique employs a wide range of sensors, including vibration sensors (accelerometers), infrared sensors, sound sensors, ultrasonic sensors, etc. This method also incorporates commonly used electrical parameter monitoring and oil analysis. Based on a range of in-the-field observations and analyses of machine performance data carried out by maintenance engineers, the condition-based maintenance plan should be implemented on manufacturing assets. When adopting CBM, common variables such the frequency and sporadic nature of failures, the necessity for repeated repairs, and the effect on quality yield should be considered.

The fundamental benefit of employing CBM is that it is a non-invasive method of maintenance, meaning that real-time data is collected while the equipment is still operating without changing how it runs. Through sensors, physical inspection, or prearranged testing, we can choose to continuously record data or do so at predetermined intervals. Vibration analysis, infrared thermography, oil analysis, electrical parameter analysis, ultrasonic analysis, and pressure analysis are a few examples of condition-based maintenance techniques. Each one has a particular use for which it is intended.

Vibration Monitoring

The method of monitoring the amplitudes and frequency of machine vibrations and using the data to assess the condition of the machines and the components is known as vibration analysis. Vibration analysis aids in the identification of defects such as imbalance, bearing failure, mechanical looseness, resonance, bent shafts, etc.

Ultrasonic Analysis

In ultrasonic analysis, high-frequency sounds are detected and converted into audio and digital data in order to identify potentially deteriorating assets. Ultrasonic analysis uses two different sorts of data collection techniques to identify different types of detectable failure. These include both direct and indirect ways. We typically employ contact methods for mechanical difficulties such bearing defects, lubrication concerns, gear corrosion, and pump cavitation. These errors all exhibit high-frequency noise. In many electrical applications as well as compressed gas systems, pressure and vacuum leaks are found using non-contact ultrasonic measuring techniques.

Vibration Monitoring

Vibration analysis can be done in 3 different ways: Time domain Analysis, Frequency domain Analysis, and Time-Frequency Analysis.

Time Domain Method

In time domain analysis, signals are examined in relation to time. A time-domain graph shows how a signal changes over time. We can identify polluted signals by separating unwanted signals from the intended signal over time using time domain signal processing.

Frequency Domain Method

In frequency domain analysis, signals are examined in relation to frequency. The frequency domain demonstrates how much of the signal is distributed across a range of frequencies within a given frequency band. However, the frequency domain provides a more detailed picture of how a signal interacts with a material. When compared to other frequencies, the medium's response to sympathetic frequency is significantly more apparent. The frequency domain of a signal is effectively converted into its many periodic and aperiodic constituents when it is decomposed or modified. This enables us to focus on each component or band of the signal spectrum and associate its behavior with a potential medium property or defect. Signal processing's crucial cognitive signal analysis can only be carried out in the frequency domain.

Time Frequency Analysis

We can simultaneously examine signal characteristics in the time and frequency domains by using the Time-Frequency Analysis technique. This method employs various time-frequency representations. The time-frequency distribution function can be expressed in a variety of ways, resulting in various time-frequency distributions such as the Short Time Fourier Transform (STFT), Wigner Ville Transform, and Wavelet Transform.

2 Literature Review

The fault detection in gear using enveloped wavelet spectrum. Using Laplace and Morlet wavelet functions, the diagnostic potential of the rapid Fourier transform power spectrum to the wavelet envelope power spectrum is investigated. The motor, single-stage gearbox, connection, loading mechanism, and bearings make up the experimental setup [1]. According to Kumar et al., 2018, the wavelet enveloped spectrum can be employed to detect fault in belt drive system. A simple transmission mechanism made out of a motor, a battery, and a cone pulley with a belt drive. First, a healthy belt's vibration data is gathered using three axial accelerometers and an appropriate data acquisition equipment (LabVIEW software and DAQ module). Flat, toothed, and V belts are intentionally flawed, and the resulting vibration data is gathered [2]. With the use of Wavelet Packet Decomposition, the vibration signals were divided. Each frequency band's energy was taken out and used as a feature in the SVM training process to determine the direction of idler faults. Based on the methodology, a belt conveyor online monitoring and fault diagnosis system that can identify issues automatically was created [3]. It has been discovered that continuous wavelet transform is a useful tool for defect diagnostics when utilized for damage identification [4]. The statistical properties of a signal are provided by wavelet coefficients and can be used as ANN inputs. The outcome of the suggested ANN's learning process demonstrates that the training with 18 iterations satisfied the MSE stopping requirement [5]. The waveform and feature characteristics were used to distinguish between damaged and healthy teeth. Support vector machines used extracted feature parameters to diagnose the state of each gear tooth [6]. The NI DAQ equipment and LabVIEW software were used to collect the vibration signals. For monitoring belt conveyors with a focus on idler fault identification, a defect diagnostic method integrating wavelet packet decomposition (WPD) and support vector machines (SVM) is proposed [7]. Techniques for processing vibration signals are categorized and examined. A literature review has been conducted to analyze the potential of each technique, underlying concepts, benefits and drawbacks, and practical applications for gear failures detection [8].

Time domain, frequency domain, and time-frequency domain vibration signal analysis is performed for monitoring and analysis. This paper is just a survey / collection of works done by various authors in the vibration analysis domain [9]. The performance of neural networks is then investigated in relation to variables such as the number of hidden layers, the number of neurons in the final hidden layer, and data augmentation. Using neural networks and data augmentation, faults are categorized. The outcomes of deep neural networks are then contrasted with those of the other techniques. [10].

3 Methodology with Experimental Setup

Through this project work an approach is being worked out to develop advanced signal processing techniques to automatically detect faults in belt drive system in its incipient stage to avoid sudden breakdown of machine (Fig. 1).

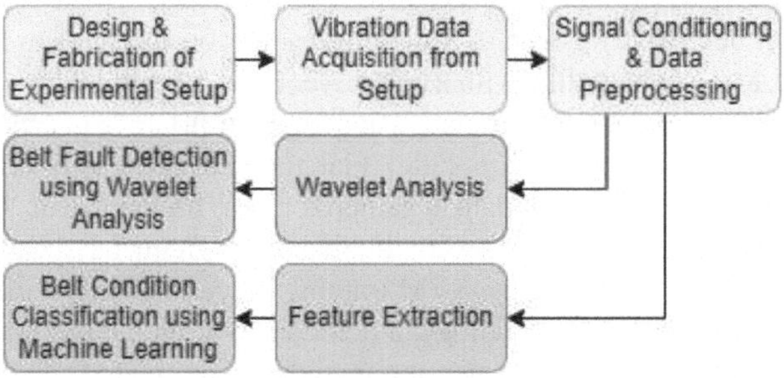

Fig. 1. Methodology

Project work necessitated the use of an experimental setup consisting of a simple belt transmission system for data collection. The system consists of a motor as the prime mover, a battery as the power source and a cone pulley with belt drive from which the vibration data is obtained. A variable frequency drive is also included in the setup to adjust the rotational speed. A 3-axial accelerometer is used as a sensor to collect vibration data from the setup. The accelerometer sensor is mounted on the revolving belt drive shaft, and the data acquisition module is connected to the computer via a USB port. The LabVIEW software is utilized to obtain data from the DAQ module and convert it into a processing-ready format (Tables 1 and 2).

Motor

- Capacity: 3-Phase, 440 V, 50 Hz
- Power: 0.38K W, 0.5 HP
- Rated speed: 1440 RPM
- Rated current: 1.4 Amps

Motor Starter

- Phase: 3
- Coil Voltage: 280/440
- Amperes: 15

Sensor Specifications

After connecting accelerometer to the DAQ module, the DAQ module is connected to the computer having the LabVIEW software installed and configured for the DAQ hardware access (Fig. 2). The accelerometer sensor is placed on its mounting on the setup. The motor is the powered ON through the switch. This procedure is done for three conditions

Table 1. Pulley specifications

Parameters	Cone Pulley		
	Small	Medium	Large
Belt Material	Low carbon steel		
External dia	50 mm	75 mm	100 mm
Internal dia	16 mm	16 mm	16 mm
Category	'V'	'V'	'V'

Table 2. Sensor specifications

Sensor used	Accelerometer
Manufacturer	IMI Sensors
Model Number	621B40
Sensitivity	1.02 $mV/(m/s^2)$
Frequency Range	up to 18 kHz
Axis covered	3 axial

Fig. 2. Experimental setup

of belt viz., healthy belt, belt with 2 mm cut in, belt with 6 mm cut in. The cut in, here, represents an underside cut of respective depth. First, the readings for healthy belt are taken for specific time period and the data is exported into an excel sheet. Same procedure in used for other two conditions of belt as well and the data is again exported into excel sheets. The generated excel file containing 3 sheets for the data of three belt conditions is used as an input for the MATLAB program. In MATLAB software, a program is prepared for outputs like Time Domain Signal, Frequency Domain Signal (FFT Spectrum) and Wavelet Enveloped Power Spectrum. For this project work, Morlet Wavelet and Mexican Hat Wavelet are used in convolution with the actual vibration signal. The belt frequency is analyzed in FFT spectrum and the Enveloped Wavelet Power Spectrum (Fig. 3). The belt frequency can be calculated as below:

$$BF = \frac{\pi D_1 N_1}{L} = \frac{\pi D_2 N_2}{L}$$

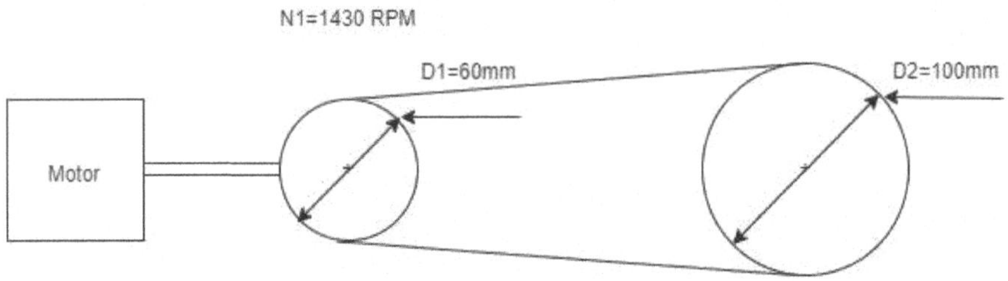

Fig. 3. Belt Drive System Dimensions

where,

BF = Belt frequency.

D1 = Drive sheave diameter D2 = Driven sheave diameter N1 = Drive sheave speed.

N2 = Driven sheave speed L = Length of Belt.

From the diagram, we can calculate the belt frequency and it comes 518.36 RPM which is equal to 8.64 Hz.

So, the behavior of the amplitude of vibration is observed at this belt frequency in the FFT spectrum as well as the Enveloped Wavelet Spectrum and results are compared.

The amplitudes of the enveloped wavelet spectrum are considered for feature extraction task. The amplitude data of all the three conditions of belt is stored in a separate excel sheet. This file is used as an input for feature extraction algorithm developed on Google Colab platform. Six statistical features are considered for this task; they are Mean, Peak, Root mean square value, Standard deviation, Variance and Kurtosis.While the Peak value indicates the highest value from the data, the Mean is the average of all the data. RMS is the data's room mean square value, as the name suggests. The standard deviation shows how much a set of values in a dataset vary or are dispersed. Data points are more closely spaced from the mean value if the standard deviation is low, and the opposite is true if the standard deviation is high. The degree to which a group of numbers deviates from the average value is also measured by variance. Kurtosis is a statistical measurement that indicates how significantly a distribution's tails deviate from those of a normal distribution. One excel sheet is produced by the Feature Extraction Algorithm for each feature, giving us six excel files for a single belt condition. Here, these files must be combined into a single Excel sheet. Similar to this, we must follow the same process for the other belt circumstances. In order to feed the classification algorithms, the feature data is coupled with each belt condition. I used the Support Vector Machine technique, Naïve Bayes and Decision Tree as three classification models in this research assignment. Both the classification job and the confusion matrix are produced using these algorithms. Additionally, the current belt status is projected based on the feature values.

4 Results and Discussions

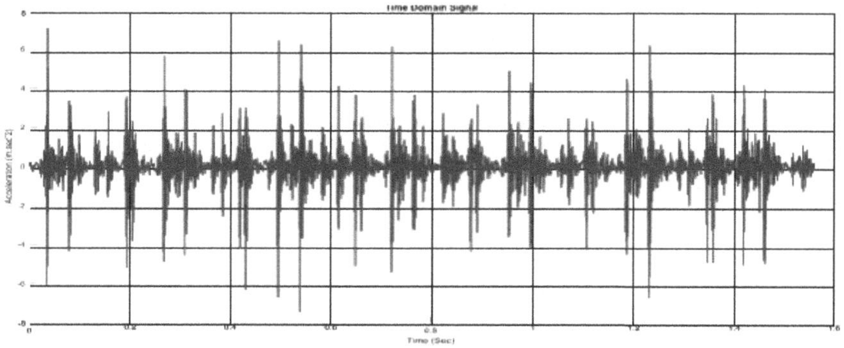

(a)

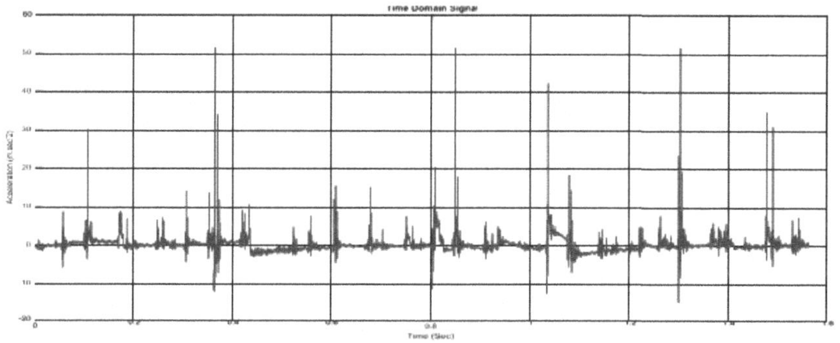

(b)

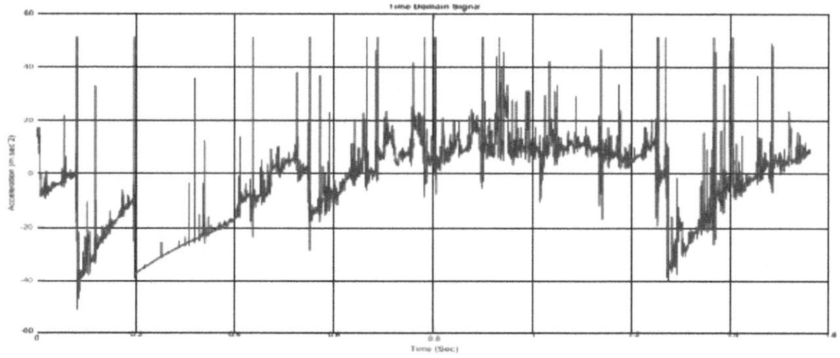

(c)

Fig. 4. Time domain spectrum for (a) healthy belt (b) belt with 2mm cut in (c) belt with 6mm cut in

MATLAB software was used to develop the Time domain signal, FFT spectrum Mexican Hat, Morlet Wavelet Enveloped Power Spectrum to analyze the vibration data (Fig. 4).

Time domain plots for different belt condition:

FFT Plots for different belt conditions:

The power spectrum shows that the belt frequency amplitude increases in proportion to the extent of the defect condition. FFT power spectrum, which is employed as a tool to gauge the extent of damage, does not, however, provide a clear indicator of induced fault (Fig. 5).

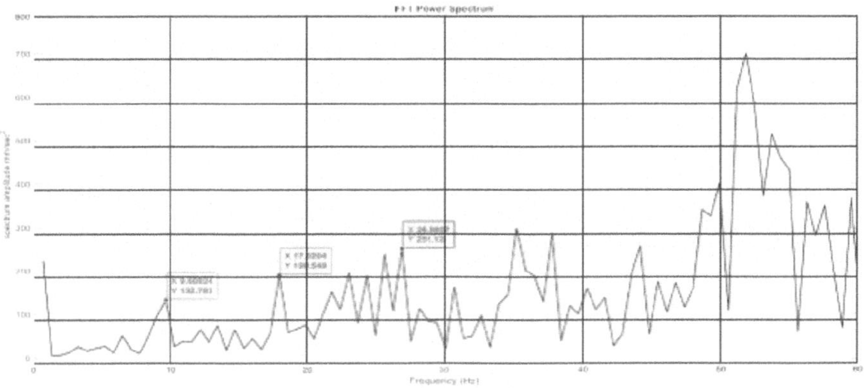

(a)

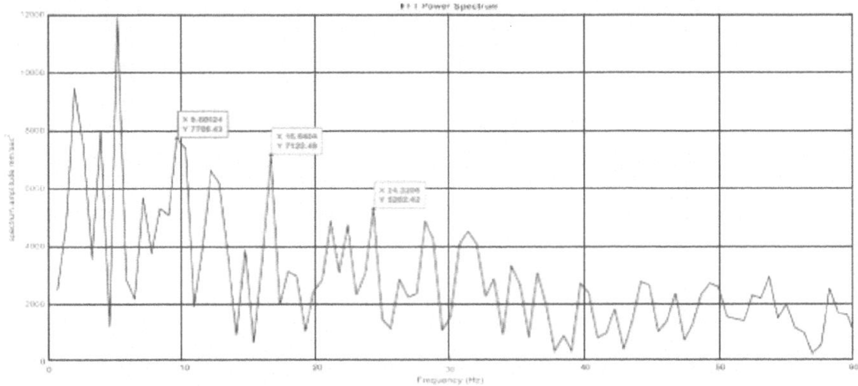

(b)

Fig. 5. FFT Power Spectrum for (a) healthy belt (b) belt with 2mm cut in (c) belt with 6mm cut in

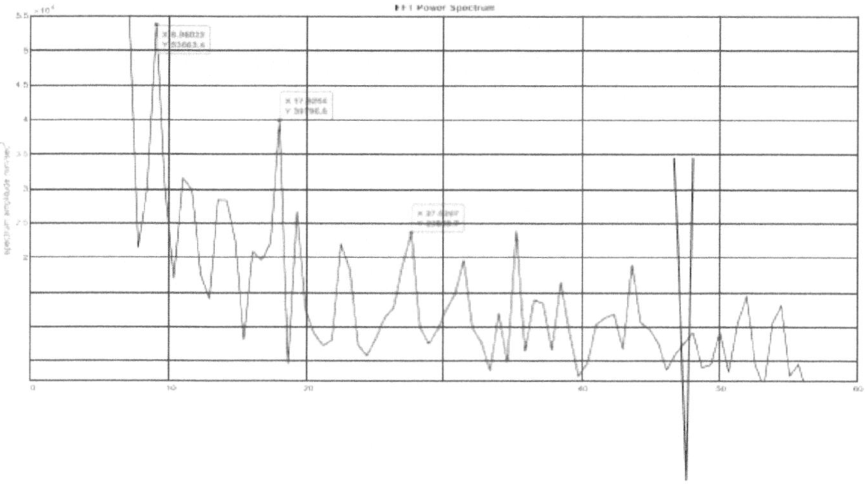

(c)

Fig. 5. (*continued*)

Enveloped Morlet wavelet Power Spectrum:

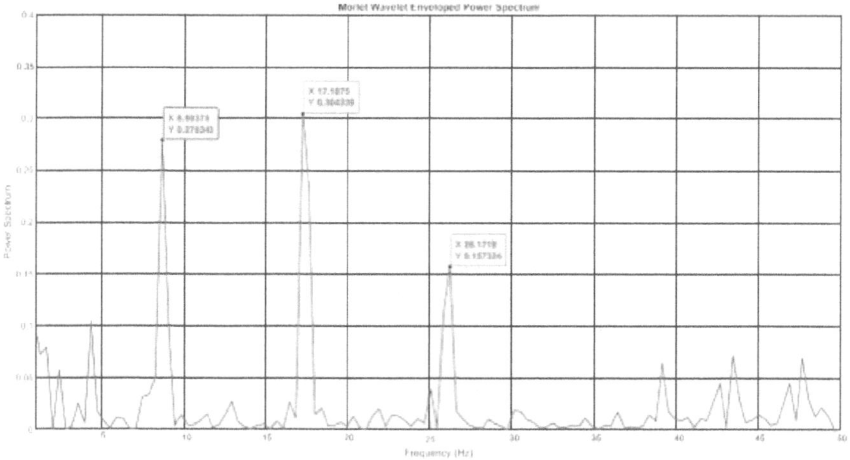

(a)

Fig. 6. Enveloped Morlet wavelet power spectrum for (a) healthy belt (b) belt with 2mm cut in (c) belt with 6mm cut in

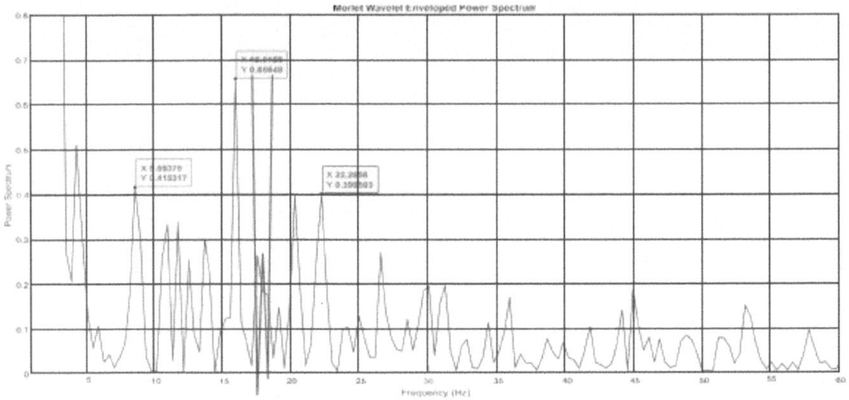

(b)

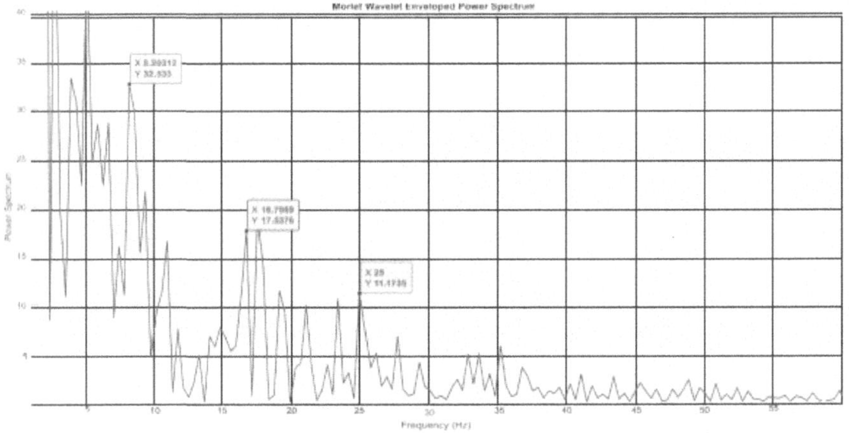

(c)

Fig. 6. *(continued)*

Figure 6(a), 6(b) and 6(c) show the Morlet wavelet enveloped power spectrums for the healthy belt and two induced stages of faults, respectively. When comparing healthy belt and faulty belt, the belt frequency amplitude increases from 0.27 to 0.41 and to 32.5 mm/s^2 in 2 separate stages.

Mexican Hat Wavelet Enveloped Power Spectrum for different belt conditions: Figure 7 (a), 7(b), and 7(c) show Mexican hat wavelet enveloped power spectrums for the healthy belt and the damaged belt at two distinct stages of fault. According to Mexican Hat wavelet-based enclosed power spectrums, vibration amplitude at belt frequency increases with fault severity from 0.36 to 0.54 to 42.57 mm/s2. Figure 7 depicts the modest amplitude of a well-maintained belt.

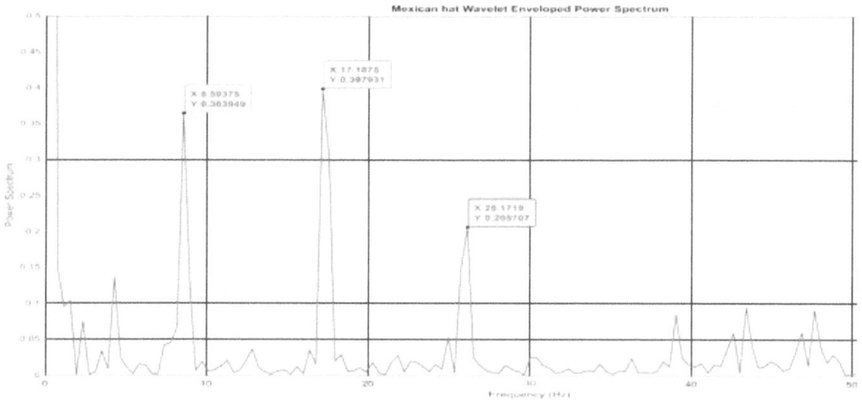

(a)

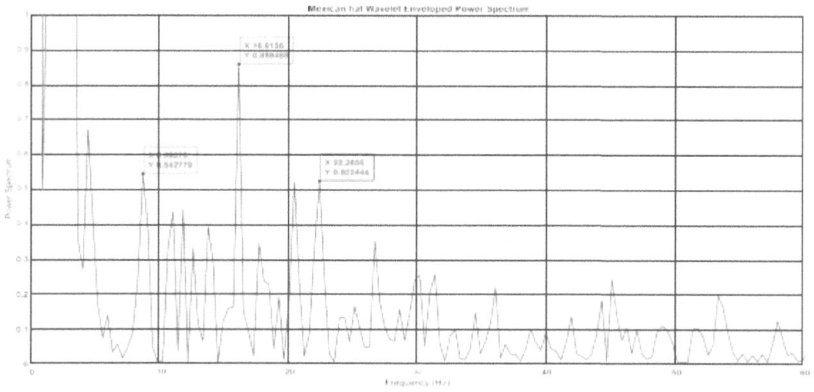

(b)

Fig. 7. Enveloped Mexican hat wavelet power spectrum for (a) healthy belt (b) belt with 2mm cut in (c) belt with 6mm cut in

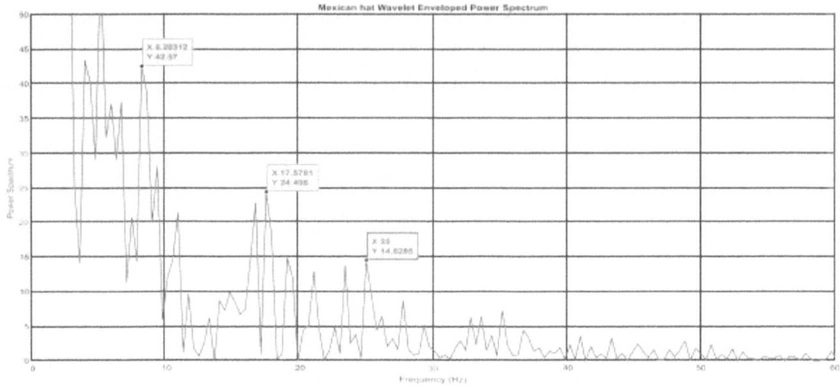

(c)

Fig. 7. (*continued*)

Based on the studies and results, we can conclude that the Enveloped wavelet power spectrum provides a more accurate assessment of the vibration amplitude, making it easier to compare the faulty and healthy states. The degree of the fault is directly inversely correlated with the vibration amplitude, which means that a 6 mm cut in the fault produces a greater vibration amplitude than a 2 mm cut in the fault in the belt.

Two wavelets viz., Morlet and Mexican hat wavelet are compared. The sensitiveness of the wavelet is responsible in measuring the amplitude for respective frequency. Form the above graphs, the Mexican hat Wavelet Spectrum is observed showing higher amplitude as compared to Morlet wavelet power spectrum. Hence, it can be concluded that Mexican hat wavelet enveloped power spectrum is more sensitive to the different

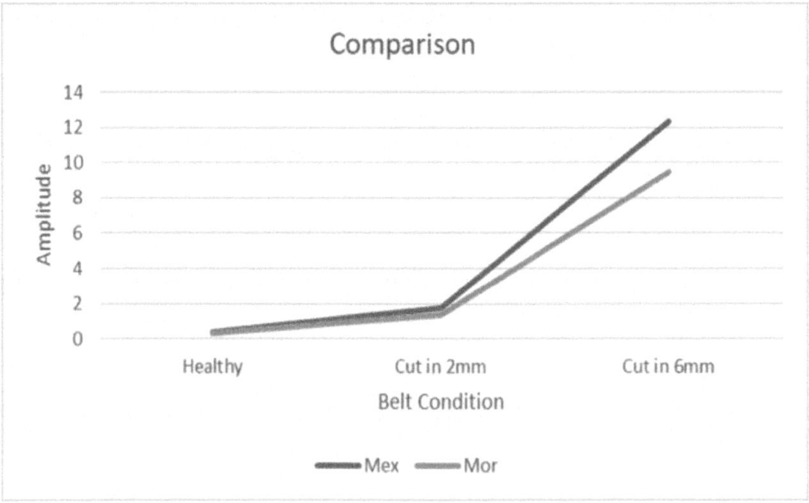

Fig. 8. Comparison of performance of Morlet and Mexican hat wavelet

types of faults present in the belt drive, as compared to the Morlet wavelet enveloped power spectrum (Fig. 8).

Machine Learning Classification
Six characteristics were extracted from the signal data: mean value, peak value, root

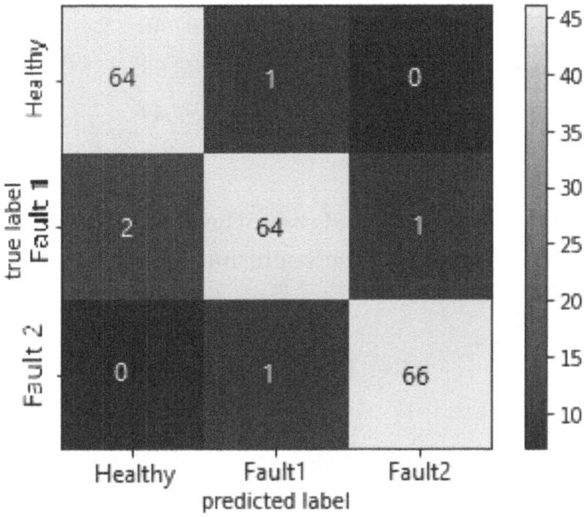

(a)

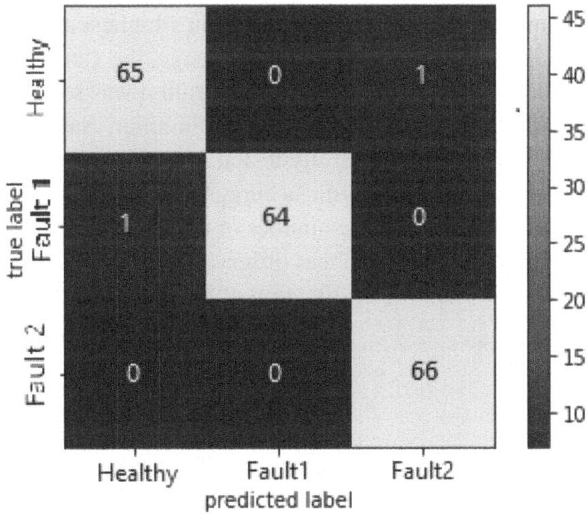

(b)

Fig. 9. Confusion Matrix for (a)- SVM classification model (b)- Naïve Bayes classification model (c)- Decision Tree classification model

mean square value, standard deviation, variance, and kurtosis (from healthy belt and from damaged belt).

Three classification models are compared for the task of belt fault detection. The confusion matrix and accuracy score observed are mentioned below (Table 3).

Table 3. Different classifier and their accuracy

Classifier	accuracy (in%)
Gaussian Naïve Bayes Classifier	97.46
Support vector Classification	98.98
Decision Tree Classification	99.49

As the project work is based on three classes of health condition of belts, it has become a multiclass classification problem. The confusion matrix for different classification models is shown below (Fig. 9).

5 Conclusion

Signal processing methods including the FFT and Mexican hat, Morlet wavelet enveloped power spectrum is used to process the vibration signal. The outcomes are contrasted for two separate belt defects. The outcome demonstrates that, in comparison to the FFT and Morlet wavelet power spectrum the Mexican hat wavelet enveloping power spectrum is more fault sensitive. The vibration's highest amplitude was discovered to be a 6 mm cut in fault. For a belt with less damage, the vibration level was lower. With increasing belt deterioration, an increase in vibration was seen.

Three classification models viz., Naïve bayes classifier, Support vector machine algorithm and Decision tree are successfully tested for real time operational data from belt drive system and observed with specific accuracies mentioned above. The confusion matrix was also generated indicating the number of correctly classified samples. In this work, Decision tree classification model was observed with maximum accuracy. Hence, we can use classification model for classification of belt defect in the rotating machines.

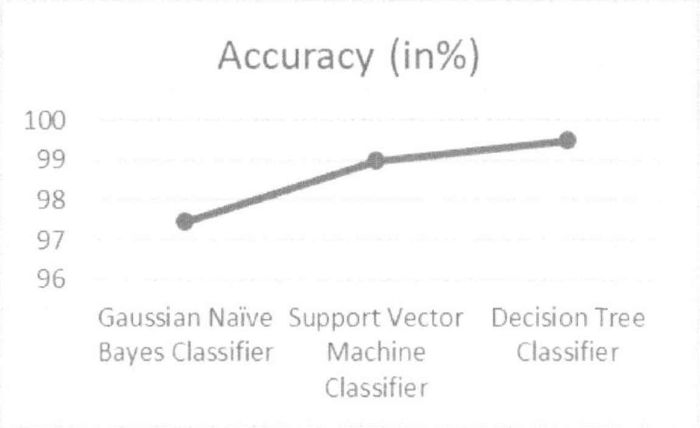

Graph 1. Comparison of performance of different classification models

References

1. Lokesha, M., Majumder, M.C., Ramachandran, K.P., Raheem, K.F.: Fault diagnosis in gear using wavelet envelope power spectrum. Int. J. Eng. Sci. Technol. **3**(8), 156–167 (2011)
2. Kumar, S., Lokesha, M., Kumar, M.K. and Ramachandra, C.G.: Fault diagnosis in belts drive transmission using wavelet spectrum. Recent Trends Mech. Eng. Appl. Sci. **7**, 5−18 (2018)
3. Li, W., Wang, Z., Zhu, Z., Zhou, G. and Chen, G.: Design of online monitoring and fault diagnosis system for belt conveyors based on wavelet packet decomposition and support vector machine. Adv. Mech. Eng. **5**, 1–10 (2013)
4. Bajrić, R., Zuber, N., Isić, S.: Recent advances in vibration signal processing techniques for gear fault detection-a review. Appl. Mech. Mater. **430**, 78–83 (2013)
5. Chandran, P., Lokesha, M., Majumder, M.C., Fathi, K., Raheem, A.: Application of laplace wavelet kurtosis and wavelet statistical parameters for gear fault diagnosis. Int. J. Multi.Sci. Eng. **3**(9), 1–7 (2012)
6. Bulushi, A., Rameshkumar, G.R. and Lokesha, M.: Fault diagnosis in belts using time and frequency-based signal processing techniques. Int. J. Multi. Sci. Eng. **6**(11), 12–19 (2015)
7. Fan, Q., Zhou, Q., Chaoqun, W., Guo, M.: Gear tooth surface damage diagnosis based on analyzing the vibration signal of an individual gear tooth. Adv. Mech. Eng. **9**(6), 1–14 (2017)
8. Vernekar, K., Kumar, H., Gangadharan, K.V.: Gear fault detection using vibration analysis and continuous wavelet transform. In: Presented at International Conference on Advances in Manufacturing and Materials Engineering, AMME (2014)
9. V.Arunkumar, K.M., Manjunath, T.C.: A brief review/Survey of vibration signals in time domain. SSRG Int. J. Electron. Commun. Eng. (SSRG-IJECE) **3**(3), 12−15 (2016)
10. Heo, S., Lee, J.H.: Fault detection and classification using artificial neural networks. IFAC Papers Online **51**(18), 470–475 (2018)
11. Zhou, F., Gao, Y., Wen, C.: A novel multimode fault classification method based on deep learning. J. Control Sci. Eng. **2017**, 1–14 (2017)
12. Rafiee, J., Rafiee, M.A., Tse, P.W.: Application of mother wavelet functions for automatic gear and bearing fault diagnosis. Expert Syst. Appl. **37**, 4568–4579 (2010)
13. Bajrić, R., Zuber, N., Skrimpas, G.A., Mijatovic, N.: Feature extraction using discrete wavelet transform for gear fault diagnosis of wind turbine gearbox. Shock Vibr. **2016**, 6748469 (2016)

14. Dekys, V.: Condition monitoring and fault diagnosis. Procedia Eng. **177**, 502–509 (2017)
15. Shenga, C., Lia, Z., Qinb, L., Guoa, Z., Zhanga, Y.: Recent progress on mechanical condition monitoring and fault diagnosis. Procedia Eng. **15**, 142–146 (2011)

Machine Learning Techniques Based Chronic Kidney Disease Detection with Performance Analysis of Fuzzy Rough Set and Correlation Attribute Selection

Ch. Ravindra Babu[1](✉), E. V. N. Jyothi[1], Kunchala Teja[1], Annem Tharun Kumar Reddy[1], Mutyala Venkata Gopi Chand[1], Peram Rajesh[1], and R. J. Shwetha[2]

[1] Pace Institute of Technology and Sciences, Ongole, AP, India
ravindra.chimmiri@gmail.com
[2] Software Developer, Global Payex Technologies Pvt. Ltd., Mumbai, MH, India

Abstract. Due to its increasing prevalence, chronic kidney disease (CKD) has a significant financial impact on the healthcare system, poor prognoses for death and health problems as well as a high probability of progressing to end-stage kidney disease. It is quickly developing into a global health emergency. This disease is largely caused by unhealthy eating habits and a lack of water consumption. The average lifespan of a person without kidney function is only 18 days, which requires dialysis and kidney transplantation. In order to correctly predict CKD in its early stages, reliable methodologies are essential. The ability to forecast CKD using Machine learning (ML) approaches is significant. Additionally, technology advancements typically have an advantageous effect on our everyday lives, particularly in the health. It is simpler to diagnose diseases presently than it was in the past because to improved technologies. In a analysis to identify The use of categorization algorithms for chronic kidney disease (CKD), this analysis's main goal is to assess that attribute selection strategies affect the algorithms performance. The CKD data set utilised in the experiments was obtained from the machine learning repository at the University of California, Irvine (UCI). The process of designing classifier models involves a number of pre-processing, normalization, and attribute selection steps. Both the fuzzy rough set based attribute selection (FRSBAS) approach and the correlation based attribute selection (CBAS) method were utilized to identify the characteristics that have a significant effect on the classification outcomes. For performance analysis, the following metrics are used: accuracy, sensitivity, precision, and specificity.

Keywords: Chronic Kidney Disease (CKD) · Fuzzy Rough Set Based Attribute Selection (FRSBAS) · Machine Learning (ML) · Correlation Based Attribute Selection (CBAS)

T. Shreekumar et al. (Eds.): ISCComm 2023, CCIS 2231, pp. 145–157, 2025.
https://doi.org/10.1007/978-3-031-75605-4_10

1 Introduction

About 10% of the worldwide population suffers from chronic kidney disease (CKD), making it one of the most prevalent illnesses worldwide [12]. Chronic kidney disease [13], it is distinguished by decreased excretory renal production or proteinuria increasing more than three months and is caused by kidney injury or kidney function damage. To filter waste from the blood and eliminate it from millions of small blood vessels fill the kidneys. When this filtration system fails, the kidneys are unable to eliminate waste, which causes kidney disease. Although CKD has several underlying causes, its progression is typically irreversible and can lead to major health issues [11].

With the severity of chronic kidney disease, such as acute renal fever, the possibility of developing new diseases increases, poor oral absorption, and diarrhoea [6]. Early-stage CKD is recoverable, but kidney damage develops in the later stages. 753 million individuals died from CKD in 2016, making it a risky and potentially fatal illness. While plant and environmental toxins can also contribute to chronic kidney disease, diabetes mellitus is generally considered to be the major toxins. The primary issue with CKD is the lack of early-stage symptoms, which can cause a 25% reduction due to a delayed diagnosis and kidney function has effected [7]. Early diagnosis is the most effective method to stop CKD from changing and leading to renal failure [1]. In addition to high diagnostic expenses, the majority of countries that acquire experience an absence of specialized physicians, particularly in remote regions. As a result, there is a increased demand for the development of innovative technologies for the early diagnosis of chronic renal disease by doctors.

Maintaining up with technological advancements is becoming more important. It is nearly difficult to be uninformed about technological developments considering the benefits they provide for human health. This study uses data from people with kidney issues to diagnosis chronic kidney disease. These include inherited polycystic kidney disease, urinary system issues, infections, and disorders associated to kidney inflammation. 500 million individuals worldwide have kidney problems, making it the most common chronic disease. However, chronic kidney disease (CKD) affects around one in six people.

Large amounts of data are being produced as a result of the substantial technological advancement in computer science. There are a lot of medical databases due to improvements in healthcare database management systems. Data mining is an increasingly popular field of study that deals with the extraction and management of knowledge from massive volumes of different information. In the current analysis, they have concentrated on the implementing of classification algorithms in the area of bioinformatics and medical science. This paper's major focus is on the categorization of CKD utilizing a data mining method in the field of medical bioinformatics. The dataset is first classified the optimal algorithm for chronic kidney disease prediction diagnosis is then selected. Recognizing symptoms in patients is the initial stage in prediction, which is followed by identifying ill people among a large population of healthy and sick people. Therefore, this analysis primary objective is to correctly predicted class in each instance by utilizing classification algorithm to analyze data from a dataset on chronic kidney disease.

The goal of this model is to develop and verify prediction models for problems with the kidneys that are chronic. Evaluation of damage to the kidneys, which indicates a need for kidney dialysis or transplantation initially, will be the major objective. These models help the doctor's assess the danger and seriousness of the problem as well as the best strategy for the treatment moving ahead. They also provide the patient lifestyle recommendations for leading a healthy life. Support Vector Machine, data mining techniques, and the ability to predicted the future incidence of specific illnesses that may have negative effects or make it possible to identify patterns in data collection.

Based on the patient's present lifestyle, the model's objective is to predict whether or not they will eventually acquire chronic renal disease. When selecting a course of treatment, with the use of this information, the glomerular filtration rate (eGFR) may be affected by kidney problems, which the doctor can evaluate. The estimated Glomerular Filtration Rate (eGFR), which measures kidney function, is used to determine the amount of damage to the kidneys. Filtration of the blood is the kidney's main function. The "silent killer" is kidney failure since it often progresses without any preceding symptoms or warning signs. A reduction in the kidney's function that occurs over months or years is known as chronic kidney disease. A common cause of kidney damage is diabetes, along with high blood pressure. A serious health issue that affects individuals all around the world is chronic kidney disease. If they do not receive help, those with limited financial resources risk suffering significantly the proper therapy for chronic kidney diseases. The most effective test to determine the severity of CKD and the condition of the kidneys is a measurement of glomerular filtration rate (GFR). It might be determined by utilizing information on age, gender, blood creatinine levels, and other variables. It's usually preferable to get sick early. As a result, major sickness is preventable [14].

Models that are suggested to predict renal disease especially utilize machine learning techniques. Compared to the current chronic kidney failure model, the database may be expanded with more data. That is, the accuracy of the predict will increase more data collected from people with chronic kidney disease's are added (even though the data must be trustworthy). In order to improve the prediction's accuracy, this approach can also be done with the support of experts to identify other characteristics that contribute to chronic kidney disease. These components might be added to the fabric paper as features.

The primary purpose of this investigation is to contrast several methods of intelligent machine learning-based kidney damage prediction. For a large number of investigations, the accuracy rate was still maintained. With this eGFR measurement, they can determine the patient's present condition. Following is the structure of the remaining portion of the paper: In Sect. 2, the Literature Survey is introduced, Sect. 3 explains Machine Learning Techniques based Chronic Kidney Disease Detection. The analysis of the experimental results is presented in Sect. 4, and Sect. 5 presents the paper to a conclusion.

2 Literature Survey

C. Bemando, E. Miranda, M. Aryuni (2021) et al. [2] used classifiers like Bernoulli Naive Bayes (NB), Random Forest, and Gaussian NB to investigate the relationship between such factors and the characteristics of blood-related disorders.

In a number of different ways, these three algorithms predict and provide statistical results. They observed this analysis, the assessed accuracy of the Naive Bayes technique was more stronger than that of different calculations.

R.S. Walse, S.D. Khamitkar, G.D. Kurundkar, P.U. Bhalchandra, A.A. Muley, S.N. Lokhande (2020) et al. [3] a statistical analysis method was developed. The authors used a range of machine learning techniques to examine the CKD dataset. To improve prediction, among the methods they examined, were decision trees, naive bayes, and random forest; they showed in terms of classification accuracy, random forest performed better than the other methods.

A. Nithya, D.R. Ramji, N. Venkatadri, A. Appathurai C.A. Palagan (2020) et al. [4] created a strategy for utilizing a dataset to forecast chronic kidney disease. Created a technique for classification and cluster-based analysis on the dataset for kidney problems. To acquire the images, the researchers used K-Means clustering on a number of image sets that were most similar to one another. They evaluated classification accuracy using artificial neural networks for prediction of kidney disease from image.

R.P. Ram Kumar, Sanjeeva Polepaka (2020) et al., [5] produced a method in the medical industry for diseases prediction. In addition to other machine learning techniques, they used Random Forest and CNNs. Regarding classification, recall, accuracy, and F1-score for datasets on illnesses, these methods offer better results. In terms of statistical performance and accuracy, Random Forest performed better in this experiment than other algorithms.

Singh H., Navaneeth N.V., Pillai G.N. (2019) et al. [8] created a method for estimating diseases datasets relevant to medicine. A support vector machine classifier was utilized to enhance the accuracy of the prediction. Over time, the author and accuracy both improved. Developed a method for predicting disease in the medical field. In this study, the author used classification methods based on back-propagation Neural networks (NN) and Logistic regression (LR). These two approaches provide different results, with logistic regression and statistical analysis providing a more precise model than previous methods.

D.D. Patil, V.M. Thakare, R.P. Singh, A.K. Gulve (2018) et al. [9] generated a database of medical conditions related to Electrocardiogram (ECG) arrhythmias. On a dataset of diseases, the authors used three machine learning techniques: a Cuckoo search-optimized neural network, a support vector machine evaluated enhanced accuracy. Dataset for statistical analysis of observed disease. They used support vector machines and other machine learning techniques to produce better outcomes in terms of negative predictive value, positive predictive value, sensitivity, and specificity.

Al Imran Abdullah, Nur Amin Md, and Tuj Johora Fatema (2018) et al. [10] evaluated datasets for chronic kidney illness were analyzed using machine learning techniques. To carry out statistical studies, the researchers used logistic regression and feed-forward neural networks such as sensitivity, precision, Area under the curve (AUC), accuracy, and compared to past methods, they provided better outcomes. Convolutional neural networks (CNN) and hybrid Support Vector Machines (SVM) were two machine learning approaches are utilized. The prediction indicated higher results of accuracy, sensitivity, and specificity.

Arasu, S. D., & Thirumalaiselvi (2017), R et al. [15] similar estimations demonstrate that the creation of missing values in the dataset, a unique technique called Weighted Average Ensemble Learning Imputation (WAELI) lowers the accuracy of CKD. To complete the missing numbers and get eliminated of incorrect information, the data cleaning procedure is required since the current approaches utilize a data preprocessing methodology. In various phases of CKD, a recalculation mechanism is included the missing values are computed and added the important. Although the current approaches are efficient, it is necessary to examine the values for CKD using a dataset created by a healthcare professional. The Feature selection (FS) procedure, it is essential for data categorization and is used to take a more focused collection of rules with particular objectives out of the training datasets.

Rajendra Acharya U., Shu Oh. Lih, Tan Y., Adam M., Hagiwara, Gertych J. H. A., Tan R. S. (2017) et al. [16] Review the medically connected disease dataset for improved statistical analysis results. They applied a number of machine learning approaches and algorithms on the ECG dataset, including Convolutional neural networks (CNN), effectively classifying data. For a chronic kidney disease dataset, they utilized a system or approach. The classification accuracy for the disease dataset was obtained by the authors using CNN machine learning.

P. Swathi Baby, T. Panduranga Vit (2015) et al. [17] developed a study on the application of machine learning to cancer diagnosis and prognosis. This study should provide an overview of many contemporary medical techniques that are used to forecast the discovery of cancer. To recognize the progress made in cancer detection up to this point, new printed information has to be evaluated. The study in this case focused on machine learning algorithms including K Star, J48, Random Forest, and Naive Bayes. Thus, for the given dataset, K-Star and random forest (RF) are the most simple approaches.

Chetty N., Vaisla K. S., and Sudarsan S. D (2015) et al. [18] three stages are used to propose a wrapper technique for CKD identification: a framework is developed using data mining, characteristics are selected using the best first search methodology the wrapper subset attribute evaluator, classification methods are then used. The results of the experiment showed that the accuracy was higher for the smaller dataset than it was for the larger dataset. Created a framework for improving CKD quality. Three processes are included in this framework: FS, ensemble learning, and classification. High classification accuracy is obtained by combining the correlation-based FS (CFS) with the K-nearest neighbor (KNN) classifiers. Developed a new technique for CKD detection using filter and wrapper methods. The simulation's results showed that better classification performance does not always equivalent to reduced characteristics.

Mahfuzah Mustafa, Mohd Nasir Tab (2012) et al. [19] provided the analytical effort to forecast internal organ disorders using Random Forest (RN) and Support vector machines (SVM). The tasks aims to compare the accuracy and execution times of two algorithms to determine which performs better. Finally, it is determined that the SVM's performance is better than that of the alternative formula.

Andrew Kusiak, Bradley Dixonb, Shital Shaha (2005) et al., [20] chemical examination with the aim of predicting the survival of urinary organs, data processing technologies are assessed for predicted performance. Several data processing methods (including logical regression, call trees, and artificial neural networks) were used in this investigation to gather data about how these variables relate to patient survival. By comparing the efficiency of data processing procedures, information is retrieved. Using information collected from multiple chemical analysis places, the concepts presented in this analysis are tested and to use in experiments. Results have been reported. Finally, SVM is advised for chemical analysis of urinary organs to provide more accurate and effective results.

3 Machine Learning Techniques Based Chronic Kidney Disease Detection

Machine learning techniques based chronic kidney disease detection is presented in this analysis. The presented model's architecture is shown in Fig. 1.

The CKD dataset was used for the investigation. Datasets for CKD patients.

The aim of the research presented in this paper is to use intelligent machine learning techniques to a comparative comparison of renal disease prediction. The block diagram of the suggested system was shown in Fig. 1. Three categorization algorithms are used in this work to predict the existence of chronic renal disease in individuals. The classifiers included the Hybrid SVM-RF classifier, Decision Tree (DT), and Area Unit Logistic Regression (LR). The chronic illness data set was acquired and utilised by each classifier to forecast the disease; therefore, the classifier's efficacy is assessed according to its specificity, precision, sensitivity, and accuracy. The following is how the design functions: The dataset for patients with chronic kidney disease (CKD) is gathered and put into the DT, LR, and Hybrid SVM-RF classifiers. CKD prediction is carried out using the implemented techniques. Because the data set for this paper's prediction was taken from the University of California, Irvine (UCI) machine learning repository. There are values and characteristics in the dataset. The CKD dataset was utilized in the study. When the number is "1," it signifies that the patient has chronic kidney disease (CKD); when it is "0," it suggests that the patient does not.

Removing undesired noise and outliers from a dataset can help avoid the need for data preparation before model development to prevent the model from differing from the training set that was planned. This stage addresses the model's efficiency. For the purpose of creating a model, data must first be collected, cleaned, and available. Next, a null value search is performed on the dataset. There are no null values in this datasets. The output values "False" and "0" in this case signify that there are no null values. While processing the uneven dataset and finishing data preparation, next is the model development phase. The task's accuracy and efficacy are increased by splitting the data into training and testing parts with an 80/20 training to testing ratio. After that, a number of classification techniques are used to train the split model. The decision tree, logistic regression, and support vector are the classification techniques utilized in this analysis.

White blood cell count, albumin, blood pressure, blood urea, serum creatinine, and sugar, potassium, and other characteristics are all positively connected with one another, based on the absolute values of the correlations between the class label and the attributes in the heat map. All elements that are favourably correlated are taken into consideration for the implementing prediction. Simply stated, there are five different sets of values for each albumin molecule. Using a urine protein test, the amount of albumin is determined When there is a high concentration of protein in the urine, it signifies that a disease, fever, or physical activity has damaged the kidneys' filtration mechanisms. A number of tests need to be performed during the period of many weeks in order to make the diagnosis.

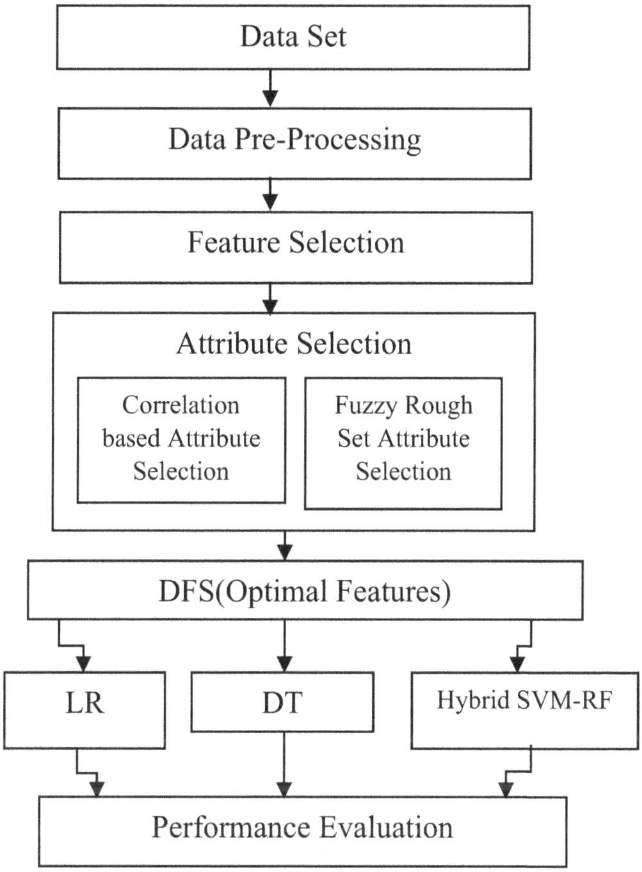

Fig. 1. The architecture of presented model.

Blood creatinine and creatinine are interchangeably for serum creatinine. Muscle breaks down the molecule creatine to produce creatinine as a byproduct.

Through way of the kidneys, creatinine is removed from the body. Blood's level of creatinine will be determined by this tests. A component of creatine is the metabolic cycle that produces the energy needed for muscle contraction. It is the same rate that the body makes creatine and creatinine. Problems with diabetes, congestive heart failure, a high-protein diet, other factors that might raise the blood level of creatinine include dehydration. The optimal range for serum creatinine levels in men and women is 0.7–1.3 mg/dL and 0.6–1.1 mg/dL, respectively. Moreover, an increase in blood pressure on blood vessel walls results in the development of hypertension, or high blood pressure. If blood pressure is not adequately controlled or treated, it can result in heart attacks, strokes, and CKD. However, people with CKD may experience hypertension.

The practise of attribute selection helps data mining systems performs better. These procedures include identifying the important qualifiers that affect the analysis and removing unimportant characteristics from the data collection. The correlation technique, which is often used in the literature as an attribute selection strategy, implements a fuzzy rough set-based method of attribute selection in this study.

Correlation-based attribute selection (CBAS) is a technique, two random variables are evidence that a linear connection. The correlation coefficient determines if a linear link exists and the powerful. The R sign represents the correlation coefficient. The range of the R coefficient ranges from -1 to $+1$. When $R = -1$, the linear connection is completely negative, and a linear connection that is only positive when $R = +1$. There's no connection if $R = 0$.

For usage in data mining applications, several studies have been done to choose whether to produce fuzzy rough sets. Studies on quality damage in fuzzy rough set applications are less numerous than those on learning algorithms. The a-interrupts of fuzzy sets and fuzzy sets with looser fuzzy equivalency features have been used to create various fuzzy rough set definitions in this context. To cross-breed fuzzy and rough sets, the two fundamental approaches are cognitive and constructive. Lower and higher methods are built on the founding approach's essential foundation, which is fuzzy relations in the universe. Only fuzzy equivalence connections were utilized at first, and later generalization studies added bilateral relations additionally.

A heuristic way for assessing the quality of features is the Density-based feature selection (DFS) method. When a feature is good, there is less overlap between every class and the other classes, according to the basic hypothesis. The DFS method takes into consideration the distribution of characteristics across all classes as well as their connection while exploring and assigning rankings. DFS's first step is to separately calculate each feature's probability density function (PDF) for every class. The ranking of the characteristics based on the overlap area is the subsequent stage. It can happen that categorize the conventional approaches for calculating PDF into parametric and non-parametric categories. The earlier technique makes the assumption density estimation's single objective is to select suitable values for the distribution's mean and variance given that the data has a Gaussian distribution. In contrast, non-parametric techniques compute the density directly from the instances rather than making any assumptions about the density function. It should be noted that many pattern recognition programmes failed to a set standard for calculating the primary data's density. As a result, the recommended solution, which serves as a replacement utilizes the parametric method.

$$\mathrm{P(x)} = \frac{K}{NV} \tag{1}$$

where N is the overall number of instances, p(x) is the number of instances in V (k), and the value of the computed PDF for example x. Increased N and lowered V can be discovered to determine the precise PDF. After estimating the PDF for each class, the next step is to examine the feature's value using the estimated PDFs for the classes.

The structure of the proposed system is shown in Fig. 1. The CKD prediction dataset is used by the framework. The DT, hybrid-SVM-RF, and logistic regression techniques have been utilized after preprocessing and feature selection.

Predictions for both discrete and continuous characteristics is possible with the Decision Tree (DT) approach, a classification and regression methodology. The technique estimates discrete qualities using a dataset's input column connections. Using the states or, more specifically, the states of a column that you designate as predictable based on the values of those columns. The method finds the input columns that are related to the column that is predicted. Because it precisely replicates the phases an individual goes through when making a decision in real life, it's simple to understand the decision tree. When dealing with decision-making issues, it could be highly helpful. The most effective strategy of action is to operate through every option for resolving a problem. Data cleaning is not as crucial it is with other techniques.

The widely used statistical technique of logistic regression is used to model binary outcomes. Utilizing a number of learning strategies, logistic regression is used in statistical research. The Logistic Regression (LR) algorithm was created using a neural network variation. However, easier to set up and utilize, this strategy is quite similar to neural networks. A categorical dependent variable's output is predicted using logistic regression. A discrete or categorical output is required. It may be 0 or 1, true or false, yes or no, etc. However, probabilities ranging from 0 to 1 are provided. The applications of linear regression and logistic regression are extremely similar. Instead of using linear regression to handle regression issues, classification issues are handled using logistic regression.

A supervised learning technique called a support vector machine classifier is helpful for both regression and classification. The fundamental principle of hybrid-SVM-RF is that separates the data points according to a binary classification approach in order to find a hyperplane when there are several possible contributions. It functions perform well in the situation of high dimensional spaces and effectively manages the outliers. It does classification using decision functions, commonly referred to as support vectors. In persons aged 18 to 75 with a diagnosis of diabetes, the kidney health evaluation measure evaluates the effectiveness of a thorough kidney evaluation that includes eGFR. Leo Breiman invented it in order to outperform Decision Trees in terms of performance. The attribute is chosen at random in this approach for building training sets. The attribute is determined for each level in develop the tree. The most prevalent characteristic is selected the features have been integrated.

4 Result Analysis

Machine Learning Techniques based Chronic Kidney Disease Detection with Performance analysis of Fuzzy rough set and correlation attribute selection is presented in this analysis. A few parameters are calculated and examined in order to evaluate performance and measure system stability. To assess the model's performance, accuracy, sensitivity, specificity, and precision were taken into consideration. The accuracy is measured by dividing the total number of occurrences in the dataset by the number of cases that were successfully recognized. Accuracy is the number of properly recognized samples divided by the overall sample count.

Sensitivity: The sensitivity indicates the model detects kidney disease. Sensitivity provides the method's capacity to recognize abnormal instances.

Specificity: The specificity measures a method's capacity to recognize typical instances (Table 1).

The development of an intelligent expert system that can categorize kidney illness utilizing LR, DT, and hybrid SVM-RF is the main goal of this analysis.

Table 1. Performance Analysis

Classifiers	Hybrid-SVM-RF	LR	DT
Accuracy (%)	98.89	85.57	76.87
Specificity (%)	93.56	80.56	75.89
Sensitivity (%)	88.67	77.67	70.35
Precision (%)	80.67	70.45	65.78

Precision: Positive prognostic value is another name for it's information. The average possibility of it is thorough since a successful retrieval is required. Precision = variety of true positives/Number + False positives.

In Fig. 2, the accuracy of the hybrid SVM-RF method and the LR, DT-based methodology are compared.

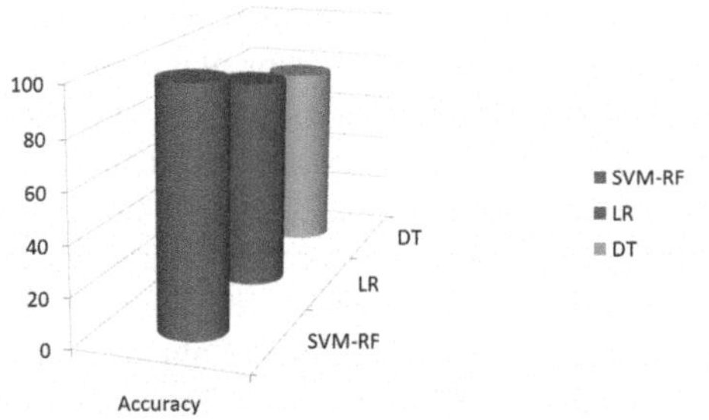

Fig. 2. Accuracy Comparison Graph

Figure 3 displays the specificity comparison between the hybrid SVM-RF method and the LR, DT-based strategy.

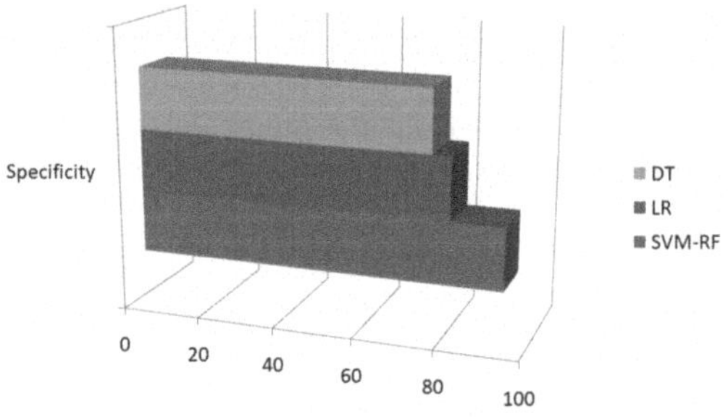

Fig. 3. Specificity Comparison Graph

Figure 4 compares the hybrid SVM-RF technique's sensitivity to that of the LR, DT-based approach.

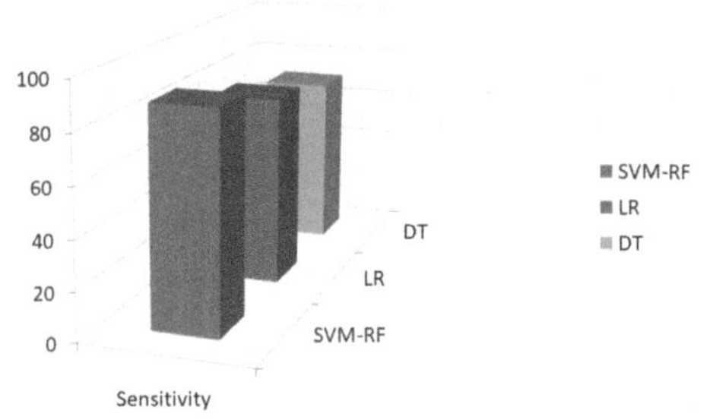

Fig. 4. Sensitivity Comparison Graph

Figure 5 compares the hybrid SVM-RF method's precision to that of the LR, DT-based technique.

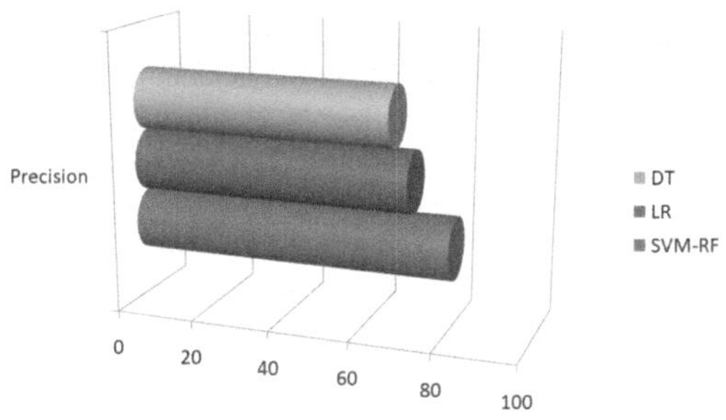

Fig. 5. Precision Comparison Graph

5 Conclusion

Machine Learning techniques based chronic kidney disease detection with Performance analysis of Fuzzy rough set and correlation attribute selection is presented in this analysis. Two data sets are produced after using the fuzzy rough set-based attribute selection (FRCBAS) and correlation-based attribute selection (CBAS) methods on the UCI chronic kidney dataset, with the objective to evaluate the way attribute selection techniques affect the effectiveness of classification systems. Then, including the raw data set, these three data sets are affected to four categorization methods. The significance of applying investigations using data mining and machine learning show the use of attribute selection approaches. It is clear from the methods used to choose the attributes for the research that FRSBAS outperformed CBAS in terms of effectiveness. Both the patient's health and their feeling of confidence in medical institutions are lost as a result of misunderstanding in the health sciences. Therefore, in particular studies related to human health, the exact and crucial data attributes must be chosen in order to get the most accurate results. According to this study, while examining the intermediate values, particularly in the domain of health analysis, it is important to keep considerations for the fuzzy logic data analysis that must be consider. It is intended to enhance the quantity of data sets and records in future research and to analyze the data using various advanced data mining techniques. The analysis's results suggest that a hybrid SVM-RF technique can more precisely forecast chronic kidney disease. Performance in terms of Specificity, Precision, Sensitivity, and Accuracy are evaluated for the provided technique. Hybrid SVM-RF outperforms other based designs in terms of performance and prediction accuracy.

References

1. Senan, E.M., et al.: Diagnosis of chronic kidney disease using effective classification algorithms and recursive feature elimination techniques. J. Healthc. Eng. **2021**, 004767 (2021)
2. Bemando, C., Miranda, E., Aryuni, M.: Machine-learning-based prediction models of coronary heart disease using naive Bayes and random forest algorithms. In: 2021 International Conference on Software Engineering and Computer Systems and 4th International Conference on Computational Science and Information Management (ICSECS-ICOCSIM), pp. 232–237. IEEE (2021)
3. Walse, R.S., Kurundkar, G.D., Khamitkar, S.D., Muley, A.A., Bhalchandra, P.U., Lokhande, S.N.: Effective use of naive Bayes, decision tree, and random forest techniques for analysis of chronic kidney disease. In: Senjyu, T., Mahalle, P.N., Perumal, T., Joshi, A. (eds.) International Conference on Information and Communication Technology for Intelligent Systems. Springer, Singapore (2020)
4. Nithya, A., Appathurai, A., Venkatadri, N., Ramji, D.R., Palagan, C.A.: Kidney disease detection and segmentation using artificial neural network and multi-kernel k-means clustering for ultrasound images. Measurement (2020)
5. Ram Kumar, R.P., Polepaka, S.: Performance comparison of random forest classifier and convolution neural network in predicting heart diseases. In: SrujanRaju, K., Govardhan, A., Padmaja Rani, B., Sridevi, R., Ramakrishna Murty, M. (eds.) Proceedings of the Third International Conference on Computational Intelligence and Informatics. Springer, Singapore (2020)
6. Hussein, N.R., Saleem, Z.S.M., Ibrahim, N., Musa, D.H., Naqid, I.A.: The impact of COVID-19 pandemic on the care of patients with kidney diseases in Duhok City, Kurdistan Region of Iraq. Diabet. Metab. Syndr. **14**(6), 1551–1553 (2020)
7. Qin, J., Chen, L., Liu, Y., Liu, C., Feng, C., Chen, B.: A machine learning methodology for diagnosing chronic kidney disease. IEEE Access **8**, 20991–21002 (2020). https://doi.org/10.1109/ACCESS.2019.2963053
8. Singh, H., Navaneeth, N.V., Pillai, G.N.: Multisurface proximal SVM based decision trees for heart disease classification. In: 2019 IEEE Region 10 Conference (TENCON), pp. 13–18. IEEE (2019)
9. Patil, D.D., Singh, R.P., Thakare, V.M., Gulve, A.K.: Analysis of ECG arrhythmia for heart disease detection using SVM and Cuckoo search optimized neural network. Int. J. Eng. Technol. **7**(217), 27–33 (2018)
10. Al Imran, A., Nur Amin, M., Johora, F.T.: Classification of chronic kidney disease using logistic regression, feedforward neural network and wide & deep learning. In: 2018 International Conference on Innovation in Engineering and Technology (ICIET), pp. 1–6. IEEE (2018)
11. Aljaaf, A.J., et al.: Early prediction of chronic kidney disease using machine learning supported by predictive analytics. In: 2018 IEEE Congress on Evolutionary Computation (CEC), pp. 1–9. IEEE (2018)
12. Subasi, A., Alickovic, E., Kevric, J.: Diagnosis of chronic kidney disease by using random forest. In: Badnjevic, A. (eds.) CMBEBIH 2017. IFMBE Proceedings, vol. 62. Springer, Singapore (2017). https://doi.org/10.1007/978-981-10-4166-2_89

13. Webster, A.C., Nagler, E.V., Morton, R.L., Masson, P.: Chronic kidney disease. The Lancet **389**(10075), 1238–1252 (2017)
14. Borisagar, N., Barad, D., Raval, P.: Chronic kidney disease prediction using back propagation neural network algorithm. Proc. Int. Conf. Commun. Netw. **19–20**, 295–303 (2017)
15. Arasu, S.D., Thirumalaiselvi, R.: A novel imputation method for effective prediction of coronary Kidney disease. In: 2017 2nd International Conference on Computing and Communications Technologies (ICCCT), pp. 127–136 (2017)
16. Rajendra Acharya, U., et al.: A deep convolutional neural network model to classify heartbeats. Comput. Biol. Med. **89**, 389–396 (2017)
17. Swathi Baby, P., Panduranga Vital, T.: Statistical analysis and predicting kidney diseases using machine learning algorithms. Int. J. Eng. Res. Technol. **4**(07), 206–210 (2015)
18. Chetty, N., Vaisla, K.S., Sudarsan, S.D.: Role of attributes selection in classification of Chronic Kidney Disease patients. In: 2015 International Conference on Computing, Communication and Security (ICCCS), pp. 1–6 (2015)
19. Mustafa, M., Tab, M.N.: Comparison between SVM and ANN classification in brain balancing application via spectrogram image. J. Comput. Sci. Comput. Math. **2**(4), 17–22 (2012)
20. Kusiak, A., Dixonb, B., Shaha, S.: Predicting survival time for kidney dialysis patients: a data mining approach. Comput. Biol. Med. **35**, 311–327 (2005). Elsevier Publication

Segmentation and Classification of Unharvested Arecanut Bunches Using Deep Learning

R. Dhanesha[1(✉)], D. K. Umesha[2], Gurudeva Shastri Hiremath[3], G. N. Girish[4], and C. L. Shrinivasa Naika[5]

[1] Department of Studies in Computer Science, Davangere University, Davangere, Karnataka, India
dhanesh025@davangereuniversity.ac.in

[2] Shivasharane Akkamahadevi Governament Womens Polytechnic, Shiralakoppa, Shivamogga, Karnataka, India

[3] Department of Computer Science and Engineering, SJEC, Mangalore, Karnataka, India

[4] Computer Science and Engineering Group, Indian Institute of Information Technology, Sricity, Chittor, India
girish.gn@iiits.in

[5] Department of Studies in Computer Science and Engineering, University B.D.T. College of Engineering, Davangere, Karnataka, India
naika2k6@vtu.ac.in

Abstract. One of the most profitable commercial palms in India is the arecanut. The market price of arecanuts, which is determined by the arecanuts maturity level, is the only factor that determines the profitability of arecanut plantation farmers. To reduce the financial loss of the farmer, knowing maturity level of the unharvested arecanut bunch is a tedious job due to lack of expertise. We presented an automated technique to determine the maturity level of the unharvested arecanut bunches in order to address this problem. The proposed method works on two phases: first, an optimized U-Net model was used for semantic segmentation of the arecanut bunch images then the classification of the segmented arecanut bunches was performed using transfer learning approach. Experiments were conducted on RGB and saturation channel of HSV color space images, and a comparative study of RGB and saturation channel of HSV color space is presented. Result of the experiments show that the segmentation and classification of the arecanut bunches from input image is efficient using RGB color space.

Keywords: Arecanut · U-Net · Semantic segmentation · Convolutional Neural Network · maturity · MobileNetV2 · Transfer learning technique

1 Introduction

Areca palm (Areca catechu L.) is a very useful commercial crop of the most South-east Asian and South Asian countries such as Myanmar, India, Bangladesh, Indonesia, China, Malaysia, Philippines, Thailand, Vietnam etc. The fruit or seed (endosperm) of this palm is called arecanut [1]. Arecanut farming yields high profits and it is used in

T. Shreekumar et al. (Eds.): ISCComm 2023, CCIS 2231, pp. 158–170, 2025.
https://doi.org/10.1007/978-3-031-75605-4_11

the production of betel nut or supari, and also used in some Ayurvedic and traditional Chinese medicine. Because of its many applications, arecanut became one of the major commercial crops along with having religious significance in Indian subcontinent [2]. India is the largest producer of arecanut in the world with Karnataka and Kerala states are the major contributors in the production [3]. Due to its various application and high demand based on maturity level, identifying and harvesting matured arecanut is essential for the farmers to get better profit [4]. Knowing maturity level of unharvested arecanut bunches to the farmer becomes paramount importance and it should be ensured that only matured arecanut bunches were harvested to maximize their profitability. As a result, technological assistance is required for farmers to automatically identify the maturity level of unharvested arecanut bunches in order to increase profitability while lowering labour expenses.

Nowadays, image processing and deep learning techniques are being used to assist farmers in obtaining high yields and profit by grading, classifying and finding the maturity level of fruits and vegetables prior to harvesting [5–7]. Because the background of the arecanut bunch may contain tree leaves, occlusions with leaves or flowers and sky, the classification of the arecanut bunches is dependent on the proper region of interest of the bunch in the image. The undesired backdrop in the arecanut bunch images obtained from the tall plants makes processing these images quite challenging. In the literature few attempts were made on automatic segmentation and classification of arecanut images. To segment the arecanut from the image, [8] have employed three sigma control limits. The blue color part is generally suppressed in the segmented region, and only the red and green color parts are used to classify the arecanut. [9], presented a method to classify the diseased arecanut using image texture features such as Local Binary pattern (LBP), Haar Wavelets, GLCM (Gray level co-occurrence matrix), and Gabor features. Dhanesha, et al. [10] presented a segmentation method using HSV color model to automatically segment arecanut bunches before harvesting and obtained results show that HSV color model-based segmentation gave better segmentation accuracy. [11] proposed a method to segment the arecanut bunch from a given image using YCgCr color space.

It is evident from the literature survey that very few research work have been done on the segmentation of unharvested arecanut bunch images using traditional image processing techniques. Convolutional Neural Networks (CNNs) have great ability of extracting hierarchical features for better image semantic segmentation [12]. To the best our knowledge none of the deep learning based methods were proposed in the literature for automatic segmentation and classification of arecanut bunches.

In this proposed work, convolution neural network architectures were employed to automatically identify the maturity levels of the arecanut bunches. This paper presents two contributions. First, segmentation of both RGB and saturation component of arecanut bunch images using optimized U-Net model which removes unwanted background data from the images. Second, classification of the segmented arecanut bunches using MobileNetV2 [4] model, also a comparative study of classification was performed on both RGB and saturation component input images.

The rest of this paper is organized as follows. In Sect. 2, the proposed methodology is explained. In Sect. 3, the experimental setup has been discussed. Results and discussions are presented in Sect. 4 and finally, in Sect. 5, the conclusion is drawn and several recommendations for future work are indicated.

2 Methodology

The three steps in our proposed method are image pre-processing, semantic segmentation of the arecanut bunch regions followed by the classification of segmented arecanut bunches based on the maturity level. These steps are shown in Fig. 1.

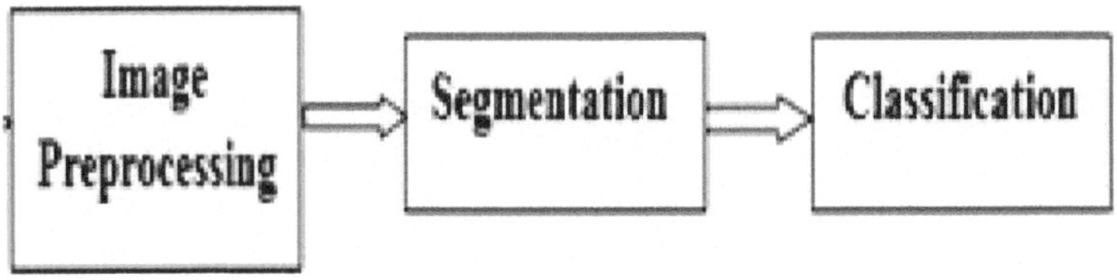

Fig. 1. Phases of the proposed method.

2.1 Dataset Description

Dataset used in [4] is utilized in this proposed method. Images of the dataset were collected from different arecanut plantations of Shivamogga, Davanagere and Chitradurga districts in Karnataka. The arecanut bunches of the considered dataset are evaluated and categorized by the expert human grader into two classes as matured arecanut bunch (which are ripened properly and ready to harvest) and immature arecanut bunch (not suitable for harvesting). The prepared dataset consists a total of 1017 images of arecanut bunches, in which 629 images are from immature and 388 are from matured arecanut bunch classes.

2.2 Image Pre-processing

Since, images in the dataset are 3–4 MB size, which consumes lot of memory and time for processing. So, each image was resized in to 224 × 224 pixels as per our model requirements.

The captured arecanut bunch images usually in RGB color space. Earlier works done on the different color models infer that HSV color space has a good capability of representing properties of human vision [13]. HSV color space image is a combination of hue, saturation and value (intensity).Hue stands for the fundamental color. Saturation, on the other hand, relates to the white light mixed with the hue and measures the color's purity. Value denotes the image's brightness or intensity.

The arecanut maturity levels are associated with the colorless of the fruit. In this work, we have considered the saturation component channel image for the input to the deep learning model. Because the saturation component of the image provides the colorfulness of a stimulus relative to its own brightness [14], saturation component can be computed from RGB image using Eq. (1)

$$Sat = \frac{3}{1 - (P1 + P2 + P3)}[min(P1, P2, P3)] \tag{1}$$

where P1, P2 and P3 are red, green and blue primary color components respectively. Figure 2 shows some of the sample images of both RGB and saturation component images used in this proposed work.

(a) (b)

Fig. 2. Sample images in the used dataset. (a) RGB images (b) Saturation component images

2.3 Segmentation with Proposed U-Net Model

Pre-processed arecanut bunch images (both RGB and saturation component) are segmented to exclude the backdrop so that aracanut images are localized. An optimized U-Net model which is inspired by Ranneberger's U-Net architecture [15], is employed for the semantic segmentation of arecanut bunches.

Figure 3, shows the architecture of proposed model. Optimized U-Net model consists of a contracting path (encoder) on the left side, bottleneck at the bottom and an expansive path (decoder) on the right side which makes it a 'U' shaped architecture. As shown in Fig. 3, all the open boxes corresponds to a multi-channel feature map, shaded box in encoder indicates the input layer and shaded box in decoder part represents the copied feature map from the encoder. The number of channels of the feature map were denoted on top of each box. Proposed U-Net model contains total 15 convolution layers.

Encoder: It has three depth levels, with two 3 × 3 convolutions layers, ReLU activation, Batch-Normalization [16], and a 2 × 2 max pooling operation with stride 2 at the end of each depth level. Number of feature channels in the first depth level is 16 and these feature channels were doubled for each further depth level of encoder as shown in the Fig. 3. The operation in the encoder is called down-sampling performed using max pooling layer, in which image size is decreased by half in each depth level of the encoder.

Bottleneck: The bottleneck contains two 3 × 3 convolution layers followed by ReLU activation and Batch-Normalization operation. Number of feature channels in bottleneck are 128.

Decoder: The Decoder block aims to revert the activations of the encoder to get the probability feature map of the size equivalent to the original input image. It uses transposed convolution to permit localization and this operation is called up-sampling. Each depth level of the decoder unit consists of a 2 × 2 up-sampling convolution, concatenation with corresponding feature map from encoder using skip connection, two 3 × 3 convolution layers which are of the same size feature map equivalent to the encoder block depth level, each followed by ReLU and Batch-Normalization operation. The final layer maps each 16-component feature vector to the appropriate number of classes using a 1 × 1 convolution followed by sigmoid activation. Table 1 shows the main differences in the proposed optimized U-Net model and Ranneberger's U-Net model.

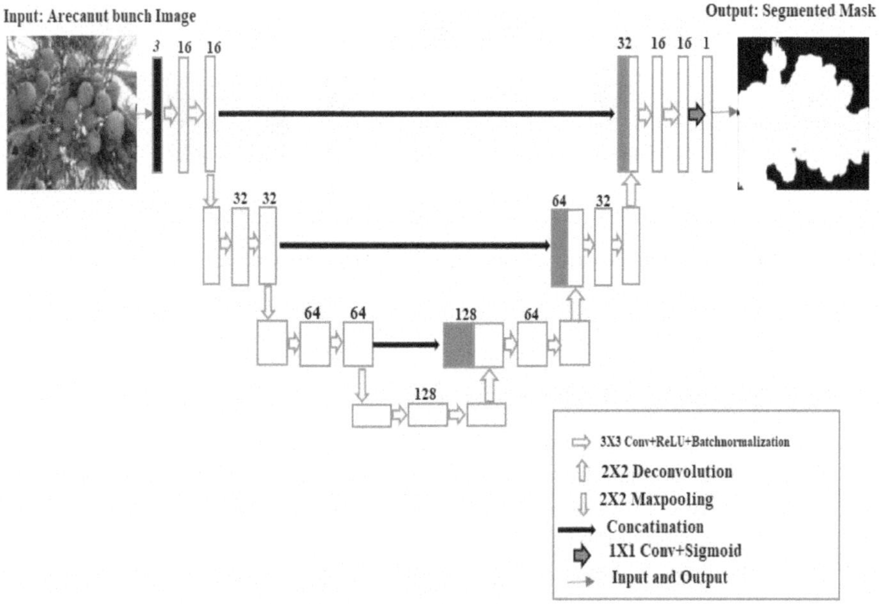

Fig. 3. Proposed U-Net model architecture.

The saturation component image will be in 224 × 224 × 1 size and these images are converted to 224 × 224 × 3 size as per our proposed U-Net model requirement. The model takes RGB and saturation channel of HSV arecanut images (size of the both inputs should be 224 × 224 × 3) as input and outputs its segmented binary mask. Corresponding segmented images can be obtained by multiplying the input image and binary mask obtained by the model.

2.4 Classification with Proposed MobileNetV2 Model

Once all the images are segmented and unwanted background is removed in the previous step (refer Subsect. 2.3), the classification of the segmented arecanut bunch images were done to classify input image into one of two classes namely "Mature" and "Immature"

Table 1. Comparison of original U-Net and proposed optimized U-Net.

Characteristic	Original U-Net	Proposed Optimized U-net
Number of Depth Level in encoder and decoder part	4	3
Each Depth level in encoder part contains	Two 3 × 3 convolutions + ReLU and a 2 × 2 max pooling	Two 3 × 3 convolutions + ReLU + Batch-Normalization and a 2 × 2 max pooling
Number of feature channels in each depth level	64,128,256,512	16,32,64
Padding in convolution operation	Padding is not used. Because of that size of the feature maps will be reduced after each convolution operation	Padding is used so that size of the feature maps before and after the convolution is same
Input and Output size	Input: 572 * 572 * 1 Output: 388 * 338 * 2	Input: 224 * 224 * 3 Output 224 * 224 * 1
Data augmentation during the model training	Used	Not used

depending on their maturity level. Recent studies in the literature have showed that for image classification problems with the smaller data sets, transfer learning techniques with pre-trained models have given better results [17]. In transfer learning a pre-trained model is fine-tuned to work on new image dataset. We used transfer learning technique for classification of segmented arecanut images because it makes model training efficient with few resources.

In the proposed method, MobileNetV2 [12] model, which is trained on more than a million images (with 1000 classes) from the ImageNet (ILSVRC-2012-CLS) Large-Scale Visual Recognition Challenge [18], is used as base model for feature extraction. On top of the base model, we have added flatten layer, dense layer and SoftMax layer [19] as a classifier layer to classify the input arecanut bunch images into 2 classes. The flatten layer is used to covert the multidimensional features extracted by MobileNetV2 base model to single dimensional feature vector. The output of the flatten layer is given to dense layer with 512 nodes and then the feature are classified into two classes using softmax classifier.

3 Experimental Setup

The models in the proposed method were built using Kera's (version-2.4.0) [20] framework with TensorFlow (version-2.4.1) [21] on Google Collab [22] using python programming language.

The proposed optimized U-Net model is tuned for hyper-parameters like learning rate, batch-size, dropout etc., during training process. The "Random Search" algorithm was used to search for best hyper-parameters configuration for the proposed model.

The hyper-parameters used while training the optimized U-Net model are tabulated in Table 2. The proposed model takes image with size 224 × 224 × 3 as input and produces a segmented mask of size 224 × 224 × 1 as output. Batch-Normalization [18] is used to speed up training and dropout of 0.05 is used in both down-sampling and up-sampling operation to reduce the over-fitting. While training the model dice coefficient is used as loss function to measure the model loss. Adam optimizer [23] with learning rate 0.006 is used to update the weights.

In the proposed method, experiments were conducted on two different dataset, RGB and saturation images to analyze the effect of segmentation. To evaluate the model, each dataset is divided into three parts as described in Table 3. Using the training dataset's images as input and the corresponding ground truth masks as labels, the model is trained. Table 4 shows the training and validation dice coefficient of the proposed method on RGB and Saturation images. Figure 4, shows the training and validation dice coefficient of the proposed model on RGB and saturation images for 40 epochs.

Table 2. Hyper-parameter setting for the proposed U-Net model.

Hyper-parameter	Number of depth levels	Learning Rate	Dropout	Optimizer	Batch size
Values	3	0.006	0.05	Adam	16

Table 3. Description of number of images in training, validation and test datasets.

Color space	Training dataset	Validation dataset	Test dataset	Total
RGB Images	732	183	102	1017
Saturation Images	732	183	102	1017

Table 4. Comparison of training and validation dice coefficient of the proposed U-Net model on RGB images and saturation component images for 40 epochs.

	Training Dice coefficient	Validation Dice coefficient
RGB-Images	94.68%	90.67%
Saturation-Images	94.03%	90.24%

After the segmentation, arecanut bunch images are classified using proposed MobilenetV2 model. Experiments were conducted on five different datasets to analyze the effect of classification on different color space images and segmented images. The five datasets on which experiments were conducted are described in Table 5. The first two datasets are the same datasets used for segmentation, which contains RGB images collected from arecanut plantation and corresponding saturation component images. The third and fourth datasets contains automatically segmented arecanut bunch images

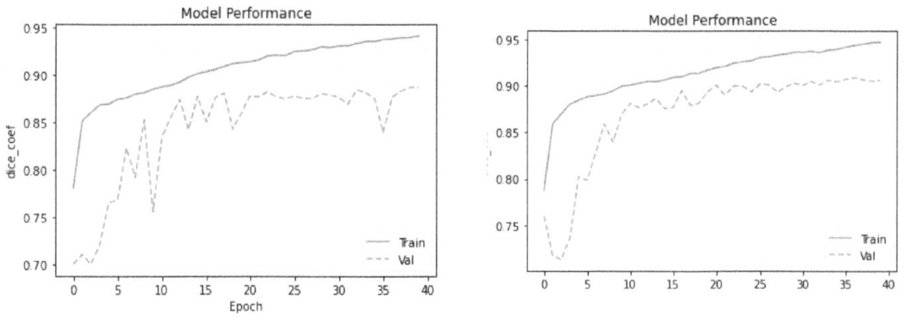

(a) (b)

Fig. 4. Dice coefficient vs. Epoch graphs of the proposed U-Net model for 40 epochs. (a) RGB images (b) Saturation image

obtained from proposed U-Net segmentation model for the RGB images and Saturation component images respectively. To compare the classification performance on segmented images we have created fifth dataset, which contains manually segmented images of the RGB images using photo shop tool. Each of these five datasets consists a total of 1017 images, which includes 388 mature and 629 immature arecanut bunch images.

Table 5. Description of different datasets used for classification using proposed MobileNetV2 model.

Name	Description
RGB images	Contains original RGB arecanut bunch images collected from arecanut plantation
Saturation images	Original RGB images converted to HSV images and then saturation channel is extracted from HSV image
Segmented-RGB images	This dataset contains segmented RGB images produced by proposed segmentation Model
Segmented-saturation images	Segmented images of the saturation images using proposed U-Net model
Manually segmented RGB images	RGB images are segmented manually using photoshop tool

Experiments were done on these 5 datasets mentioned in Table 5. Using proposed MobileNetV2 model. The base model in the proposed MobileNetV2 contains pre-trained weights from ImageNet and these weights are freeze during the training. Only the classier layer, which is added on top of the base layer, is trained on the datasets. A Dropout of 0.1 is used to reduce overfitting and model is compiled using "binary cross entropy" as loss and "Adam optimizer" [23] with learning rate 0.0001. Batch size 32 is used for training and validation datasets while training. Model is trained for 40 epochs using early stopping [24] strategy (which stops the training when validation accuracy is not changed for 10 epochs). Validation and training accuracy of the model for five different datasets is tabulated in Table 6. The training curves of the above method on five different datasets are shown in Fig. 5.

Table 6. Comparison of training and validation accuracy of the proposed MobileNetV2 model on different dataset for 40 epochs.

	Saturation Images	Segmented Saturation Images	RGB Images	Segmented RGB Images	Manually Segmented Images
Training	77.09	74.39	83.66	84.17	96.82
Validation	70.29	70.29	81.88	84.78	95.75

4 Results and Discussion

The segmented output of the segmentation model (U-Net) model on test data set is compared with the corresponding expert graded ground truth images to evaluate the model. There are many metrics available in literature to evaluate the Segmentation performance [25]. In semantic segmentation each pixel values are classified as two classes (Positive-mature and Negative-immature). Depending on this classification we can use Recall, Precision and Dice Co-efficient as evaluation metrics for segmentation.

Table 7, shows the obtained results of the proposed model on the test dataset with RGB and saturation component subsets. It can be inferred from the results of the Table 7 that proposed model can segment RGB images more accurately than the saturation images.

Qualitative results using the proposed segmentation method on RGB and saturation component images are shown in Fig. 6, first column shows the original image taken from arecanut plantation, second column shows manually created ground truth image, third column shows segmented binary mask obtained from proposed optimized U-Net architecture, and the fourth column shows segmented image which is produced by multiplying the binary mask with original images. The proposed classification model (MobileNetV2) is tested for their performance with five different test datasets and the results are tabulated in Table 8. From the results, it is observed that segmentation of arecanut images for localization of arecanut bunches improves classification accuracy compared to original non-localized input images.

The prediction of the proposed MobileNetV2 model on Segmented-RGB image dataset is shown in Fig. 7. The model is able to classify the test images into one of the mature and immature classes with an accuracy of 88.24%. In Fig. 7, predicted label and predicted confidence is displayed on top of each image. The bar graph which is right to the images shows the prediction probabilities of the proposed model for two classes (mature and immature). The class with heights probability is considered for the prediction of the image.

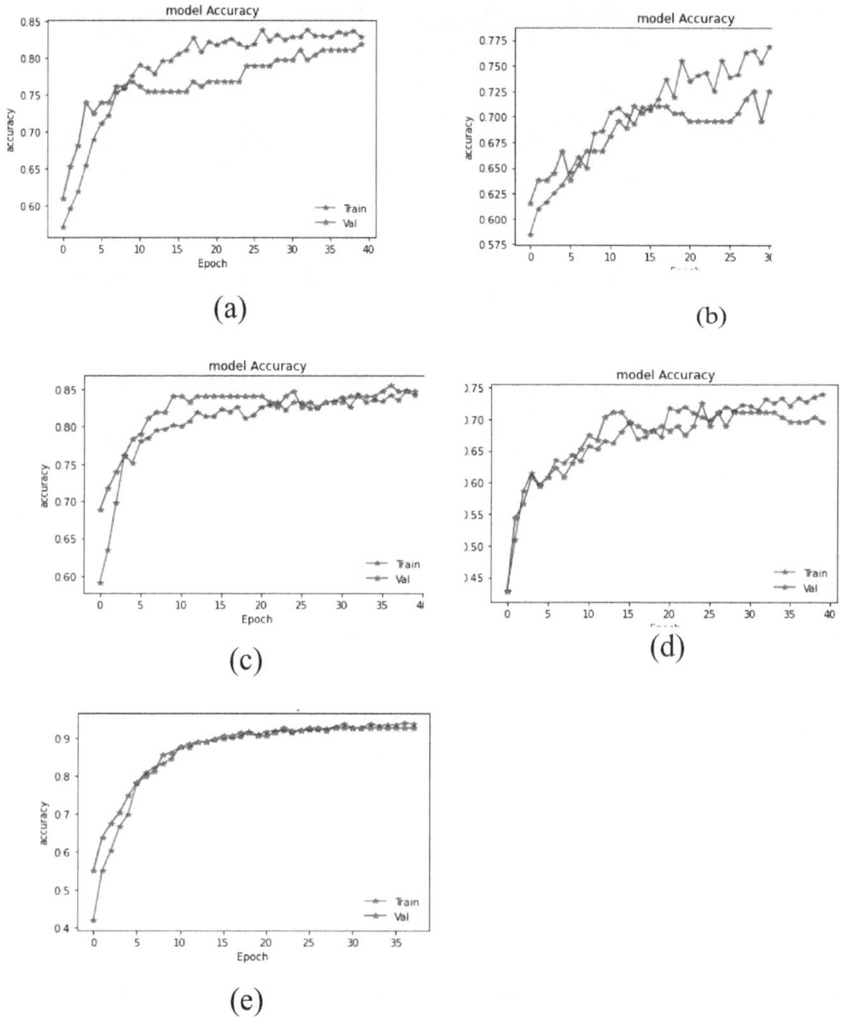

Fig. 5. Training accuracy vs. Epoch graph of the proposed MobileNetV2 model on different dataset for 40 epochs. a) RGB images, b) Saturation images, c) Segmented-RGB images d) Segmented-saturation images, e) Manually segmented RGB images.

Table 7. Comparison of different segmentation metrics of the proposed U-Net model on RGB and saturation test dataset.

	Dice coefficient	Precision	Recall
RGB-Images	91.27	93.30	76.50
Saturation-Images	90.30	91.85	77.72

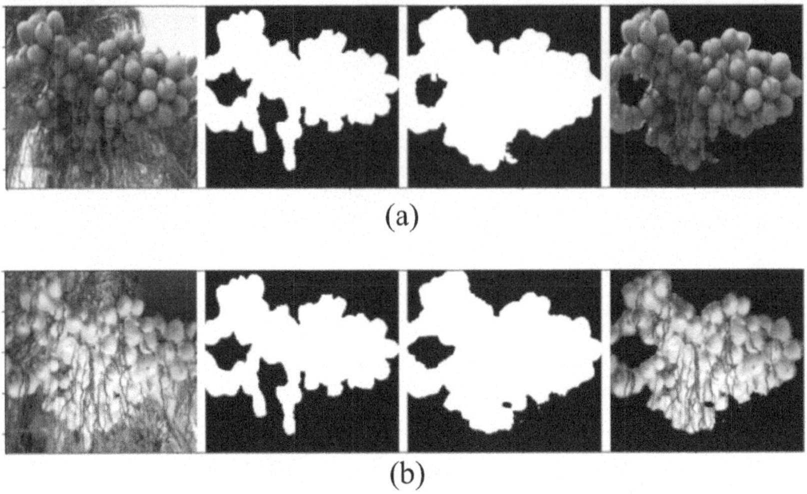

Fig. 6. Results of proposed segmentation method on arecanut bunch images of different color space (a) RGB image (b) Saturation component image.

Table 8. Comparison of classification accuracy for the proposed MobileNetV2 model on five different test datasets.

Saturation images	Segmented Saturation images	RGB images	Segmented RGB images	Manually Segmented RGB
78.43	81.37	82.45	88.24	96.38

5 Conclusions

This study described a method for automatically segmenting and classifying images into mature and immature arecanut bunches depending on their maturity level. For segmentation, an optimized U-Net model was employed on two types of color space images: RGB and the saturation channel of HSV color space. Dice coefficient, precision, and recall performance measures were utilized to evaluate the segmentation of RGB and saturation channel of HSV color space images. Experimental results show that the proposed model segments RGB images more correctly than the saturation channel of an HSV color space image, with a dice rate of 91.27%. The classification of segmented arecanut bunch images was done using a transfer learning approach with the MobileNetV2 model. The proposed model classifies segmented RGB images more accurately (88.24%).

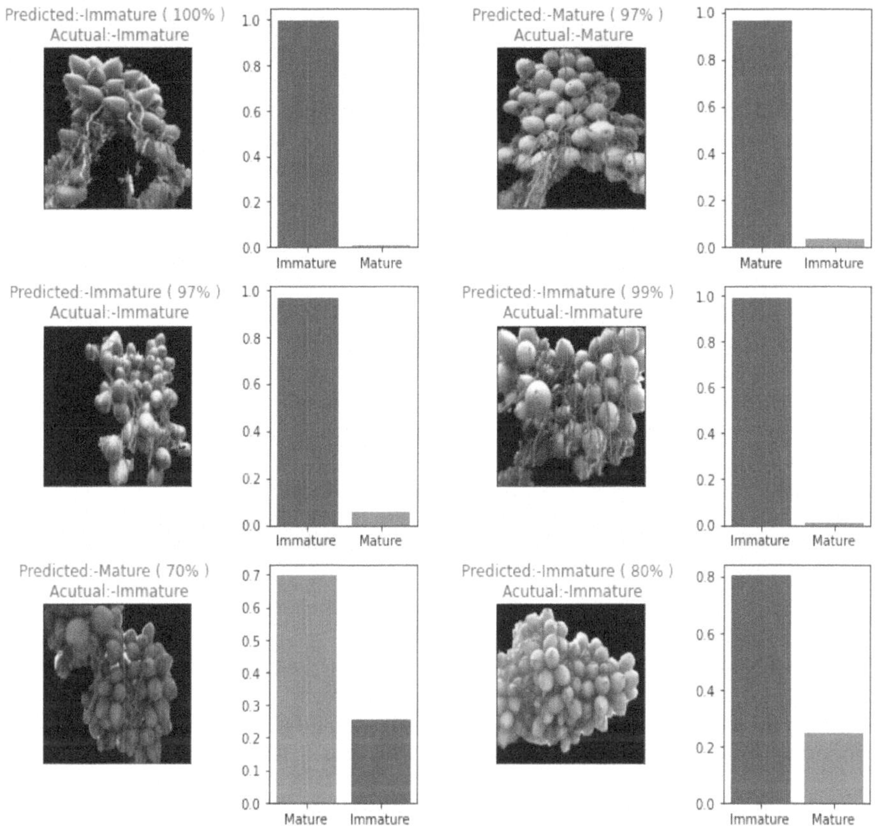

Fig. 7. Classification results of the proposed MobileNetV2 model on segmented-RGB test dataset.

References

1. Nagarajan, R., Meenakshi, R.: Analysis of arecanut production and export in India. Int. J. Bus. Soc. Sci. **3**, 67–81 (2016)
2. ICAR. ICAR-Arecanut (2021). http://agrigoaexpert.res.in/icar/category/horitculture/plantationcrops/arecanut.php
3. UAHS, S.: Areca Referral Laboratory at UAHS, Shivamogga (2021)
4. Dhanesha R., S.N.C.L.: Segmentation of arecanut bunches using YCGCR color model. In: 1st IEEE International conference on Advances in Information Technology, pp. 50–53 (2019). https://doi.org/10.1007/978-1-7281-3241-9
5. Zhu, L., Li, Z.B., et al.: High performance vegetable classification from images based on Alex net deep learning model **11**(4), 217–223 (2018)
6. Steinbrenner, J., et al.: Hyperspectral fruit and vegetable classification using convolutional neural networks. Comput. Electron. Agricult. **162**(2), 364–372 (2019). https://doi.org/10.1016/j.compag.2019.04.019
7. Gao, Z., et al.: Real-time hyperspectral imaging for the in-field estimation of strawberry ripeness with deep learning. Artif. Intell. Agricult. **4**, 31–38 (2020). https://doi.org/10.1016/j.aiia.2020.04.003

8. Ajit Danti, S.M.: Segmentation and classification of raw arecanuts based on three sigma control limits. Elsevier J. C3IT-2012 **4**, 215–219 (2012). https://doi.org/10.1016/j.protcy.2012.05.032
9. Suresha, M., et al.: Classification of diseased arecanut based on texture features. Computer Applications (0975–8887) Recent Advances in Information Technology (2014)
10. Dhanesha, R., Shrinivasa, N.C.L.: Segmentation of arecanut bunches using HSV color model. In: 2018 International Conference on Electrical, Electronics, Communication, Computer, and Optimization Techniques (ICEECCOT), pp. 37–41 (2018). https://doi.org/10.1109/ICEECCOT43722.2018.9001632
11. Girish, G.N., Thakur, B., et al.: Segmentation of intra-retinal cysts from optical coherence tomography images using a fully convolutional neural network model. IEEE J. Biomed. Health Inf. **23**(1), 296–304 (2019)
12. Sandler, M., et al.: Mobilenetv2: inverted residuals and linear bottlenecks. In: 2018 IEEE/CVF Conference on Computer Vision and Pattern Recognition, pp. 4510–4520. IEEE (2018)
13. Ganesan, P., Rajini, V., et al.: Satellite image segmentation based on YCbCr color space. Ecology **8**(1), 35–41 (2015)
14. Fairchild, M.: Color Appearance Models. Reading (1998)
15. Ronneberger, O., et al.: U-net: convolutional networks for biomedical image segmentation. In: MICCAI 2015, pp. 234–241. Springer, Cham (2015). https://doi.org/10.1007/978-3-319-24574-4_28
16. Ioffe, S., et al.: Batch normalization: accelerating deep network training by reducing internal covariate shift (2015)
17. Hussain, M., et al.: A study on CNN transfer learning for image classification (2019)
18. Russakovsky, O., et al.: ImageNet large scale visual recognition challenge (2019)
19. Wang, M., et al.: A high-speed and low-complexity architecture for SoftMax function in deep learning. In: IEEE Asia Pacific Conference on Circuits and Systems (APCCAS) (2018)
20. Kera's, A.: (2019). https://keras.io/api/
21. Machine Learning Platform. (2019). https://www.tensorflow.org
22. Collaboratory, G.: (2019). https://research.google.com/colaboratory/local-runtimes.html
23. Kingma, D.P., et al.: Adam: a method for stochastic optimization (2015)
24. Stopping, K.E.: (2019). https://keras.io/api/callbacks/earlystopping/
25. Gao, Z., et al.: Real-time hyperspectral imaging for the in-field estimation of strawberry ripeness with deep learning. Artif. Intell. Agricult. **4**, 31–38 (2020). https://doi.org/10.1016/j.aiia.2020.04.00

Enhanced Satellite Image Fusion Using Deep Learning and Feature Extraction Techniques: A Survey

Swathi Nallagachu(✉) and R. Sandanalakshmi

Department of Electronics and Communication Engineering, Puducherry Technological University, Puducherry, India
swathiprasad549@gmail.com, sandanalakshmi@ptuniv.edu.in

Abstract. This paper presents an overview and analysis of numerous research projects on image fusion methods, with a particular emphasis on deep learning-based methods. The research analyses the inadequacies of current fusion models and suggests novel methods for feature extraction, visual sensor networks, remote sensing applications, medical imaging, and multi-resolution image fusion. The study demonstrates the advantages of deep learning techniques for image fusion tasks, including Convolutional Neural Networks (CNNs), Stacked Auto encoders, and Convolutional Sparse Representation (CSR). These methods provide better fusion quality, fast processing, and better visual perception. Numerous studies provide novel methods, such as the rapid Integer Lifting Wavelet Transform (ILWT), Dual-Tree Complex Contourlet Transform (DT-CCT), and hybrid fusion methods that combine Discrete Cosine Transform (DCT) and Integer Lifting Wavelet Transform (ILWT) techniques. The review literature survey shows how these techniques can improve fusion outcomes while introducing little information distortion. In general, this paper brings together improvements in feature extraction and image fusion techniques, demonstrating the strength and potential of deep learning-based approaches. It is a useful tool for academics and professionals working in various fields who want to comprehend and advance these fields of study.

Keywords: Convolutional Neural Networks · Convolutional Sparse Representation · Stacked Auto encoders · Dual-Tree Complex Contourlet Transform (DT-CCT) · Discrete Cosine Transform (DCT) · Integer Lifting Wavelet Transform (ILWT)

1 Introduction

A fundamental procedure in computer vision called image fusion tries to integrate numerous images of the same scene that were taken using various sensors or modalities to produce an augmented image that contains more information than the individual input images. This method has a wide range of applications in remote sensing, medical imaging, surveillance, and robotics. In these areas, it is essential for improving image quality, enhancing information extraction, and assisting decision-making.

T. Shreekumar et al. (Eds.): ISCComm 2023, CCIS 2231, pp. 171–187, 2025.
https://doi.org/10.1007/978-3-031-75605-4_12

Deep learning's capacity to automatically generate hierarchical representations from data has made it one of the most effective methods for fusing images. Particularly Convolutional Neural Networks (CNNs) [22], which have been widely used in image fusion, have shown amazing performance in a variety of computer vision tasks. The caliber and applicability of the derived features, however, strongly influences how well an image fusion system based on deep learning performs.

As it directly affects how well the fusion algorithm performs, feature extraction is a crucial phase in the image fusion process. The final fused image's clarity, spatial details, and retention of crucial information are substantially influenced by the feature extraction method that is selected. In order to find the best strategy for a certain application area and obtain the best fusion outcomes, a detailed review of feature extraction approaches in image fusion is necessary.

In this paper, we give a thorough examination of feature extraction methods used in deep learning to fuse images. We examine a variety of well-liked CNN architectures to see how well they can extract pertinent data from multispectral, hyperspectral, and thermal images [23–27]. In order to maximize the effectiveness of transfer learning and get over data constraints, we also take into account a variety of pre-trained models and fine-tuning techniques.

The following are the primary goals of this analysis:

Review and synthesis of the literature on feature extraction and image fusion methods based on deep learning.

To help academics and practitioners choose the best strategy for their unique image fusion jobs, it is important to understand the advantages, disadvantages, and trade-offs of each feature extraction technique.

The outcomes of this literature survey investigation open the door for further developments in this fascinating and fast developing area of study by offering insightful information about the best practices and cutting-edge methods for feature extraction in image fusion. The use of deep learning algorithms for feature extraction is covered in the next section. A thorough analysis of machine learning algorithms for image fusion models is provided in Sect. 3. Deep learning-based image fusion models are the main topic of Sect. 4. Various discussions about the research are offered in Sect. 5. In the final section, which summarizes the research's findings and contributions, the study comes to an end.

2 Feature Extraction Using DL

Feature extraction is crucial to deep learning, especially with CNNs. Converting raw input data into useful and non-redundant characteristics aids analysis and pattern detection. Automatic feature extraction-based CNN classifiers, MLP-based image analysis, and MLP-based feature extraction are discussed [28–36]. Automatic feature extraction uses CNN architecture to extract essential features from input photos, and then fully linked layers classify them. MLP-based image analysis trains and classifies the entire image using an MLP network. MLP feature extraction uses manually or automatically extracted features as inputs to an MLP network [28, 41–44]. The publication [1] emphasizes feature extraction's use in image retrieval, character recognition, and plant modeling. Future research should improve deep learning feature extraction and examine the benefits of integrating ensemble classifiers with CNNs.

This research uses feature extraction and machine learning-based classification techniques to identify pollen in photos. It compares 12 machine learning classifiers and 11 feature extraction/selection methods. Different feature extraction/selection methods affect classification results, according to the study [2]. The results show that principal component analysis, best-first with backward search, and greedy search fail all classifiers. However, evolutionary algorithms, reranking search, best-first with forward search, and linear forward selection yield better outcomes [37–40]. Research on feature selection's potential to reduce predictive factors is also highlighted. Reranking search, linear forward selection, evolutionary and genetic algorithms, and other feature selection methods commonly outperform the baseline. This suggests that deleting unimportant elements improves categorization. Despite the study's modest focus on pollen recognition, its findings can help extract picture classification elements. Further research should examine how deep learning architectures automatically extract features to improve classification performance, especially for tiny photo datasets. In summary, the research emphasizes the importance of selecting appropriate feature extraction methods and suggests promising algorithms for deep learning feature extraction.

Research [3] describes a deep learning-based automated multimodal brain tumor classification method. It highlights the challenges of categorizing brain tumors using different MRI modalities and proposes contrast stretching, deep learning feature extraction using trained CNN models, joint learning based on correntropy for feature selection, fusion of selected features, and classification using extreme learning machine. Based on BraTS datasets, the offered approach is accurate. The feature selection procedure is the main focus of the research to save computational time, decrease redundancy, and improve classification accuracy.

Reference [4] introduces deep learning-based feature selection for remote sensing scene classification and identification. Consider feature selection as feature reconstruction in a deep belief network (DBN) [45–47]. Deep Belief Networks (DBNs) abstract by minimizing reconstruction error across input features. Features with reduced reconstruction errors should provide more intrinsic feature information for picture representation. The suggested technique selects reconstructible discriminative qualities. An iterative technique modifies the Dynamic Bayesian Network (DBN) and generates reconstruction weights. The technique was tested on 2800 remote sensing scene photos. To overcome pixel-level classification difficulties in high-resolution satellite images, the strategy uses feature-based picture representation. Deep learning improves classification accuracy and helps identify discriminative features in the suggested strategy. Unsupervised feature selection improves high-resolution satellite picture interpretation. The findings show that the proposed technique outperforms baseline methods and has promise for remote sensing scene classification.

The studies emphasize feature extraction in deep learning. They compare feature selection, CNN-based automatic feature extraction, and MLP-based image analysis for extracting important features. Studies suggest that feature extraction improves image retrieval, character recognition, plant modeling, pollen detection, brain tumor classification, and remote sensing scene identification. The study emphasizes the need for advanced feature extraction methods and the benefits of ensemble classifiers and deep learning architectures. The research improves feature extraction and classification in various applications (Table 1).

Table 1. Comparative analysis of Research Gap, Advantages and Disadvantages in Feature Extraction using DL.

Author	Year	Model	Research Gap	Advantages	Disadvantages
S. Dara and P. Tumma [1]	2018	CNN and MLP	Limited ability to compare and interpret feature extraction techniques	Automatic high-level representation extraction, complicated pattern recognition, and effective management of enormous datasets	Labelled data requirements, complicated computations, potential for overfitting, and poor interpretability
M. C. Popescu and L. M. Sasu [2]	2014	PCA, Greedy Search, Best-First Search with Forward Search, Linear Forward Selection, Evolutionary Algorithms, Genetic Algorithm, PSO, Reranking Search, Tabu Search	The lack of exploration and comparison of deep learning models for feature extraction in pollen recognition from images	The success of deep learning models in computer vision tasks, automatic feature learning, management of complicated data, potential for high accuracy, and exploitation of massive datasets are some of their benefits	Data dependency, computational requirements, overfitting, interpretability limitations, resource-intensive training

(*continued*)

Table 1. *(continued)*

Author	Year	Model	Research Gap	Advantages	Disadvantages
Khan, Muhammad Attique, et al. [3]	2020	VGG16 and VGG19	Inability to comprehend and contrast with other feature selection methods for classifying brain tumours	Automatic classification of brain tumors, high accuracy, application of deep learning models, and careful feature choice	Big Labelled datasets and a lot of computing power is needed to train deep learning models
Q. Zou, L. Ni, T. Zhang and Q. Wang [4]	2015	DBN	Deep learning-based feature selection techniques have only been lightly investigated for remote sensing scene classification	Use of deep learning approaches, effective feature selection, enhanced performance, discriminative feature identification, and feature identification	Limited benchmark datasets, dependency on labelled training data, computational complexity, and probable overuse of feature reconstruction

The compared algorithms span many feature selection and machine learning methods. CNNs and VGG models excel at visual tasks, whereas MLPs have flexibility with structured data. Greedy Search, Best-First Search, Linear Forward Selection, Evolutionary Algorithms, Genetic Algorithm, PSO, Reranking aim to find useful features.

Search, Tabu. PCA reduces dimensions. Deep Belief Networks answer unsupervised learning questions. The method you use depends on data qualities, task difficulty, and interpretability. A smart algorithm choice can improve model performance and insights in many machine learning applications.

3 Traditional Machine Learning Algorithms Image Fusion

An overview of multi-resolution image fusion is given in the survey paper [5], which also proposes a unique approach based on rapid integer lifting wavelet transform [6]. It analyses the shortcomings of several currently available models and offers fusion techniques for both high- and low-frequency sub-bands. Improved fusion quality and fast processing time are just two of the suggested algorithm's benefits. Its usefulness in real-time image fusion with improved visual quality and detail retention is demonstrated by experimental results. The report makes recommendations for further research, including the use of threshold selection algorithms and objective evaluation indices. In the end, the study advances the field of multi-resolution image fusion and provides guidance for future developments (Fig. 1).

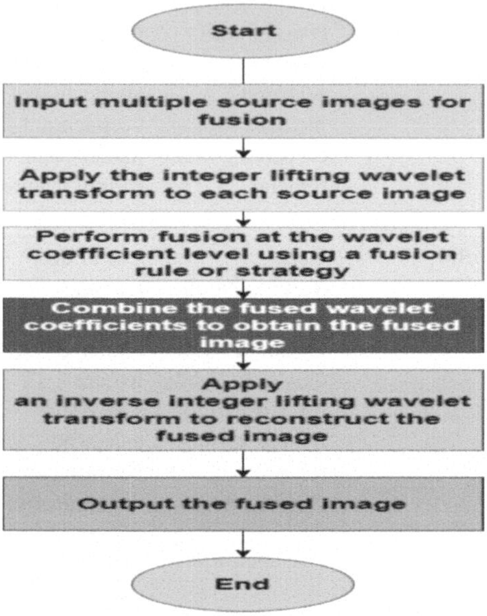

Fig. 1. Flowchart for Image Fusion using Integer Lifting Wavelet Transform

Reference [7] focuses on Visual Sensor Networks (VSNs) and multimodal data fusion of pictures. The suggested hybrid image fusion approach uses the Discrete Cosine Transform (DCT) [8] and the Integer Lifting Wavelet Transform (ILWT). The fusion technique fuses estimate coefficients in the DCT domain by variance and detail coefficients in the ILWT domain by optimum weighted average based on correlation. The suggested solution addresses information loss, computational complexity, and memory space limits using integer operations in the ILWT domain. Multiple tests show that the suggested method outperforms other fusion methods in qualitative and quantitative evaluations. The technique will eventually include hardware implementation and video fusion (Fig. 2).

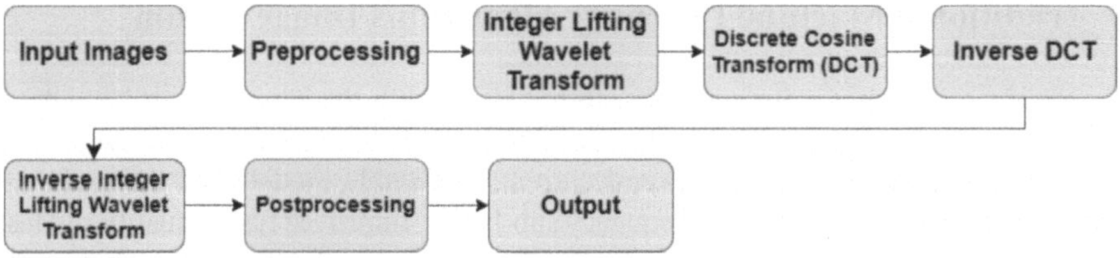

Fig. 2. Flowchart for Hybrid Image Fusion Method based on Integer Lifting Wavelet and Discrete Cosine Transform

In order to obtain superior spatial and spectral resolutions, this work [9] focuses on image fusion algorithms that combine high spatial resolution panchromatic images with low spatial resolution multi-spectral images. The research contrasts various fusion techniques, including IHS, BT, PCA, and HPF, and suggests a hybrid strategy that combines PCA and HPF [9, 48–50]. While enhancing spatial resolution, the suggested approach keeps the multi-spectral image's spectral features. Comparing the proposed fusion method to previous fusion methods, experimental results show that it is more effective at decreasing spectral distortion and improving spatial quality. The publication offers tips on how to pan-sharpen images with better spatial resolution and less spectral distortion overall.

According to [10], the fusion of high-resolution panchromatic (pan) and low-resolution multispectral (MS) satellite images with dual-resolution sensors is the subject of this research. It contrasts various fusion techniques, including IHS, BT, PCA, and HPF, and suggests a hybrid strategy that combines PCA and HPF. With the suggested technique, a pan-sharpened image should have better spatial resolution and less spectral distortion. The suggested fusion method keeps the spectral properties of the MS image while enhancing spatial resolution, according to experimental data. It reduces spectrum distortion better than other fusion methods and reaches spatial quality on par with PCA, GS, DWT, and SWT fusion methods.

Intensity-hue-saturation (IHS), principal component analysis (PCA), the Brovey-transform (BT), and the wavelet transform (WT) are some of the modern image fusion methods that are examined in this study [11] for their color distortion issues. The study emphasizes that variations in saturation are what lead to color distortion. It assesses these fusion techniques and provides a thorough analysis of their spectrum and spatial characteristics. The paper highlights the requirement for quantitative evaluations of various fusion methodologies. The color distortion issue is confirmed by experimental findings utilizing the SPOT and CARTERRA datasets, which also offer guidance for developing new fusion algorithms. The objective of the study is to improve the comprehension and advancement of image fusion methods in remote sensing applications.

In particular, the fusing of MRI and CT images is the subject of this paper's [12] discussion of the use of image fusion techniques in medical imaging. By merging data from various imaging modalities, the aim is to increase spatial resolution and boost diagnostic precision. The limitations of current fusion techniques like Wavelet Transform (WT), Principal Component Analysis (PCA), and Intensity-Hue-Saturation (IHS) are discussed in the paper. It introduces a novel method for fusion called the Dual-Tree Complex Contourlet Transform (DT-CCT), which combines PCA and local energy weighted selection. The suggested method provides fused images with better details and little information distortion, according to experimental findings. The method improves fusion approaches' dependability in a clinical setting.

Due to its insidious course, coronary artery disease is a primary cause of death and is frequently discovered too late. A semi-invasive technique called Intra Vascular Ultra Sound (IVUS) imaging gives precise details regarding plaque-induced artery blockages. Manual IVUS picture interpretation, nevertheless, takes time and is prone to mistakes. In order to precisely recognize coronary layers in IVUS pictures, this paper [13] investigates the use of deformable models and image fusion techniques. To enhance the quality of the images, pre-processing techniques like wavelet transform and edge-preserving smoothing are used. The suggested method has the advantage of being entirely automated with a respectable execution time, and it performs border detection with favorable outcomes.

The studies on image fusion approaches that have been evaluated present a variety of benefits and drawbacks. These methods seek to speed up processing, boost spatial and spectral resolutions, and enhance fusion quality. The suggested techniques show benefits like improved fusion quality, fast execution times, and the preservation of spectral information. However, difficulties include a limited applicability to various datasets, computational cost, significant color distortions, and information loss. The studies emphasize the need for additional study to enhance the field of image fusion in areas like threshold selection algorithms and objective assessment indices. In general, image fusion offers promise for applications in many fields, such as medical imaging, remote sensing, and visual sensor networks, but the best outcomes require careful analysis of the individual techniques and their constraints.

The surveyed image fusion algorithms encompass a diverse range of approaches and applications. Improved fusion quality and fast processing times are demonstrated by the novel method based on rapid Integer Lifting Wavelet Transform (ILWT). It is particularly useful in real-time image fusion with enhanced visual quality and detail retention. Another hybrid approach combining Discrete Cosine Transform (DCT) and ILWT demonstrates superior performance in multimedia data fusion and visual sensor networks. However, fusion techniques that combine low-resolution multi-spectral images with high spatial resolution panchromatic images show promise in maintaining spectral details while enhancing spatial resolution. However, challenges persist, such as color distortion issues and limited applicability to various datasets.

4 Deep Learning-Based Image Fusion Model

4.1 Deep Convolutional Neural Network

This study [14] introduces deep CNN-based multi-focus image fusion. The preferred method involves directly mapping source photographs to a focus map to provide clear and useful data. The Convolutional Neural Network (CNN) model generates the activity level measurement and fusion rule simultaneously to solve modern fusion problems. Experimental data shows that the suggested method improves visual quality and objective evaluation of fusion. The trained CNN model may do further image fusion tasks.

A Pixel-Wise Convolutional Neural Network (p-CNN) [15, 51–53] is introduced for multi-focus image fusion [15]. The p-CNN uses contextual information to identify source image pixels as focused, defocused, or unknown. The suggested technique partitions source images into overlapping patches and uses a well-trained p-CNN to evaluate pixel focus intensity. Focus levels help create score matrices and decision maps. These matrices and maps are improved and utilized to create the merged picture. The study report also proposes a strategy for converting the parallel convolutional neural network (p-CNN) into an image-wise CNN to improve computational efficiency. According on empirical data, the suggested strategy beats complex fusion procedures in subjective and objective assessment measures. Further research is needed to minimize processing and expand image fusion applications. Convolutional Neural Networks (CNN) is the foundation of the IFCNN framework, a broad image fusion framework presented in this research [16]. The system uses two Convolutional layers to separate out important characteristics from a variety of input images before merging them according to one of three fusion rules: element wise-max, element wise-min, or element wise-mean. The fusion image is created by reconstructing the fused features using two Convolutional layers. The suggested model can be trained end-to-end without post-processing because it is completely convolutional. The model shows higher generalization capability and delivers comparable or better outcomes than existing state-of-the-art fusion techniques. A large-scale multi-focus picture dataset is developed for training. In addition to picture alignment, deeper networks, larger datasets, more complicated fusion modules, and customized architectures for particular image types, the paper makes suggestions for topics for further investigation.

4.2 Convolutional Sparse Representation (CSR)

In lieu of Sparse Representation (SR), Convolutional Sparse Representation (CSR) is presented in this paper [17]. The majority of SR-based fusion techniques currently in use have issues with misregistration sensitivity and detail retention. These issues are resolved by the CSR model, which depicts the full image as opposed to isolated regions. For the merging of multi-focus and multi-modal images, a CSR-based framework is suggested. In terms of objective evaluation and visual quality, experimental results show that CSR-based methods are preferable to SR-based methods. Future work will examine how algorithm parameters affect performance and create better fusion strategies based on CSR. Future image fusion models are anticipated to increasingly use CSR.

This research [18] presents CSCFuse, a deep learning-based image fusion technique that applies Deep Convolutional Sparse Coding (CSC) networks to three distinct image fusion tasks: the fusing of infrared and visible images, the fusion of images with multiple exposures, and the fusion of images with multiple spectral components. Dictionary convolution units use the CSC model as well as an iterative shrinkage and thresholding technique, enabling all hyperparameters to be learned from the data. Numerous tests and comparisons show that CSCFuse is superior in terms of quantitative assessment and visual examination. The suggested networks do better than cutting-edge techniques and are very repeatable. A potential method for picture fusion in numerous applications is presented by CSCFuse [18, 54, 55].

In order to improve signal representation and scale back computational complexity, this work [19] introduces a unique method for image fusion using Online Convolutional Sparse Coding (OCSC) [19, 56]. The suggested technique divides source images into low-frequency and detail components, and then applies fusion principles to produce fused components. The proposed method's advantage in terms of quantitative evaluations and visual perception are demonstrated by results from experiments on the fusion of infrared-visible and medical images. The contributions include the application of ADMM for effective Convolutional Sparse Coding, the addition of OCSC to multimodal image fusion, and the attainment of encouraging results in structure feature transmission and detail retention. Future research focuses include the investigation of OCSC in additional signal processing tasks, problem-oriented medical fusion techniques, and successful and cost-effective infrared-visible fusion.

4.3 Stacked Auto Encoders

Using the Non Sub Sampled Contourlet Transform (NSCT) and Stacked Sparse Auto Encoders (SSAE) [20, 57, 58], the researchers offer a new method for combining IR and VI pictures [20]. The suggested approach efficiently integrates the infrared item into the fused picture while maintaining fine details. The NSCT method divides source pictures into low-frequency and high-frequency subbands. The Spectral Subtraction Amplitude Estimation (SSAE) approach on the low-frequency sub band of the infrared (IR) image is used to determine object reliabilities (OR). A logical operator OR-based adaptive fusion rule fuses low-frequency sub bands. Use a choose-max fusion rule to fuse high-frequency subbands. Experimental results support the proposed method. Using stacked autoencoders to obtain feature representations from unlabeled data improves fusion strategies.

A novel feature fusion approach for Synthetic Aperture Radar (SAR) target detection is presented in [21]. This method uses Stacked Auto-Encoder (SAE), as explained in references [20, 59–61]. Baseline and Three-Patch Local Binary Pattern (TPLBP) characteristics characterize synthetic aperture radar (SAR) pictures globally and locally. After unsupervised learning, the SAE network combines these features. Target classification accuracy is improved by using a softmax classifier. The suggested technique achieves 95.43% classification accuracy on a 10-class SAR target dataset. The method improves recognition efficiency and accuracy while simplifying the network.In conclusion, the research papers under consideration demonstrate important developments in the area of image fusion, with a special emphasis on deep learning-based methods. These works cover the gaps in the literature and problems with computational complexity, feature fusion, multi-focus image fusion, and multi-modal image fusion. Convolutional Sparse Representation, Stacked Auto-Encoders, and Deep Convolutional Neural Networks all demonstrate higher fusion performance in terms of visual quality and objective assessment criteria. Deep learning models can be used to generate fusion rules, extract features more efficiently, and enhance recognition.

The surveyed deep learning-based image fusion algorithms cover a wide range of methodologies. Convolutional Neural Networks (CNNs) are utilised in the first two methods for multi-focus image fusion. They outperform conventional techniques in terms of fusion performance, producing advanced visual quality and impartial assessments. The CNN-based models overcome problems with conventional fusion methods by simultaneously generating activity level data and fusion rules.

Convolutional Spare Representation (CSR) and Deep Convolutional Spare Coding (CSC) approaches, which outperform conventional sparse representation-based algorithms, bring sparse coding to fusion challenges. They provide enhanced detail retention and decreased sensitivity to misregistration, leading to more accurate assessments of visual quality in multi-focus and multi-modal picture fusion.

For the purpose of detecting SAR targets, Stacked Auto Encoders (SAE) offers an effective feature fusion technique. From SAR images, they extract baseline and local features, and then combine them using the SAE network. The target classification accuracy is increased, and the network structure is made simpler as a result of the fused features. All things considered, deep learning-based algorithms like CNN, CSR, CSC, and SAE show its potential for image fusion problems. They appear to have the potential to improve fusion quality, boost computational effectiveness, and boost performance for a range of image fusion tasks, including multi-focus, multi-modal, and infrared-visible image fusion.

The research papers discuss compromised techniques aimed at addressing certain image fusion problems. Convolutional Sparse Representation (CSR) improves fusion results by addressing misregistration sensitivity and detail retention concerns. Deep Convolutional Neural Network (CNN) improves fusion performance and efficiency, opening up new opportunities in image fusion tasks. Stacked Auto Encoders (SAE) improves target identification efficiency and accuracy in synthetic aperture radar. By merging Convolutional Sparse Coding with online learning, Online Convolutional Sparse Coding (OCSC) presents a novel fusion strategy. Image Synthesis Deep CSC networks are used in Convolutional Sparse Coding (CSCFuse) for various image fusion tasks.

In all, these future developments aim to further enhance and broaden the compromised algorithms' capabilities, allowing them to tackle complicated fusion scenarios, improve efficiency, and attain even better performance levels. Collaborative research efforts and cross-disciplinary inquiry can help to evolve these algorithms and broaden their effect in the field of picture and data fusion (Fig. 3 and Table 2).

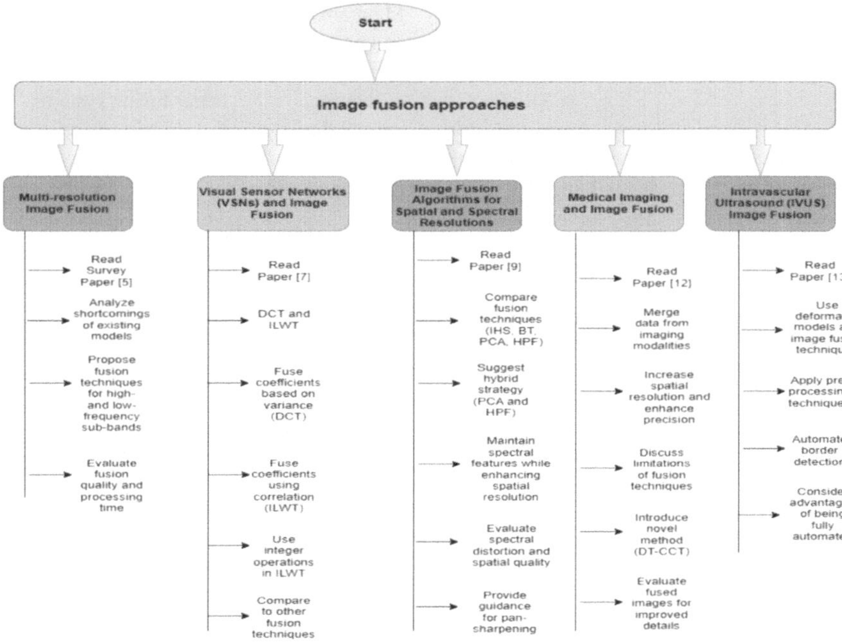

Fig. 3. Flowchart for Image Fusion Techniques and Applications

Table 2. Comparative analysis of Research Gap, Advantages and Disadvantages in Deep Learning-based Image Fusion Model.

Author	Year	Model	Research Gap	Advantages	Disadvantages
Liu, Yu, et al. [14]	2017	CNN	Limited research of deep learning for fusion rule and activity level	Potential for additional tasks, superior fusion, and simultaneous generation	Need for extensive data, computational speed constraints
Tang, Han, et al. [15]	2018	CNN	Limited exploration of pixel-wise CNNs for multi-focus fusion	Efficient learning, superior performance over existing method	Data requirements, computational efficiency can be improved
Zhang, Yu, et al. [16]	2020	CNN	Insufficient investigation of CNN-based fusion frameworks for various image types	High-resolution dataset, fully convolutional, end-to-end training, and competitive performance	Minimal dataset; room for network depth and module enhancement
Liu, Yu, et al. [17]	2016	CSR	Convolutional Sparse Representation (CSR) for picture fusion has only been lightly investigated	Maintains details and performs better in evaluating quality than sparse techniques	The necessity for parameter analysis and the opportunity for better fusion systems
Zhao, Zixiang, et al. [18]	2023	CSR	Deep CSC networks have not been fully explored for image fusion	The cutting-edge, flexible hyper parameter learning performs better by CSCFuse	Limited architecture analysis and training computational resources

(continued)

Table 2. *(continued)*

Author	Year	Model	Research Gap	Advantages	Disadvantages
Zhang, Chengfang, et al. [19]	2022	CSR	Online Convolutional Sparse Coding (OCSC) for picture fusion has received limited attention	Superior depiction, simplicity, promising structural transmission, and preservation of detail	Medical fusion presents difficulties, requires optimisation, and makes efficiency-performance trade-offs
Luo, Xiaoqing, et al. [20]	2018	SAE	Inability to integrate IR objects while maintaining details	Integration that works, preservation of details, and unsupervised feature learning	Limited labelled data and difficulty in designing an activity level measure
Kang, Miao, et al. [21]	2017	SAE	SAE-based fusion for SAR target recognition has received limited attention	Distinct fusion, lessening redundancy, raising precision and productivity	Complex network complicated pre-processing, and limited progress in some methods

5 Discussions

The research papers under discussion offer insightful information on deep learning feature extraction and image fusion methods. They discuss a variety of topics, including feature selection, multi-modal fusion, and multi-resolution fusion, highlighting the benefits and drawbacks of various approaches.

In the final analysis, the papers suggest new methods for fusing images, including the use of Discrete Cosine Transform, Rapid Integer Lifting Wavelet Transform, and deep learning-based methods like Convolutional Neural Networks (CNNs), Convolutional Sparse Representation (CSR), and Stacked Auto Encoders. In comparison to earlier algorithms, these show enhanced fusion quality, decreased computing complexity, and superior classification efficiency.

According to the research's interpretations, feature extraction using deep learning architectures improves classification accuracy and visual quality. The effective extraction of useful features from complicated picture data is made possible by the employment of cutting-edge algorithms like CNNs, CSRs, and Stacked Auto encoders.

The results have implications for the prospective use of these approaches in numerous industries, including medical imaging, remote sensing, and multimedia data fusion. Improved diagnosis, analysis, and interpretation of images in various fields may be a result of developments in feature extraction and image fusion. It's crucial to take into account the studies' limitations. The use of certain datasets, which could not accurately reflect real-world situations, and the need for additional study to investigate how the proposed approaches might be generalised across various datasets and image kinds are some drawbacks. Further research is needed since it's possible that using these strategies in time-sensitive or resource-constrained situations would provide difficulties.

Several recommendations can be made in light of the research's findings. Future research should investigate how ensemble classifiers and deep learning models might be combined to improve classification performance. Investigations on how algorithm parameters affect feature extraction and fusion approaches can also shed light on how to enhance their performance. Additionally, the creation of hardware implementations and evaluation on huge datasets can confirm the scalability and efficiency of these methods in practical applications.

In conclusion, the research papers under discussion promote deep learning-based picture fusion and feature extraction algorithms. These works demonstrate the advantages of using deep learning architectures, suggest fresh ideas, and offer insightful suggestions for future research directions. The results show promise for enhancing image analysis, pattern identification, and decision-making processes and have ramifications for numerous industries.

6 Conclusions

In conclusion, the research papers covered here highlight important developments in image fusion methods, with a focus on deep learning-based approaches. In terms of fusion quality, processing speed, spatial and spectral resolutions, detail retention, and colour distortion reduction, these research address the drawbacks of conventional fusion systems and provide superior alternatives. Deep learning models like Convolutional Sparse Representation, Stacked Auto Encoders, and Deep Convolutional Neural Networks regularly outperform more traditional approaches, according to experimental results. The relevance of feature extraction in numerous image analysis tasks across diverse domains is also highlighted, underscoring the significance of feature extraction in deep learning. These publications shed light on feature extraction methods and how they affect data representation, pattern recognition, and classification precision. Even though the research papers under discussion make a substantial contribution to the discipline, further research is still needed. In order to increase classification accuracy, future research might investigate the integration of ensemble classifiers with deep learning models, improve feature extraction approaches, and assess how well deep learning architectures perform in feature extraction.

In the end, these works expand image fusion, deep learning, and feature extraction, laying the groundwork for later developments and applications in a variety of industries, including remote sensing, medical imaging, multimedia data fusion, and more.

References

1. Dara, S., Tumma, P.: Feature extraction by using deep learning: a survey. In: 2018 Second International Conference on Electronics, Communication and Aerospace Technology (ICECA), Coimbatore, pp. 1795–1801 (2018). https://doi.org/10.1109/ICECA.2018.8474912

2. Popescu, M.C., Sasu, L.M.: Feature extraction, feature selection and machine learning for image classification: a case study. In: 2014 International Conference on Optimization of Electrical and Electronic Equipment (OPTIM), Bran, pp. 968–973 (2014). https://doi.org/10.1109/OPTIM.2014.6850925
3. Khan, M.A., et al.: Multimodal brain tumor classification using deep learning and robust feature selection: a machine learning application for radiologists. Diagnostics **10**(8), 565 (2020)
4. Zou, Q., Ni, L., Zhang, T., Wang, Q.: Deep learning based feature selection for remote sensing scene classification. IEEE Geosci. Remote Sens. Lett. **12**(11), 2321–2325 (2015). https://doi.org/10.1109/LGRS.2015.2475299
5. Hu, G., Zheng, Y., Qin, X.: Image fusion based on integer lifting wavelet transform. In: Image Fusion and Its Applications. IntechOpen (2011)
6. Seyyedi, S.A., Sadau, V., Ivanov, N.: A secure steganography method based on integer lifting wavelet transform. Int. J. Netw. Secur. **18**(1), 124–132 (2016)
7. Latreche, B., Saadi, S., Kious, M., et al.: A novel hybrid image fusion method based on integer lifting wavelet and discrete cosine transformer for visual sensor networks. Multimed. Tools Appl. **78**, 10865–10887 (2019). https://doi.org/10.1007/s11042-018-6676-z
8. Li, X., et al.: Discrete CosinTransFormer: image modeling from frequency domain. In: Proceedings of the IEEE/CVF Winter Conference on Applications of Computer Vision (2023)
9. Metwalli, M.R., et al.: Image fusion based on principal component analysis and high-pass filter. In: 2009 International Conference on Computer Engineering & Systems. IEEE (2009)
10. Metwalli, M.R., et al.: Satellite image fusion based on principal component analysis and high-pass filtering. JOSA A **27**(6), 1385–1394 (2010)
11. Tu, T.-M., et al.: A new look at IHS-like image fusion methods. Inf. Fusion **2**(3), 177–186 (2001)
12. Al-Azzawi, N., Abdullah, W.A.K.W.: Medical image fusion schemes using Contourlet transform and PCA bases. Image Fusion Appl. 93–110 (2011)
13. Ashoori, A, Moshiri, B., Setarehdan, S.K.: Detecting coronary layers in IVUS pictures using image fusion approach. Image Fusion Appl. IntechOpen (2011)
14. Liu, Y., et al.: Multi-focus image fusion with a deep convolutional neural network. Inf. Fusion **36**, 191–207 (2017)
15. Tang, H., et al.: Pixel convolutional neural network for multi-focus image fusion. Inf. Sci. **433**, 125–141 (2018)
16. Zhang, Y, et al.: IFCNN: a general image fusion framework based on convolutional neural network. Inf. Fusion **54**, 99–118 (2020)
17. Liu, Y., et al.: Image fusion with convolutional sparse representation. IEEE Signal Process. Lett. **23**(12), 1882–1886 (2016)
18. Zhao, Z., et al.: Deep convolutional sparse coding networks for interpretable image fusion. In: Proceedings of the IEEE/CVF Conference on Computer Vision and Pattern Recognition (2023)
19. Zhang, C., Zhang, Z., Feng, Z.: Image fusion using online convolutional sparse coding. J. Ambient Intell. Human. Comput. 1–12 (2022)
20. Luo, X., et al.: Infrared and visible image fusion based on NSCT and stacked sparse autoencoders. Multim. Tools Appl. **77**, 22407–22431 (2018)
21. Kang, M., et al.: Synthetic aperture radar target recognition with feature fusion based on a stacked autoencoder. Sensors **17**(1), 192 (2017)
22. Shao, Z., Cai, J.: Remote sensing image fusion with deep convolutional neural network. IEEE J. Select. Topics Appl. Earth Observ. Remote Sens. **11**(5), 1656–1669 (2018)
23. Nevavuori, P., Narra, N., Lipping, T.: Crop yield prediction with deep convolutional neural networks. Comput. Electron. Agric. **163**, 104859 (2019)

24. Liu, L., et al.: Multi-dataset hyper-CNN for hyperspectral image segmentation of remote sensing images. Processes **11**(2), 435 (2023)
25. Wang, Q., et al.: GETNET: a general end-to-end 2-D CNN framework for hyperspectral image change detection. IEEE Trans. Geosci. Remote Sens. **57**(1), 3–13 (2018)
26. Chen, Y., et al.: Deep feature extraction and classification of hyperspectral images based on convolutional neural networks. IEEE Trans. Geosci. Remote Sens. **54**(10), 6232–6251 (2016)
27. He, X., Chen, Y., Ghamisi, P.: Heterogeneous transfer learning for hyperspectral image classification based on convolutional neural network. IEEE Trans. Geosci. Remote Sens. **58**(5), 3246–3263 (2019)
28. Shaheen, F., Verma, B., Asafuddoula, M.: Impact of automatic feature extraction in deep learning architecture. In: 2016 International Conference on Digital Image Computing: Techniques and Applications (DICTA). IEEE (2016)
29. Sharma, R., Kaushik, B.: Offline recognition of handwritten Indic scripts: a state-of-the-art survey and future perspectives. Comput. Sci. Rev. **38**, 100302 (2020)
30. Dara, S., et al.: Feature extraction in medical images by using deep learning approach. Int. J. Pure Appl. Math. **120**(6), 305–312 (2018)
31. Valls, J.M., et al.: Supervised data transformation and dimensionality reduction with a 3-layer multi-layer perceptron for classification problems. J. Ambient Intell. Humaniz. Comput. 1–13 (2021)
32. Ebrahim, M., Al-Ayyoub, M., Alsmirat, M.A.: Will transfer learning enhance imagenet classification accuracy using imagenet-pretrained models? In: 2019 10th International Conference on Information and Communication Systems (ICICS). IEEE (2019)
33. Bharathi, P.T., Subashini, P.: Optimization of image processing techniques using neural networks: a review. WSEAS Trans. Inf. Sci. Appl. **8**(8), 300–328 (2011)
34. Xiao, B., Nguyen, M, Yan, W.Q.: Fruit ripeness identification using transformers. Appl. Intell. 1–12 (2023)
35. Chowdhury, S.P., et al.: Genetic algorithm based global and local feature selection approach for handwritten numeral recognition. In: Metaheuristics in Machine Learning: Theory and Applications, pp. 745–769. Springer, Cham (2021)
36. Dalai, R., Das, P.: Effective fine-grained feature extraction and classification of solid materials using hybrid region convolutional neural networks. Multim. Tools Appl. **80**, 32171–32196 (2021)
37. Manikandan, G., Abirami, S.: A survey on feature selection and extraction techniques for high-dimensional microarray datasets. Knowl. Comput. Appl.: Knowl. Comput. Specific Domains **I**, 311–333 (2018)
38. Sasikala, S., Appavu alias Balamurugan, S., Geetha, S.: Multi Filtration Feature Selection (MFFS) to improve discriminatory ability in clinical data set. Appl. Comput. Informat. **12**(2), 117–127 (2016)
39. Balogun, A.O., et al.: Software defect prediction using wrapper feature selection based on dynamic re-ranking strategy. Symmetry **13**(11), 2166 (2021)
40. Deng, X., et al.: Feature selection for text classification: a review. Multim. Tools Appl. **78**, 3797–3816 (2019)
41. Hoq, M., NazimUddin, M., Park, S.-B.: Vocal feature extraction-based artificial intelligent model for Parkinson's disease detection. Diagnostics **11**(6), 1076 (2021)
42. Lerner, B., et al.: A comparative study of neural network based feature extraction paradigms. Pattern Recognit. Lett. **20**(1), 7–14 (1999)
43. Thomas, S., Ganapathy, S., Hermansky, H.: Multilingual MLP features for low-resource LVCSR systems. In: 2012 IEEE International Conference on Acoustics, Speech and Signal Processing (ICASSP). IEEE (2012)

44. Chatterjee, R., Bandyopadhyay, T.: EEG based motor imagery classification using SVM and MLP. In: 2016 2nd International Conference on Computational Intelligence and Networks (CINE). IEEE (2016)
45. Hamel, P., Eck, D.: Learning features from music audio with deep belief networks. In: ISMIR, vol. 10 (2010)
46. Hua, Y., Guo, J., Zhao, H.: Deep belief networks and deep learning. In: Proceedings of 2015 International Conference on Intelligent Computing and Internet of Things. IEEE (2015)
47. Li, Z., et al.: Manifold-based multi-deep belief network for feature extraction of hyperspectral image. Remote Sens. **14**(6), 1484 (2022)
48. Shahdoosti, H.R., Ghassemian, H.: Spatial PCA as a new method for image fusion. In: The 16th CSI International Symposium on Artificial Intelligence and Signal Processing (AISP 2012). IEEE (2012)
49. Wenbo, W., Jing, Y., Tingjun, K.: Study of remote sensing image fusion and its application in image classification. Int. Arch. Photogramm. Remote. Sens. Spat. Inf. Sci. **37**(B7), 1141–1146 (2008)
50. Liu, W., Huang, J., Zhao, Y.: Image fusion based on PCA and undecimated discrete wavelet transform. In: International Conference on Neural Information Processing. Springer, Heidelberg (2006). https://doi.org/10.1007/11893257_54
51. Bhalla, K., et al.: A fuzzy convolutional neural network for enhancing multi-focus image fusion. J. Visual Commun. Image Represent. **84**, 103485 (2022)
52. Mustafa, H.T., Yang, J., Zareapoor, M.: Multi-scale convolutional neural network for multi-focus image fusion. Image Vision Comput. **85**, 26–35 (2019)
53. Zhou, Y., et al.: Gradient-based multi-focus image fusion method using convolution neural network. Comput. Electric. Eng. **92**, 107174 (2021)
54. Stephanakis, I.M., Chochliouros, I.P.: Multimedia content distribution over next-generation heterogeneous networks featuring a service architecture of sliced resources. In: Artificial Intelligence Applications and Innovations: AIAI 2012 International Workshops: AIAB, AIeIA, CISE, COPA, IIVC, ISQL, MHDW, and WADTMB, Halkidiki, 27–30 September 2012, Proceedings, Part II 8. Springer, Heidelberg (2012)
55. Maheshwari, A.: Tutorial on Network Quality and IMS Services Used in LTE. (2011)
56. Li, H., et al.: A novel fusion method based on online convolutional sparse coding with sample-dependent dictionary for visible–infrared images. Arab. J. Sci. Eng. 1–11 (2023)
57. Zhang, L., Ma, W., Zhang, D.: Stacked sparse autoencoder in PolSAR data classification using local spatial information. IEEE Geosci. Remote Sens. Lett. **13**(9), 1359–1363 (2016)
58. Peng, Y., Wang, Y.: Prediction of the chlorophyll content in pomegranate leaves based on digital image processing technology and stacked sparse autoencoder. Int. J. Food Prop. **22**(1), 1720–1732 (2019)
59. Tawfik, N., et al.: Multimodal medical image fusion using stacked auto-encoder in NSCT domain. J. Digit. Imaging **35**(5), 1308–1325 (2022)
60. Liu, Y., et al.: Deep learning for pixel-level image fusion: recent advances and future prospects. Inf. Fusion **42**, 158–173 (2018)
61. Xia, K.-J., Yin, H.-S., Wang, J.-Q.: A novel improved deep convolutional neural network model for medical image fusion. Clust. Comput. **22**, 1515–1527 (2019)

Intelligent Aircraft Antiskid Braking Systems – A Review

P. Sudheendra[1](✉) and Vinayambika S. Bhat[2]

[1] Aircraft Research and Design Centre, Hindustan Aeronautics Limited, Bengaluru, India
sudhiaero1476@gmail.com

[2] Department of Electronics and Communication Engineering, Mangalore Institute of Technology and Engineering, Moodabidri, India

Abstract. This paper presents the evolution of application of Artificial Intelligence (AI)techniques in the design of Aircraft Antiskid Braking Systems (AABS). AABS designs that are based on conventional heuristic algorithms fail to generate desired responses when exposed to unpredictable nonlinearities such as actuator and sensor faults and variations in aircraft dynamics. AI techniques such as Artificial Neural Networks (ANN), Fuzzy logic and Machine Learning (ML) can make control systems adapt in response to changes in both internal and external disturbances. This work begins with an introduction on evolution of AABS and limitations of conventional AABS designs. Then an adapted aircraft landing gear and 'tyre-runway' dynamic models are presented. This is followed by the review of AABS that are based on the three common AI techniques namely ANN, Fuzzy logic, and ML and intelligent PID controllers. A comparison of the AI based AABS is provided in the concluding section.

Keywords: Aircraft Antiskid Braking System (AABS) · Artificial Neural Networks (ANN) · Fuzzy logic systems · Coefficient of friction · slip

1 Introduction

AABS is a closed loop controller that modulates the brake pressure to generate the right amount of brake toque in response to varying tyre to runway coefficient of friction. It reduces the chances of skidding or slipping, avoids drifting of the aircraft from the runway, thereby providing directional stability and ensures safety of both aircraft and passengers. AABS uses the wheel speed output from the sensors embedded within the wheels as a feedback signal. In most of the controllers, wheel slip is used to control the braking effort. A threshold slip value is defined for each class of aircraft based on the landing energy and dynamics.

This paper begins with a brief historical perspective on the evolution of AABS. Limitations of the conventional AABS are highlighted in the subsequent section. This section concludes by how AI techniques have penetrated the design of AABS. Then a dynamic model of aircraft and wheel are presented that are used as references for deriving AABS parameter set. The core of the work, a review of AABS that incorporate

T. Shreekumar et al. (Eds.): ISCComm 2023, CCIS 2231, pp. 188–206, 2025.
https://doi.org/10.1007/978-3-031-75605-4_13

AI techniques, is presented in the subsequent sections. Review categorizes the AABS into four main AI platforms namely Fuzzy logic, ANN, Machine Learning in general and Deep Learning in particular and Intelligent PID Controllers. An analysis of the AABS is presented along with the comparison of the techniques described. The paper concludes by summarizing the review and highlighting the challenges that lie ahead in the design of AI based intelligent controllers.

2 Historical Perspective

Antiskid Braking System (ABS)was first introduced on aircraft in the decade that followed 1950. ABS prevents wheel skid by application of required levels of braking force to the wheels, provides directional control and ensures shortest possible stopping distance under varying runway conditions.

The earliest ABS were hydraulic systems that modulated the braking pressure based on mechanical wheel speed sensor input and prevented aircraft wheels from locking and thus providing directional control and vehicular stability. The first commercial aircraft to be equipped with an antiskid system was the Boeing 707, which became operational in the late 1950s. Subsequently, other commercial aircrafts began to adopt ABS.

The North American F-100 Super Sabre was the first military aircraft on which an ABS was implemented. The F-100 Super Sabre was a supersonic jet fighter that was put into combat use by the United States Air Force (USAF) during later part of 1950.

Since then, antiskid systems have become one of the most essential equipment on both commercial and military aircrafts and have evolved continuously from simple on-off controllers to present day advanced electronic controllers operating on continuous pressure modulating systems. These systems play a critical role in ensuring safe and efficient braking under various runway conditions.

3 Limitations of Conventional Design

Rule-based or heuristic antiskid control systems have several drawbacks, as listed below, which limit their performance and adaptability.

3.1 Limited Adaptability

Heuristic ABS rely on set of rules that are designed beforehand and for a specific type of aircraft or vehicle. Such systems have limited adaptability and cannot be used on a different aircraft platform. Also, changes either in the aircraft dynamics or external environment leads to reduced braking efficiency.

3.2 Poor Handling of Uncertainty

Heuristic rule-based ABS is designed based on input parameters whose excursions are limited to a predefined range. Nonlinearities exhibited by the parameters that excite the system are generally approximated to piecewise linear models. Actuator and sensor faults are generally not considered to name a few.

3.3 Absence of Learning

Rule-based systems lack the ability to self-learn from experience and make changes to the control algorithm in response to such changes. They require expert intervention and manual reprogramming every time a new set of parameters need to be incorporated into the design.

3.4 Lack of Scalability

Aircraft mass configuration could undergo incremental changes during design phase and during testing or certification phase. Changes in mass configuration could lead to changes in associated systems and aircraft subsystems need to evolve along with the changes. In case of conventional ABS, frequent tuning of control algorithm is required in response to changes in the aircraft configuration making them untenable for scalability.

It is quite evident that here has been a steady transition from rule-based design to model-based or learning-based approaches that can adapt to changing conditions, learn from data, and make more informed and context-aware control decisions. These approaches leverage on the advantages offered by artificial intelligence and control theory to improve overall system performance and safety.

4 Aircraft and Wheel Dynamics Models

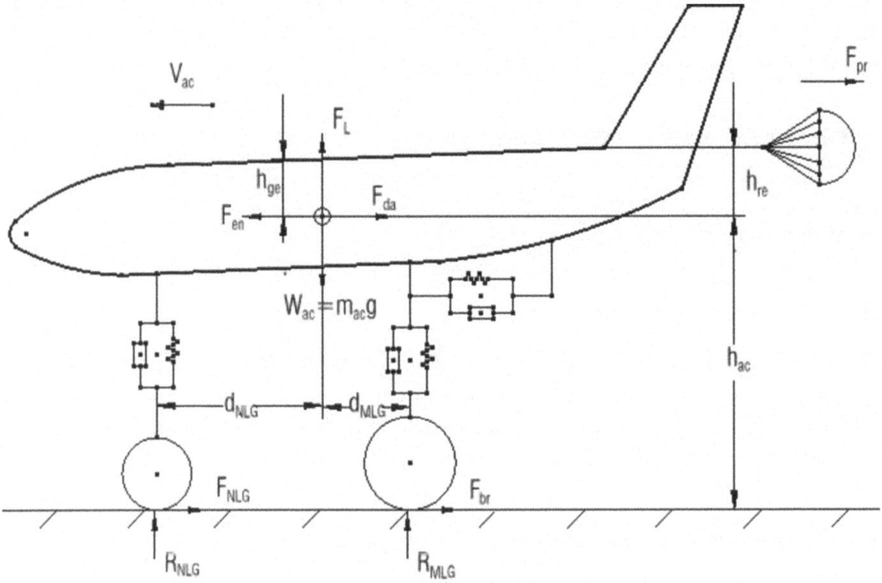

Fig. 1. Aircraft dynamic model

Adaptation of AI techniques for AABS requires an adequate model of aircraft and wheel dynamics during braking. Simplified force balance equations and forces acting on the wheel during braking [3, 13], as illustrated in Figs. 1 and 2, are adapted from the existing literature for review of AABS. Nomenclatures used for definition of aircraft and wheel dynamics are listed in Table 1.

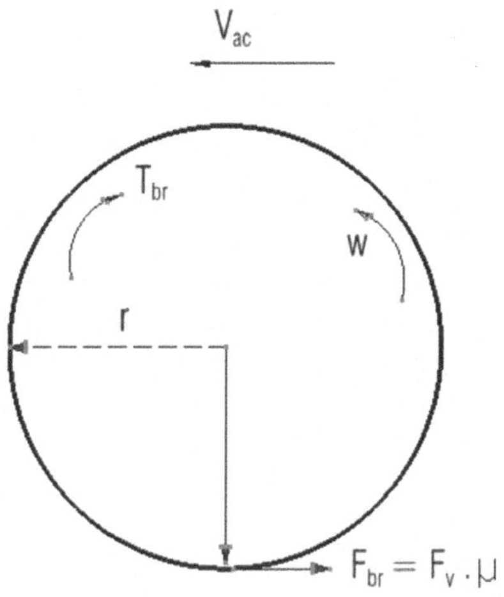

Fig. 2. Forces acting on the wheel during braking

Table 1. Nomenclatures used in aircraft and wheel modeling

Name	Description
FL	Aircraft lift
W_{ac}	Aircraft weight
F_{da}	Aircraft drag force
F_{en}	Aircraft thrust
F_{NLG}	Nose wheel frictional force
F_{br}	Main wheel braking force
F_{pr}	Brake parachute drag force
R_{NLG}	Nose wheel reaction
R_{MLG}	Main wheel reaction
vac	Aircraft forward velocity
d_{NLG}	Distance between aircraft centre and nose wheel
d_{MLG}	Distance between aircraft centre and main wheel

(*continued*)

Table 1. (*continued*)

Name	Description
H_{prc}	Distance between F_{pr} and aircraft Centre of Gravity C.G
H_{gc}	Distance between aircraft thrust F_{en} and C.G
I_{wl}	Wheel inertia
ω	Main wheel angular velocity
F_b	Main wheel frictional force
r	Main wheel radius
T_{br}	Brake torque
P_{br}	Brake pressure
F_v	Vertical force at the main wheel axle
μ	Coefficient of friction or adhesion coefficient
S	Wheel slip
rdl,	Deflected tyre radius
C_r	Surface or runway conditions label

5 Fuzzy Logic AABS

Fuzzy logic developed by Lotfi Zadeh is a multivalued logic that can adapt to complex nonlinear nature of control systems. Uncertainties are effectively modelled by employing fuzzy logic. This logic is used to build robust controllers that easily adapt to nonlinearities in system parameters and uncertainties such as external disturbances, faults in actuators and sensors and modelling inaccuracies.

A review of the existing Fuzzy logic AABS indicates that two groups of parameters are considered for design of controllers. First group of parameters refer to the aircraft dynamics or the force balance equations and the wheel dynamics or forces acting on the wheel during braking is the other group. Runway conditions are also included in the wheel dynamics. Based on the tyre to runway/surface coefficient of friction, three types of fuzzy membership functions, namely Dry, Wet, and Icy are identified. Fuzzy variables pertaining to wheel dynamics such as coefficient of friction 'μ', Slip 'S', main wheel angular velocity'ω', brake pressure 'P_{br}', braking force or the frictional force 'F_{br}', braking torque '$T_{br,}$deflected tyre radius $r_{dl,}$ and vertical load at the point of contact of the wheel $V_L$areusedfor defining the fuzzy sets.

Fuzzy set consisting of variables such as landing gear to aircraft stiffness coefficients, actuator performance, aircraft velocity, engine thrust, aircraft drag, parachute drag etc. are associated with aircraft performance parameters.

Application of Artificial Intelligence techniques to AABS started with use of Fuzzy Logic based self-tuning controllers. H. Chris Tseng and CharlieW.Chi[1, 2] proposed a simplified self-tuning fuzzy logic PI controller based on nonlinear interpolation technique. A limited number of data points pertaining to AABS nonlinearities namely coefficient of friction μ, vertical load V_L, tyre deflection$d_r$andwheel slip S are assigned as

Gaussian fuzzy membership functions. Tyre deflection radius r_{dl} varies with vertical load V_L and μ is a function of wheel slip S. Brake torque T_{br} is input to the fuzzy controller based on the fuzzy rules formulated by taking the difference between actual wheel speed and desired wheel speed ε and also the rate of $\dot{\varepsilon}_e$ A simplified ABS model equivalent to a constant self-point tracking algorithm in which desired aircraft slip velocity V_{dv} determined from the aircraft dynamics, is used as reference and wheel speed is controlled to achieve maximum coefficient of friction. Desired wheel velocity is then defined as

$$\omega_{\mathrm{d}S} = \frac{v_{ac} - v_{\mathrm{d}v}}{r_{\mathrm{d}l}} \tag{1}$$

Researchers have tried to supplement the conventional AABS with rule based Fuzzy Expert Supervision systems that estimate reference slip values in association with the stable adhesion characteristics. Two different models have been suggested for the design of supervisory systems by B. Ewers, J. Bordeneuve-Guibe, J.-P. Garcia and J.Piquin [3]. In the first approach, a new reference slip value is synthesized directly using aircraft system variables. The second approach uses fuzzy identification of runway conditions to estimate the reference slip. A simplified model of the aircraft as depicted in Fig. 2 is proposed by neglecting the aerodynamic drag and reverse engine thrust. Coefficient of friction μ is a function of wheel slip S.

$$S = \frac{V_{ac} - \omega R_r}{V_{ac}} \tag{2}$$

Only two forces vertical force F_v and frictional force F_{fr} are considered in the braked wheel model of Fig. 1.

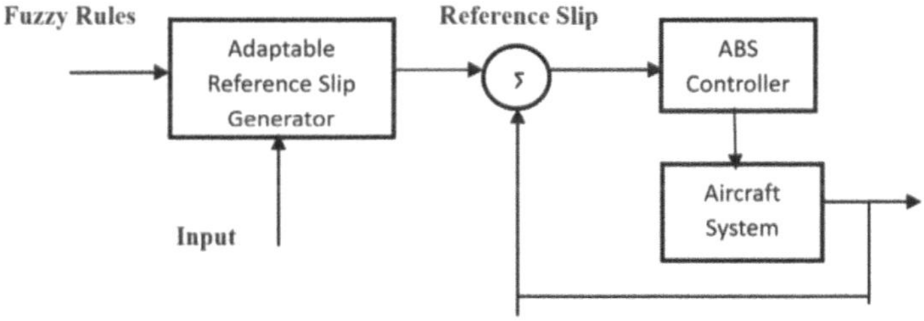

Fig. 3. Reference Slip Supervisor based on aircraft data

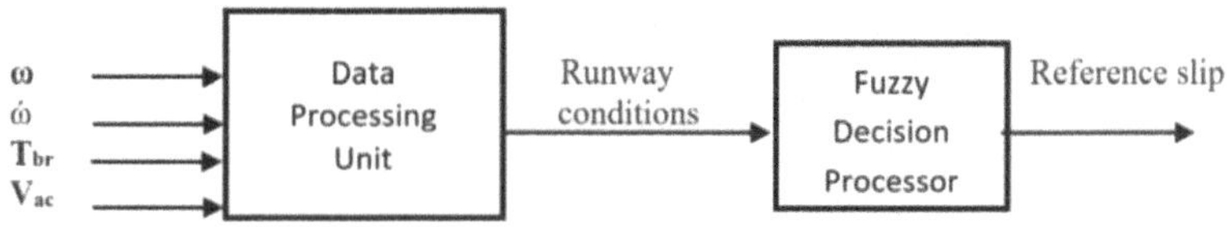

Fig. 4. Reference Slip Supervisor based on aircraft data

A Fuzzy AABS Controller for a military aircraft that incorporated both Nose and Main wheels models was proposed by Ioan Ursu and Felicia Ursu [4]. A simplified aircraft model that neglected the forces or moments generated due to vertical movement of landing gear, moments due to pitch and roll, was adapted to generate the required fuzzy framework. Nose wheel's angular velocity was taken as the reference for calculation of main wheel slip S. A fuzzy controller based on "Mamdani" model was built comprising of a fuzzifier, fuzzy rules and a de-fuzzifier. Membership functions called the "road labels" that were based on road conditions, namely, dry, wet and icy, were defined. Coefficient of friction μ is a function of wheels slip S, aircraft velocity V_{ac} and road conditions C_r. Thus $\mu = f(S, V_{ac}, C_r)$. Nine threshold values for road conditions were defined as inputs to fuzzy controller. Control variable, the brake pressure supplied to the rear wheels, is generated by genetic algorithms. Triangular membership functions were employed for building the fuzzy logic. A singleton fuzzifier for the control variable P_{br} and a centre-average defuzzyfier is used for modeling the control system (Figs. 3 and 4).

Generally, only longitudinal stability is considered while designing AABS effects of lateral drift due to differential braking and differences in adhesion coefficients experienced by Starboard and Port side wheels, are neglected. An attempt was made by Yang Li, Zongxia and Yaoxing Shang to model the lateral disturbances and propose an ABS [5] that takes care of lateral stability along with longitudinal stability. Lateral disturbances, the drift angle and drift angular velocity are input to a fuzzy logic-based controller in which triangular functions were are used for fuzzy membership definitions. The proposed model addresses the limitations of differential braking whose effectiveness decreases with increase in lateral disturbances (Fig. 5).

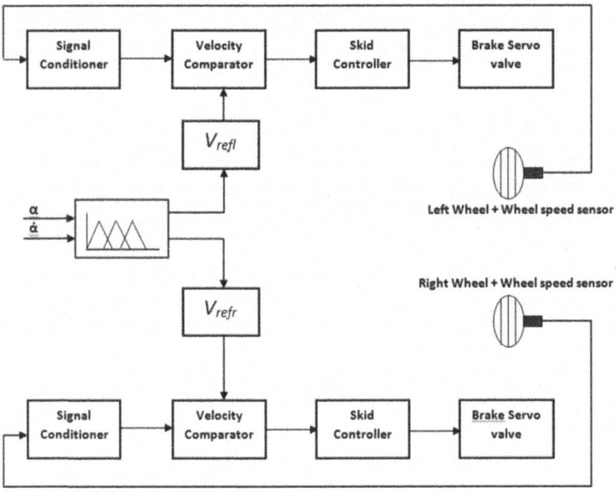

Fig. 5. Lateral Fuzzy Controller

The lateral disturbance caused by factors such as crosswind or runway surface issues can affect the anti-skid braking system's performance. In case of lateral disturbances, different decelerations of the two main wheels lead to aircraft drift. The lateral control system is designed to compensate for this drift angle and ensure the lateral stability of the taxiing aircraft. The compensation is achieved through the adjustment of left and right wheel reference linear velocities dependent on the drift angle and drift angular velocity. Left and right main wheel antiskid control reference velocities are modelled as

$$v_{\text{rl}}(\text{t}) = \begin{cases} v_{\text{wl}}(\text{t}) & v_{\text{wl}}(\text{t}) \geq v_{\text{rl}}(\text{t - } \Delta \text{ t}) \\ v_{\text{rl}}(\text{t}- \Delta \text{ t}) - \text{dv} - (1 - k_{vr})\ v_{\text{wl}}(\text{t}) \leq v_{\text{rl}}(\text{t - } \Delta \text{ t}) \end{cases} \tag{3}$$

$$v_{\text{rr}}(\text{t}) = \begin{cases} v_{\text{wr}}(\text{t}) & v_{\text{wr}}(\text{t}) \geq v_{\text{rr}}(\text{t - } \Delta \text{ t}) \\ v_{\text{rr}}(\text{t}- \Delta \text{ t}) - \text{dv} - V(1 + k_{vr})\ v_{\text{wl}}(\text{t}) \leq v_{\text{rl}}(\text{t - } \Delta \text{ t}) \end{cases} \tag{4}$$

The input and output functions are modelled respectively as

$$\text{Input}\, E = \frac{\alpha}{\dot{\alpha}} = \frac{E}{\text{d}E} \tag{5}$$

E is the error signal ratio, α is the wheel speed acceleration

$$\text{Output}\, u = k_{vr} \tag{6}$$

k_{vr} is a factor determined by the aircraft dynamics.

6 Artificial Neural Network (ANN) Based AABS

ANNs are inspired by human brain. ANNs consist of nodes or artificial neurons like neurons contained in the human brain, interconnected, and organized into hierarchal layers. Information is transmitted from the inner layer to the outer layer through one or many hidden layers. Each neuron in an inner layer receives and processes the data using an activation function and produces an output. Neurons are interconnected with associated weights that determine its strength and biases that control the activation threshold. Activation functions define the nonlinearities of the system. Neurons associated with a layer compute their output by addition of weighted inputs and activation control until an output is generated. This process is called feedforward propagation. ANNs predicted output is compared with target output and the errors that are generated are optimized using various algorithms and gradient decent being one of the most important one. This process called backpropagation adjusts the weights and biases associated with the network to minimize the error. Thus, ANNs are trained until the proposed model converges to a predefined accuracy value.

ANNs are used in natural language processing, image processing, speech recognition, autonomous vehicle control, computer vision and intelligent controllers. ANNs are best suited for highly nonlinear systems and therefore find immense use of design of AABS. In the subsequent sections, a review of the AABS designs based on ANNs are presented.

An intelligent ABS was implemented using a combination of ANN and fuzzy logic by L Bareilli, G.Bidini and Bonucci [6]. Main objectives of the controller were maximization of friction coefficient under varying runway conditions, optimization of antiskid controller response time, minimization of input parameters to fuzzy controller and design of an efficient adaptive controller. A bi-directional aircraft model that focuses on the balance of forces acting on the aircraft during braking was used to describe the aircraft dynamics.

Three different types of runway conditions, namely dry, wet, and icy were input to the fuzzy controller. A total of nine fuzzy membership functions such as "Zero Dry," "Medium Dry", "Large Dry" etc. were defined for slip evaluation. Triangular and trapezoidal functions were incorporated for definition of fuzzy membership functions. Aircraft deceleration was the second input to the fuzzy based controller and here also membership functions "very large", "medium large" and "large" were specified.

Outcome of simulation runs show that fuzzy based anti-skid system provides better results on dry tracks with almost constant deceleration of aircraft speed. Performance of the controller in medium dry and icy tracks is comparable with that on the dry track.

Sensitivity analysis of the fuzzy logic ABS showed that the slip factor, wheel angular speed, aircraft velocity and deceleration are not significantly affected by changes in aircraft mass. It was observed that braking torque was a function of the aircraft mass, increasing with heavier loads.

Application of ANNs to AABS that can work under uncertain runway conditions was the highlight of the work carried out by R Somakumar and J Chandrasekhar [7]. A simplified aircraft model was considered that neglected lateral forces and fuselage to landing gear coupling.

Aircraft wheel slip S was monitored along with aircraft velocity and deceleration. Nonlinear mapping ability and pattern recognition were used to build an ANN for identification of nature of runway namely dry, wet, and icy. Neural network adjusts the brake pressure P_{br} to achieve maximum coefficient of friction μ. Neural network continuously monitors the slip S and brake torque is varied in response to wheel slip so that maximum μ is achieved regardless of runway conditions.

A two-layer network with three input neurons namely wheel speed ω, V_{ac} and aircraft deceleration, five hidden neurons and one output layer with one neuron was designed to produce the required command to the brake system. Hidden and output layers were realized using sigmoid functions. A representation of the proposed ANN controller is shown (Fig. 6).

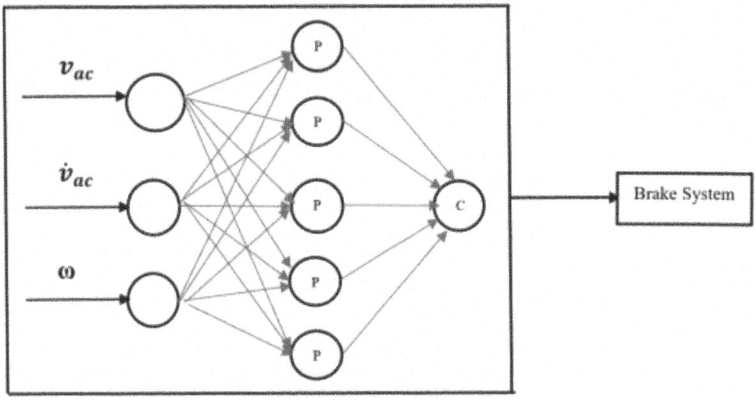

Fig. 6. Neural network AABS

A direct use of PID controllers in aircraft brake control applications may not yield better response due its limited capability when exposed to uncertainties or nonlinearities present in aircraft system dynamics. Samuel John designed a feedforward optimized PID controller [8] that makes use of inverse model neural network was designed to overcome the disadvantages of a simple PID controller. V_{ac} and ω were the state variables. An auto-regressive moving average neural network model was employed for simulation of the controller. Desired slip S_d was tracked using the inverse neural network model as shown in the representation (Fig. 7).

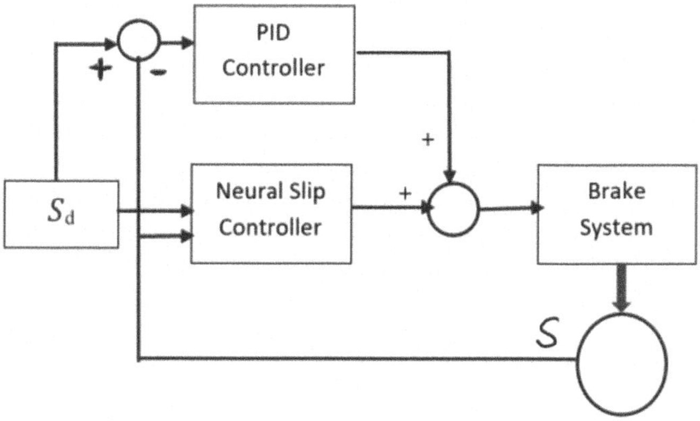

Fig. 7. Inverse neural network model PID AABS

Traditional ANNs use sigmoid functions at the hidden layers for computation. Sigmoid functions use complex exponentials and could result in neuron outputs being more computationally intensive especially when number of neurons is more. On the contrary, Radial Basis Function (RBF) ANNs use Gaussian functions or radial functions and are much simpler in comparison with ANNs using sigmoid functions.

A complete definition of the aircraft dynamics requires a vast set of nodes and if traditional ANNs are used could lead to slower response and increased complexities. AABS are required to respond faster to mitigate wheel locking. It therefore becomes necessary to use RBF ANNs that use Gaussian or radial functions. RBF neurons are always upfront and weights can be adjusted using least square method or other simple regression techniques resulting in faster response.

One such RBF neural network-based AABS was proposed by Hui Sun, Jianguo Yan, Yaohong Qu and Yuan Liu [9] that determined the type of runway and optimized the braking based on RBF network response. In the present work, aircraft is assumed to be a ideal rigid body and lateral disturbances are neglected. Tyre deformations are also neglected. AABS is designed to regulate wheel slip S. .An assumption is made that the function $f(x) = f(S) = \{x_1, x_2, x_3, \ldots .x_n\}$ is nonlinear and uncertainties along with external disturbances are bounded. An ideal weight vector w^* is assumed such that RBFNN could be approximated to a function $f(\Theta)$ defined by a asymmetrical barrier Lyapunov function.

A nonlinear back stepping algorithm was used to predict the runway conditions precisely Aircraft dynamics were modelled using minimal-learning-parameter technique and uncertainties were capped at a predefined upper bound.

Backward Propagation (BP) or Feed Forward Neural (FNN) network techniques have meticulously been used in the design of PID based AABS. BPNNs have certain limitations when applied to complex nonlinear systems like AABs due to the large number of uncertainties. Weights tend become very small during backpropagation and this could result in slow convergence or a sub-optimal convergence value. Sometimes, during training, gradients tend become very large leading to nonconvergence. Also, it is very important in a BPNN to define the initial values of the weights and biases corresponding to slip factor S and μ.

To overcome problems associated with AABS that are based on BPNN, Yuanjin Zhang, Huawei Wu and Zhigang Tan implemented a BPNN using Genetic Algorithm (GA) [10]. Here also, a simplified aircraft model neglecting the lateral effects are considered. It means that, when lateral disturbances are neglected, braking on starboard and port sides of the aircraft are equal. Training of the BPNN include ω, slip velocity v_s and $\mu.\omega$ and V_{ac} are the two input nodes. Two hidden layers were specified comprising of ten nodes each and a single output node namely wheel slip S was included. A large number of initial population number with a value of 50 was considered. Initial crossover probability and mutation were selected based on calculated guess. A representation of the controller is given (Fig. 8).

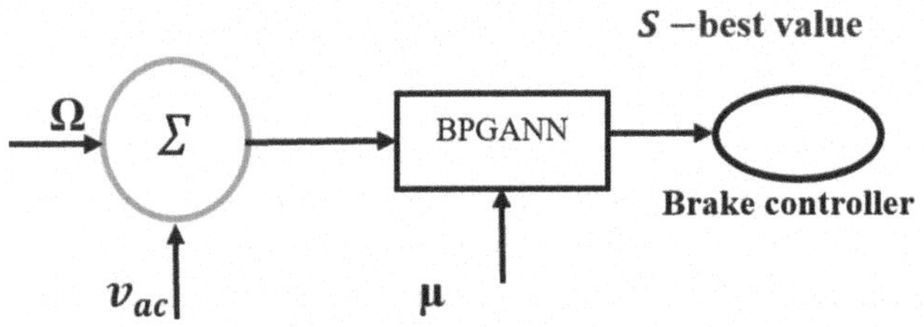

Fig. 8. BPGA antiskid controller

All the design discussed so far focused on only the nonlinearities associated with aircraft dynamics. Faults or nonlinearities associated with actuators, sensors and strong coupling between systems and aircraft structures were not considered hitherto during the design. A reconfiguration controller based Adaptive Fuzzy Active Disturbance Rejection Control (AFADRC) [11] was suggested by Zhao Zhang, Zhong Yang, Guoxing Zhou, Shuchang Liu, Dongsheng Zhou, Shuang Chen and Xiaokai Zhang, that consisted of a fuzzy Nonlinear State Error Feedback (NLSEF) and an ANN based Adaptive Extended State Observer (AESO). NLSEF compensates for the fault perturbations, external disturbances and noise in measurement systems. AESO is built to estimate the measurement noise, component faults and external disturbances (Figs. 9 and 10).

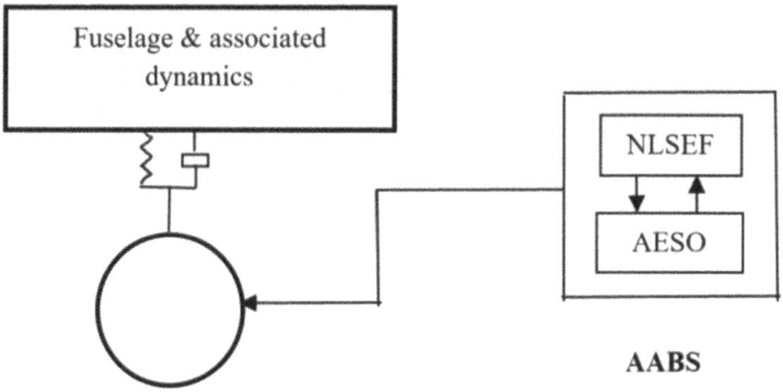

Fig. 9. Simplified representation of AFADRC

7 Machine Learning Based (Deep Learning) AABS

ML techniques have made deep inroads into the engineering application environment and have not left any field of engineering untouched by their influence. Aircraft systems were one of the first to adapt the ML techniques starting from passenger and cargo booking, logistics, environmental control systems inside the aircraft, air traffic control and many more. Desing of critical controllers that are part of the aircraft started evolving gradually due to the passenger and pilot safety norms and very stringent certification guidelines. It is therefore evident from the survey of the available literature that only a handful of ML based AABS designs have been attempted and hitherto only simulation studies have been carried out.

As discussed in the previous sections, control strategies in all the AABs were derived from a simplified aircraft model. Brake system failures such as actuator failure and sensor failures were not considered in the design. Such failures are uncertain and highly nonlinear. Modeling such component failures using conventional PID algorithms and AI techniques described in the previous sections may not always yield the desired performance. A reconfiguration controller consisting of a linear state observer and a linear state error feedback along with Twin-Delayed Deep Deterministic Policy (TD3) gradient algorithm which is a model free reinforcement learning method deep reinforcement learning was suggested by Shuchang Liu et.al [12]to address the gaps in the control logic. In this method, control reconfiguration concept is employed that uses the results of fault diagnosis results to restructure the control loop to adapt the controller to the faulty component such as actuator or sensors. An Active Disturbance Rejection Control (ADRC) technique is used for design of AABS. ADRC is a model free technique that is well suited for AABS due to the uncertainties in the dynamics.

Uncertainties in the dynamics are modeled using this actor-centric method that searches for optimal slip ratios and maximizes chances of obtaining the desired value of slip in compliance with the variations in the dynamics. The design of the AABS starts with known aircraft models and authors have incorporated a complete description of the aircraft that includes aircraft, fuselage to landing gear coupling, landing gear, landing gear to wheel stiffness, wheel and hydraulic servo system components comprising of actuators.

The reconfiguration controller was defined based on only the longitudinal taxiing. A slip control logic was used along with aircraft velocity V_{ac}as the reference. Wheel speed V_{ω} was adjusted to achieve antiskid braking. Aircraft dynamics with only longitudinal forces considered is defined by the equation

$$V_{ac} = f\,(V_{ac,}\, V_{ac,}\, \Phi_{ed},\, \Phi_{fl}) + g_o u \tag{7}$$

where V_{ac} is the controlled parameter, ϕ_{ed} is the external disturbance, ϕ_{fl} is the fault term, g_o is the controller gain and u is the system input. Here AABSA dynamics represented by Eq. (7) is considered as state variable. A simplified functional block of the controller is represented in Fig. 11.

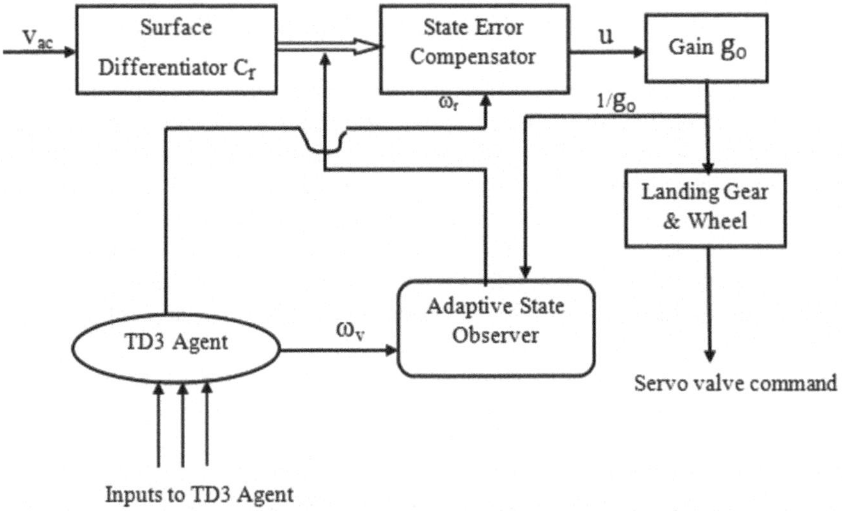

Fig. 10. Reconfiguration controller based on TD3 Deep learning

8 Intelligent PID Controllers

Robust controllers built around PID algorithms are an effective means of dealing with aircraft nonlinearities. But such PID controllers are prone to slow learning due to unbounded uncertainties.

A self-learning robust PID controller based on swarm optimization technique was proposed by Fengrui Xu, Xuelin Liang, Mengqiao Chen and Wen sheng Liu [13] to address the strong nonlinearities and external disturbances during braking.

This paper introduces a robust self-learning PID algorithm that does not depend on the conventional mathematical model of the controlled object. Robust controller is modeled keeping in view the aircraft nonlinearities and the self-learning function adapts to the varying runway conditions. A Particle Swarm Algorithm (PSA) is used to optimize the self-learning process. PSA comprises of a fitness function that is designed by taking the tracking error between the actual slip rate and the optimal slip rate. An exponential fitness function is defined with fitness values between 0 and 1. Here 1 represents best fitness value and 0 represents no fitness.

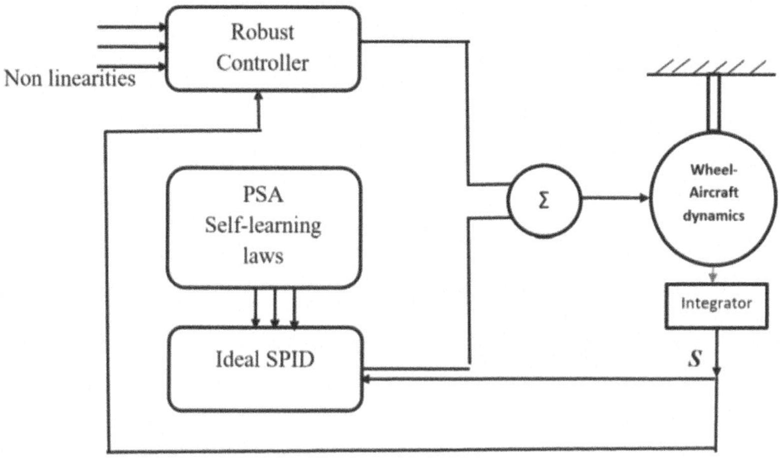

Fig. 11. Self-learning robust controller

It is a well-known fact simple PID controllers are quite effective when disturbances are known beforehand. But traditional PID controllers fail to generate the optimal performance when subjected to unknown disturbances such as aircraft nonlinearities and component failures. It is always desirable to design a controller that is immune to external perturbances. Haibin Song, Bin Fang and Pu Wang built an Immune PID controller (IIPD) [14] inspired by human immunity system. This controller assumed a simplistic aircraft model without lateral drift velocities caused due to differential braking and also side loads due to cross winds. Control function is defined as

$$r(t) = k_p \mathrm{e}(t) \tag{8}$$

Here $r(t)$ is the command to the AABS and $e(t)$ is the error signal generated based on the difference between aircraft velocity and main wheel velocity. k_p is a proportionality constant determined by the aircraft dynamics.

A discrete form of ideal PID controller is added to the controller represented by the nonlinear function represented by Eq. (8). Slip S is controlled by modulating he brake pressure which in turn varies the brake torque to achieve optimum wheel deceleration under various runway conditions.

9 Conclusion

Evolution of AI based AABS started with application fuzzy logic followed by ANN. AABS is highly nonlinear in nature and modeling of aircraft dynamics requires factoring uncertainties in the form of external disturbances, system faults and sensor failures. Fuzzy logic and ANN techniques can be used to model AABS due their inherent nature. Uncertainties or nonlinearities defined as fuzzy membership functions offer gradual transition between parameters. Fuzzy logic AABS are robust and handle multi-valued variables. But fuzzy AABS requires expert intervention due to the lack of a solid mathematical framework. AABS based on ANNs handle nonlinearities quite well and ae scalable. ANN AABS are much faster in comparison with fuzzy logic AABS. ANN AABS are flexible and can easily integrated with reinforced and deep learning (Table 2).

Table 2. Comparison of AI AABS techniques

Controller	AI technique	Dynamic input parameters	Features
Self Tuning PI [1, 2]	Fuzzy logic Nonlinear interpolation	μ, V_L, r_{dl}, S	• Simple fuzzy PI controller that works on linear interpolation • Runway conditions are not identified • Aerodynamic drag and lateral loads are not considered • Only main landing gear is considered in the model • Only main wheel speed reference considered • Lateral drift and stability aspects neglected • Pitch and Roll moments neglected
Expert Supervision- Approach 1 [3]	Fuzzy logic Reference Slip Supervisor	μ, V_L, r_{dl}, S	• Expert Supervisory system estimated slip • A new reference slip is generated from the aircraft dynamics • Runway conditions are not considered • Aerodynamic drag and lateral loads are not considered • Only main landing gear is considered in the model • Only main wheel speed reference considered • Lateral drift and stability aspects neglected • Pitch and Roll moments neglected

(continued)

Table 2. (*continued*)

Controller	AI technique	Dynamic input parameters	Features
Expert Supervision- Approach 2 [3]	Fuzzy logic Reference Slip Supervisor	μ, V_L, r_{dl}, S, Dry, Wet, Ice	• Expert Supervisory system estimated slip • Reference slip is generated based on type of runway • Aerodynamic drag and lateral loads are not considered • Only main landing gear is considered in the model • Only main wheel speed reference considered • Lateral drift and stability aspects neglected • Pitch and Roll moments neglected
Mamdani model fuzzifier [4]	Fuzzy logic Genetic algorithm	μ, r_{dl}, S, Dry, Wet, Ice	• Reference slip is generated based on type of runway (road labels) • Aerodynamic drag and lateral loads are not considered • Main landing gear is considered in the model • Nose wheel angular speed is used as reference for main wheel slip • Lateral drift and stability aspects neglected • Pitch and Roll moments neglected
Lateral Fuzzy Controller [5]	Fuzzy logic Triangular fuzzy functions	μ, V_L, r_{dl}, S, v_{refr}, v_{refl}	• Runway conditions are not identified • Aerodynamic drag is considered • Lateral loads (drift) are considered • Lateral drift and stability aspects considered • Only main landing gear is considered in the model • Pitch and Roll moments effects are addressed
Intelligent Fuzzy ANN Controller [6]	Fuzzy logic + ANN Triangular and Trapezoidal functions ANN Adaptive Controller	μ, V_L, r_{dl}, S, α, Dry, Wet, Ice	• Wheel acceleration α and μ are input as references • Runway conditions are considered • Aerodynamic drag and lateral loads are considered • Only main landing gear is considered in the model • Only main wheel speed reference considered • Lateral drift and stability aspects neglected • Pitch and Roll moments neglected

(*continued*)

Table 2. (*continued*)

Controller	AI technique	Dynamic input parameters	Features
ANN Antiskid Controller [7]	ANN Controller Nonlinear mapping & pattern recognition Two-layer ANN with three input neurons Five hidden layers One output layer	μ, V_{ac}, r_{dl}, S, α, Dry, Wet, Ice	• Reference slip is generated based on type of runway • Aerodynamic drag and lateral loads are not considered • Only main landing gear is considered in the model • Only main wheel speed reference considered • Lateral drift and stability aspects neglected • Pitch and Roll moments neglected
Inverse Neural Network PID AABS (Feedforward optimized PID) [8]	Auto-regressive moving average ANN	μ, r_{dl}, S, V_{ac}	• Reference slip is generated based on aircraft velocity • Aerodynamic drag and lateral loads are not considered • Main landing gear is considered in the model • Nose wheel speed is used as aircraft velocity reference • Lateral drift and stability aspects neglected • Pitch and Roll moments neglected
RBF ANN AABS [9]	RBF ANN Gaussian or Radial functions	μ, $V_{L,}$, S, Dry, Wet, Ice	• Runway conditions are identified • Simplistic aircraft model • Aerodynamic drag and lateral forces not considered • Only main landing gear is considered in the model • Pitch and Roll moments effects neglected
BPNN Genetic Algorithm AABS [10]	BPNN with Genetic algorithm Two input nodes Two hidden layers with ten nodes each One output node	μ, ω, Vac , S, v_s, Dry, Wet, Ice	• Runway conditions are not identified • Simplistic aircraft model • Aerodynamic drag and lateral forces not considered • Only main landing gear is considered in the model • Pitch and Roll moments effects neglected

(*continued*)

Table 2. *(continued)*

Controller	AI technique	Dynamic input parameters	Features
Active Disturbance Rejection Control [11]	Fuzzy ADRC + ANN Observer NLSEF compensator + Adaptive State Observer	μ, V_L,V_{ac}, r_{dl}, S, α, Dry, Wet, Ice	• Reference slip is generated based on type of runway • Aerodynamic drag and lateral loads are not considered • Only main landing gear is considered in the model • Lateral drift and stability aspects neglected • Pitch and Roll moments neglected • Servovalve, sensor and hydraulic failures considered
Deep Learning -TD3 Gradient AABS [12]	Deep learning TD3 Gradient algorithm Model free reinforcement learning	μ, ω, V_L, r_{dl}, S, V_{ac}, ϕ_{ed}, ϕ_{fl}, g_o,	• Reference slip is generated based on aircraft velocity • Full Aerodynamic model of the aircraft considered • Main landing gear is considered in the model • Nose wheel speed is used as aircraft velocity reference • Lateral drift and stability aspects neglected • Pitch and Roll moments neglected
Self-learning Robust PID Controller [13]	Self-learning Robust PID Particle swarm algorithm	μ, ω, V_L, r_{dl}, S, V_{ac},	• Runway conditions are not identified • Simplistic aircraft model • Aerodynamic drag and lateral forces not considered • Only main landing gear is considered in the model • Pitch and Roll moments effects neglected
Immune PID Controller [14]	Immune PID Immune to external disturbances	μ, ω, V_L, r_{dl}, S, V_{ac}	• Runway conditions are not identified • Simplistic aircraft model • Aerodynamic drag and lateral forces not considered • Only main landing gear is considered in the model • Pitch and Roll moments effects neglected

AABS built on Robust control PIDs offer enhanced disturbance rejection when subjected to sudden perturbances in the form of runway variations and system faults. Such AABS can handle a greater mismatch between actual aircraft dynamics and controller model. AABS built on Deep learning techniques are more adaptable and learn from real time data. Table 2 summarizes the review of AABS carried out in this work.

The review also points towards the fact that most of the studies have focused on simulation of the AABS using software models. Real time experimental studies on AABS are difficult due to the challenges in logistics, certification, safety and cost involved.

Certification of onboard controllers such as AABS requires analysis of real time data from the aircraft sorties. Simulation studies provide a strong foundation for structuring the qualification process leading to certification. It is here that the utility of test rigs that are designed to test aircraft systems on ground take centre stage. Ground testing of AABS helps to test the design in real time economically. Issues like passenger and pilot safety can be addressed easily on ground.

References:

1. Chris Tseng, H., Chi, C.W.: Aircraft anti-lock brake system (ABS) with fuzzy logic control. In: IFAC Symposium in Automatic Control in Aerospace, vol.27, no.13, pp. 387–392 (1994)
2. Chris Tseng, H., Chi, C.W.: Aircraft antilock brake system with neural networks and fuzzy logic. J. Guidance Dyn. **18**(8), 1113–1118 (1995)
3. Eweres, B., Bordenneuve, J., Garcia, J.P., Piquin, J.: Expert supervision of an anti-skid control system of a commercial aircraft. In: Proceedings of International Symposium on Intelligent Control, pp. 420–425 (1996)
4. Ursu, I., Ursu, F.: An intelligent ABS control based on fuzzy logic. Aircraft application. In: Proceedings of International Conference on Theory and Applications of Mathematics and Informatics, vol. 7, pp. 355--368 (2003)
5. Li, Y., Jiao, Z., Shang, Y.: Research on aircraft taxiing lateral control based on fuzzy controller. In: Proceedings of International Conference on Fluid Power and Mechatronics, pp. 582–587 (2011)
6. Barelli, L., Bidini, G., Bonucci, F.: AN anti-skid controller for aircraft applications based on computational intelligence. Int. J. Autom. Control Eng. **2**(3), 101–112 (2013)
7. Somakumar, R., Chandrasekhar, J.: Intelligent anti-skid brake controller using a neural network. J. Int. Fed. Autom. Control, 611–621 (1999)
8. John, S.: Artificial intelligent-based feedforward optimized PID wheel slip controller. In: Proceedings of AFRICON (2013)
9. Sun, H., Yan, J., Qu, Y., Liu, Y.: RBF neural network based dynamic surface control for anti-skid braking system using barrier Lyapunov theorem. Chin. Control Decis. Conf., 1498–1495 (2016)
10. Zhang, Y., Wu, H., Tan, Z.: Design of aircraft antiskid braking system based on BP neural network and genetic algorithm. In: Conference on Advanced Design and Manufacturing Engineering, vol. 96, pp. 72-76 (2017)
11. Zhang, Z., et al.: Adaptive fuzzy active-distributive rejection control – based reconfiguration controller design for aircraft anti-skid braking system. MDPI J. Actuators **10**(8), 1–21 (2021)
12. Liu, S., et al.: Application of deep reinforcement learning in reconfiguration control of aircraft anti-skid braking system. J. MDPI Aerospace, 1–25 (2022)
13. Song, H., Fang, B., Wang, P.: Research and applications of immune PID adaptive controller in anti-skid braking system for aircraft. In: Proceedings of International Conference on Information Engineering and Computer Science (2009)
14. Xu, F., Liang, X., Chen, M., Liu, W.: Robust self-learning PID control of an aircraft anti-skid braking system. MDPI J. Math., 1–18 (2022)

Artificial Neural Networks Applied in the Detection of Breast Cancer

Deepthi Kotian(✉), Vinayambika S. Bhat, Dhanush Poojary, and V. Kavya

Mangalore Institute of Technology and Engineering, Moodabidri, Mangalore, India
deepthikotian@mite.ac.in

Abstract. Breast cancer is a serious health issue for women everywhere, and effective treatment depends on early identification. For medical imaging, automated breast cancer identification is essential to speeding up diagnosis; however, current approaches frequently have extensive analysis durations. In order to address this, we investigate the usage of neural networks (NN) and concentrate on the backpropagation technique in particular. We found deficiencies in the sensitivity and specificity of the existing NN models by conducting a thorough literature analysis. Our method uses backpropagation to train a neural network (NN) for effective categorization of benign and malignant patterns in breast cancer images. Preliminary findings demonstrate decreased processing time and increased accuracy. A comparative study with current methods demonstrates our method's effectiveness. To improve diagnostic precision even more, future improvements might include advanced NN architectures.

Keyword: Neural Networks (NN)

1 Introduction

The backpropagation technique can be used to train artificial neural networks (ANNs), which have been widely used in the diagnosis of breast cancer [3]. Artificial neural networks are used to diagnose breast cancer by applying a computer model that was influenced by the design and operation of biological neural networks in the human brain [1]. Artificial neurons (ANNs) are networks of interconnected nodes, also referred to as perceptron's, that collaborate to interpret and analyze data. For training ANNs, a key method is the backpropagation algorithm. It gives the network the ability to gain knowledge from a given dataset, enhance its functionality, and produce precise predictions or classifications [3]. Backpropagation uses supervised learning, hence it needs a dataset that has been labeled and in which each sample is linked to a known result or class. Forward propagation and backward propagation are the two primary stages of the backpropagation algorithm [2]. During the forward propagation phase, input data are delivered into the neural network, where they are processed by artificial neurons in the network's layers. Each neuron processes the inputs according to an activation function, adds up the weights, and sends the information to the layer underneath [4]. Up until the output layer generates a prediction or categorization, this process is repeated. The phase

T. Shreekumar et al. (Eds.): ISCComm 2023, CCIS 2231, pp. 207–220, 2025.
https://doi.org/10.1007/978-3-031-75605-4_14

of backward propagation starts after the forward phase. It entails figuring out the mistake, or the discrepancy between the output that was expected and what actually happened. The weights of the connections between the neurons are then updated using this error, iteratively working backward through the network [8]. The objective is to decrease the error by altering the weights in a way that increases the network's effectiveness using methods like gradient descent. In the context of breast cancer screening, ANNs trained using the backpropagation approach can look at various traits or qualities extracted from clinical data, genetic data, or medical images [18]. Tumor characteristics including size, form, texture, and other pertinent elements may be included in these criteria. By speaking with the ability of ANNs to understand complicated patterns and relationships in the data has been effectively proved by their usage in the detection of breast cancer [17]. It's critical to remember that the performance of these models is influenced by a variety of factors, including the quality and representativeness of the training data, clever feature selection, and optimized neural network architecture [15] (Figs. 1 and 2).

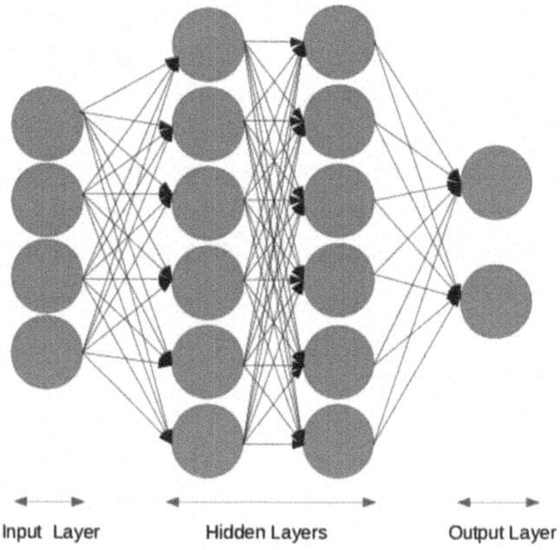

Fig. 1. CNN Layers

Breast cancer detection using artificial neural networks trained with the backpropagation algorithm offers a useful method with the potential for early diagnosis and individualised treatment plans [2]. Neural networks may be able to recognise patterns that are challenging to spot using traditional methods employing a range of information, such as imaging data, clinical evaluations, and genetic markers. By providing clinicians with crucial visual information about the breast tissue, medical imaging technologies including mammography, ultrasound, and magnetic resonance imaging (MRI) help diagnose breast cancer. By leveraging ANNs, it becomes possible to extract meaningful features from these images and train the network to recognize patterns associated with cancerous or benign tumours [7]. This can help radiologists and healthcare professionals in accurately diagnosing breast cancer and making informed treatment decisions. Furthermore, artificial neural networks can be trained to integrate various data sources, including demographic information, family history, genetic mutations, and biomarker measurements. By combining these diverse inputs, ANNs have the potential to enhance the accuracy of breast cancer detection and prediction models, leading to more personalized

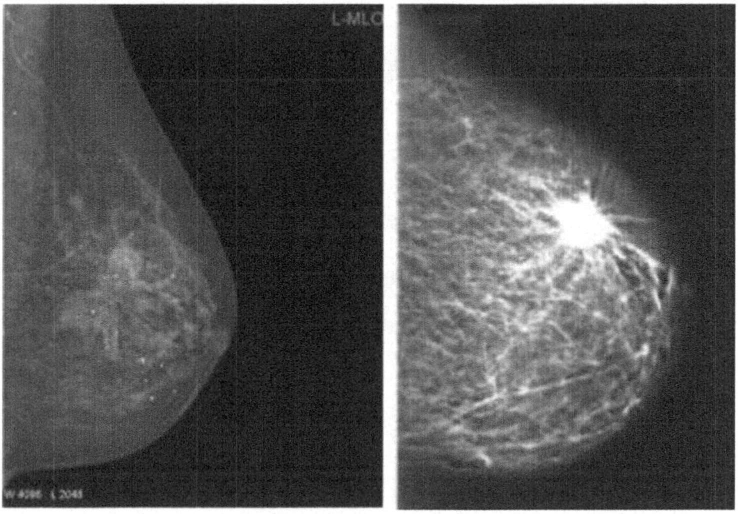

Fig. 2. Cancerous Breast

and effective healthcare interventions [12]. It's important to keep in mind that artificial neural networks do have limitations, despite having shown promise in the detection of breast cancer. The effectiveness of these models depends heavily on the availability of high-quality and diverse training data, as well as careful preprocessing and feature engineering [16]. Additionally, interpreting the decisions made by neural networks, often referred to as their "black box" nature, can pose challenges in terms of explain ability and trust in the predictions [8]. Ongoing research in the field of artificial intelligence and healthcare aims to address these limitations and further improve the performance and interpretability of breast cancer neural networks detection [10]. Researchers and healthcare professionals continue to push the limits of what is feasible in the early diagnosis and treatment of breast cancer by utilising breakthroughs in deep learning, reinforcement learning, and other related fields. Breast cancer detection using artificial neural networks trained with the backpropagation method is made possible. By learning from labelled datasets, these networks can analyze complex features and patterns to make accurate predictions on new, unlabeled cases [13]. While there are challenges and considerations associated with their implementation, ANNs offer the potential to improve the accuracy and efficiency of breast cancer diagnosis, contributing to better patient outcomes and advancements in the field of oncology [4]. The ability of artificial neural networks to handle enormous volumes of data and identify nuanced correlations between variables is one of their advantages when used for breast cancer screening. Neural networks may be able to identify patterns that are difficult to identify using conventional techniques by analyzing a variety of variables, such as imaging data, clinical assessments, and genetic markers [7]. Medical imaging methods like magnetic resonance imaging (MRI) and ultrasound offer useful visual data on breast tissue when it comes to the early detection of breast cancer (Fig. 3).

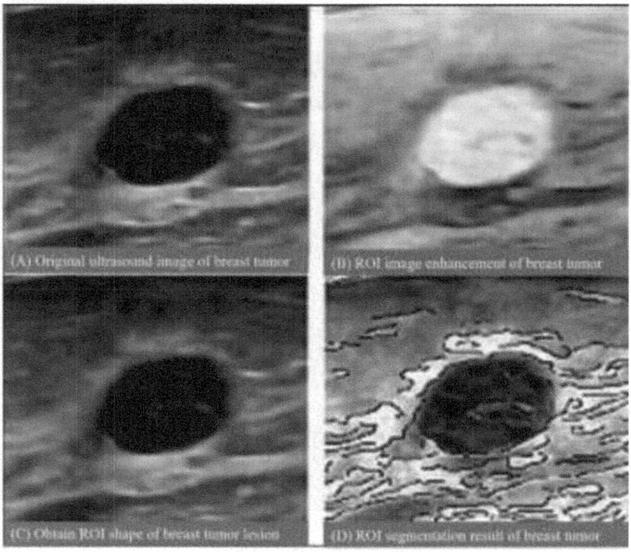

Fig. 3. ROI of the calibrated area of interest and the original breast cancer ultrasound image

Artificial neural networks (ANNs) can be trained to distinguish patterns linked to either benign or malignant malignancies by extracting pertinent data from these images [6]. Radiologists and medical professionals can use this to accurately diagnose breast cancer and make knowledgeable decisions. Artificial neural networks can also be taught to incorporate a variety of data sources, such as biomarker measurements, family history, genetic mutations, and demographic data. ANNs have the potential to improve breast cancer detection and prediction models' precision, providing patients with more precise findings, and successful healthcare solutions that are customized [4]. Despite showing potential in the identification of breast cancer, artificial neural networks do have some drawbacks that must be considered. These models' performance significantly depends on the availability of diverse, high-quality training data as well as rigorous feature engineering and preprocessing [21]. These models' performance significantly depends on the availability of diverse, high-quality training data as well as rigorous feature engineering and preprocessing. Through ongoing research in the disciplines of artificial intelligence and healthcare, neural networks for breast cancer screening are now being made more functional and comprehensible. Finally, artificial neural networks that have been trained via backpropagation are an effective method for the detection of breast cancer. These networks can analyze intricate traits and patterns to generate precise predictions on fresh, unlabeled cases by learning from labeled information [9]. Despite the difficulties and issues involved in putting them into practice, ANNs have the potential to increase the efficacy and accuracy of breast cancer diagnosis [6].

2 Literature Survey

This paper brief about the accuracy of backpropagation algorithm in predicting breast cancer is examined in this work. Using MATLAB R2016a software, the researchers developed and put into practice an artificial neural network (ANN) system. By adjusting certain training parameters, such as epochs, learning rates, goals, and hidden layers, they were able to attain an accuracy of 96.929% with an error of 3.071%. The Wisconsin breast cancer dataset is used to train the artificial neural network (ANN). To guarantee the ANN system's interoperability, the dataset is pre-processed. The outcomes of the numerical simulation show how effective the suggested strategy is. The ANN system is intended to differentiate between benign and malignant forms of cancer. They come to the conclusion that a useful technique for predicting breast cancer is the backpropagation algorithm [1].

This paper outlines how, many metrics are used to assess the effectiveness of the breast cancer categorization process, such as positive and negative predictive values, sensitivity, specificity, and accuracy. The study uses nine carefully chosen variables to classify breast cancer using a backpropagation neural network. Model training and assessment are conducted using the 683-patient Wisconsin Breast Cancer Dataset (WBCD). The diagnostic potential of machine learning techniques, specifically backpropagation neural networks, for breast cancer is impressive. The research emphasizes how important machine learning is to the field of medical science. The results support additional research into and use of machine learning algorithms for tasks related to breast cancer classification. The results of this study open new avenues for improved patient outcomes and diagnosis of breast cancer [2].

This paper provides information about a unique feature fusion method using computers A CNN approach is proposed for picture categorization and diagnosis of breast cancer. CNNs with prior training automatically extract characteristics from medical photos. For categorization, extracted features are combined using feature fusion. The technique classifies breast cancer images with 89% accuracy. CNNs are utilized for image categorization and detection related to breast cancer. CNNs are trained to automatically extract features from photos of medical conditions. Integrating retrieved features from two distinct CNN structures is known as feature fusion. The fused characteristics are classified by a classifier. Expertise and time consumption are decreased by doing away with manual feature extraction. The accuracy of classifying images of breast cancer increases dramatically [3].

This research studies the application of CNN and MLP to the diagnosis and detection of breast cancer. In terms of breast cancer diagnosis, CNN performs marginally better than MLP. The application of MLP and CNN, particularly for the diagnosis and detection of breast cancer, is the main topic of this research. CNN is a deep learning network with exceptional feature extraction capabilities, whereas MLP is a feedforward neural network. CNN can be adjusted and utilized to classify breast cancer subtypes. Based on gene signatures, deep learning methods such as CNN are highly accurate in identifying different subtypes of breast cancer. The best architecture is proposed to be found by comparing MLP and CNN on the same dataset, according to the article. Future studies can investigate CNN for early breast cancer diagnosis and enhancing the effectiveness of the healthcare system [4].

This paper addresses the differentiate between benign and malignant breast tumors, the study employed three neural networks (NN) with ATR-FTIR spectral data. Significant IR regions were found using NN-based sensitivity analysis, which agreed with the FTIR peak identification. In diagnostic metrics, neural network models surpassed six other machine learning models. The findings indicated that NN-ATR-FTIR is a potentially useful diagnostic tool for breast cancer. Financial limitations forced the study to use only basic FTIR spectrometry, highlighting the need for more sophisticated methods. Finding spectral biomarkers in plasma for treatment monitoring appeared to be possible with the integration of AI and FTIR. Lipid, nucleic acid, phospholipid, and carbohydrate regions were found to have distinct spectral features. This research is the first to apply several machine learning techniques to cancerous breast tissue [5].

This paper aims on, using artificial neural networks in image processing for early breast cancer detection, the study examines the benefits of automated classifiers over manual diagnosis in terms of efficiency and accuracy. Enhancing specificity and sensitivity is the main focus of this review of recent research on NN techniques in medical imaging. The writers stress that NN is applicable in a variety of medical fields and talk about how versatile it is in these medical imaging techniques. Furthermore, explored in the paper is the application of hybrid NN for breast cancer detection. In medical imaging applications, it describes the various types of neural networks (NNs) used and the various data feeding strategies [6].

This paper emphasizes the value of routine examinations for early breast cancer detection. Using radial basis neural networks (RBFN) and back propagation neural networks (BPNN), the study suggests an automated classification technique. Discrete Haar Wavelets improve image representation, which helps with region modeling of breast cancer. According to reports, RBFN's accuracy is 70.4%, while BPNN's is 59.0%. Using the same set of test and training images, both networks are assessed. The results suggest that RBFN performs better than BPNN in the classification of breast cancer images, indicating the usefulness of Haar Wavelets for representing image content [7].

3 Methodology

Breast cancer detection using artificial neural networks (ANNs) is a promising area of research that has shown great potential in improving the accuracy and reliability of diagnosis [1]. The backpropagation algorithm is a popular technique used for training ANNs. Here are the steps i or breast cancer detection [2, 8] (Fig. 4).

3.1 Data Preparation

Gather and preprocess the data needed for training your neural network in the first stage. This entails gathering a dataset of breast cancer cases, organizing the data, and carrying out any required preparation operations like feature scaling or data normalization. The dataset ought to contain instances with labels on it, where each example comprises of input information (such as patient demographics and mammography pictures) and corresponding output labels (such as "cancer" or "non-cancer").

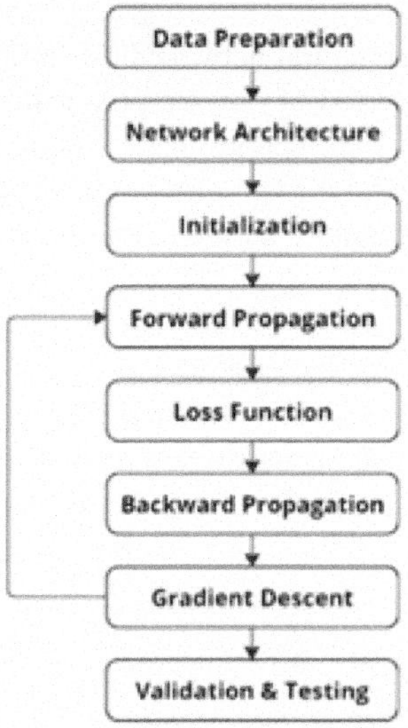

Fig. 4. Back propagation Flowchart

3.2 Network Architecture

The arrangement and configuration of an artificial neural network are referred to as its network architecture. Two popular forms of neural networks utilized for the diagnosis of breast cancer are feedforward neural networks (FNN) and convolutional neural networks (CNN). The number of layers, the number of neurons in each layer, and other architectural information.

3.3 Initialization

In this step, the neural network's weights and biases are set either at random or in accordance with a predetermined procedure. The learning process and the network's ability to converge to an ideal solution are both parameters impacted by the starting values of these.

3.4 Forward Propagation

The input data is supplied into the neural network layer by layer, going from the input layer to the output layer in the period of onward propagation. Each neuron in a layer receives its inputs as a weighted sum, applies an activation function (such as sigmoid or ReLU), and sends the outcome to the layer below. This process is repeated until the forecast or set of output data is produced by the output layer.

3.5 Loss Function

The loss function calculates the difference between the dataset's actual output labels and the neural network's anticipated output. Depending on how the problem is formulated, a potential loss function for breast cancer diagnosis can be binary cross-entropy or mean square error.

3.6 Backward Propagation

The Central concept of backpropagation algorithm's is backward propagation, sometimes referred to as backpropagation. It determines the gradients of the loss function with regard to the neural network's weights and biases. The parameters are later updated using these gradients during the optimisation phase. Applying the chain rule of calculus, backpropagation means moving the error layer by layer backward across network.

3.7 Gradient Descent

The Central concept of backpropagation algorithm's is backward propagation, sometimes referred to as backpropagation. It determines the gradients of the loss function with regard to the neural network's weights and biases. The parameters are later updated using these gradients during the optimisation phase. Applying the chain rule of calculus, backpropagation means moving the error layer by layer backward across network.

3.8 Validation and Testing

In order to prevent overfitting, it is crucial to track the neural network's performance on unlabelled data during training. After backpropagation has been used to obtain the gradients, gradient descent is used to update the neural network's parameters. This enables you to adjust the hyperparameters based on the validation findings, such as the learning rate or the network design. The final model can be evaluated for performance and generalizability using a different testing set after training. These procedures can be used to train an artificial neural network to detect breast cancer using the backpropagation methodology.

4 Result and Discussion

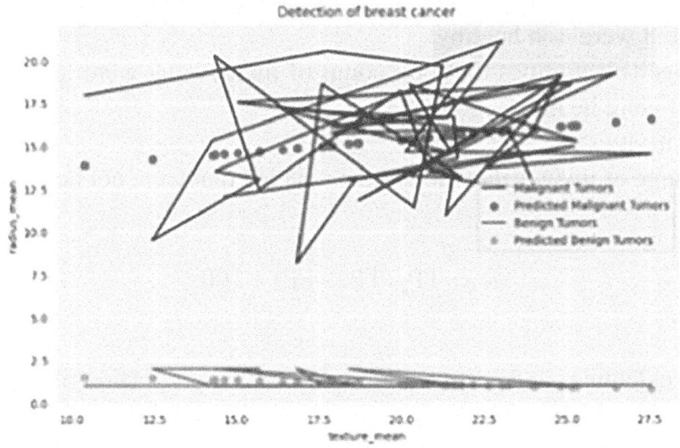

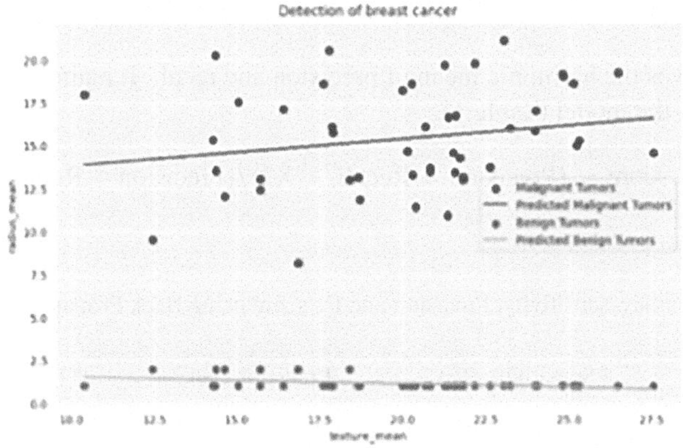

The neural network for breast cancer detection using backpropagation. Preparation of the data, description of the network architecture, initialization of the weights and biases, forward and backward propagation, gradient descent for optimization, network validation, and testing are all covered. The precision of the trained network might vary depending on the dataset and hyperparameter values. There are true positives, true negatives, false positives, and false negatives among the network's findings. The dataset and hyperparameter combinations employed have an impact on the network's accuracy. The accuracy, sensitivity, and specificity of the algorithm may be evaluated using the true positive, true negative, false positive, and false negative rates.

$$Accuracy = \frac{\mathrm{TP} + \mathrm{TN}}{\mathrm{TP} + \mathrm{TN} + \mathrm{FP} + \mathrm{FN}}$$

where,

True Positive (TP) defines the total count of images that were accurately classified as non healthy class.

False Positive (FP) defines the total count of images that were falsely classified as non healthy.

False Negative (FN) defines the total count of images that our model failed to detect as non healthy but were non healthy.

False Positive (FP) defines the total count of images that were falsely predicted as non healthy, but were healthy.

Recall/Sensitivity:-

The percentage of images that the system recalled that were not healthy is calculated as Recall R

$$R = TP/(TP + FP) * 100$$

Precision:-

Precision P calculates and graphically illustrates the proportion of people who were correctly categorised by the suggested system.

$$P = TP/(TP + FN) * 100$$

F1 score:-

F1 score gives the harmonic mean of precision and recall. It mainly tells the overall performance of the model (Table 1).

$$\text{F1 score} = (\text{Precision} * \text{Recall} * \text{X2})/(\text{Precision} + \text{Recall})$$

Table 1. Accuracy, sensitivity, Precision and F1 score using Back Propagation Algorithm

Accuracy (%)	Sensitivity (%)	Precision (%)	F1 Score (%)
95	92	94	93

5 Program Code

```
% Breast Cancer Detection using Backpropagation Algorithm

% Step 1: Data Preparation
load breast_cancer_data.mat % Assuming you have a dataset with features (X) and
corresponding labels (y)
X = normalize(X); % Normalize the features
X_train = X(train_indices,:);
y_train = y(train_indices,:);
X_test = X(test_indices,:);
y_test = y(test_indices,:);

% Step 2: Network Architecture
input_size = size(X_train, 2); % Number of features
hidden_size = 64; % Number of neurons in the hidden layer
output_size = 1; % Number of neurons in the output layer

% Step 3: Initialization
% Randomly initialize the weights and biases
W1 = randn(input_size, hidden_size);
b1 = randn(1, hidden_size);
W2 = randn(hidden_size, output_size);
b2 = randn(1, output_size);

% Step 4: Forward Propagation
% Calculate the output of the network
Z1 = X_train * W1 + b1;
A1 = sigmoid(Z1);
Z2 = A1 * W2 + b2;
A2 = sigmoid(Z2);

% Step 5: Loss Function
% Calculate the loss/error
loss = sum((A2 - y_train).^2) / size(X_train, 1);

% Step 6: Backward Propagation
% Calculate the gradients
dZ2 = A2 - y_train;
dW2 = A1' * dZ2;
db2 = sum(dZ2, 1);
dZ1 = dZ2 * W2' .* (A1 .* (1 - A1));
dW1 = X_train' * dZ1;
db1 = sum(dZ1, 1);

% Step 7: Gradient Descent
% Update the weights and biases
```

```
learning_rate = 0.01;
W2 = W2 - learning_rate * dW2;
b2 = b2 - learning_rate * db2;
W1 = W1 - learning_rate * dW1;
b1 = b1 - learning_rate * db1;

% Step 8: Validation & Testing
% Perform forward propagation on the test set
Z1_test = X_test * W1 + b1;
A1_test = sigmoid(Z1_test);
Z2_test = A1_test * W2 + b2;
A2_test = sigmoid(Z2_test);

% Calculate the accuracy
threshold = 0.5; % Threshold for classification
predictions = A2_test >= threshold;
accuracy = sum(predictions == y_test) / size(y_test, 1);

% Display the accuracy
disp(['Accuracy: ' num2str(accuracy)]);

% Sigmoid function
function output = sigmoid(x)
output = 1 ./ (1 + exp(-x));
end
```

6 Future Scope

It is exciting that ANNs may one day be used to diagnose breast cancer. As ANN technology develops, it is likely that it will become more precise and efficient. New, improved breast cancer screening tools might be developed as a consequence, potentially saving lives. New breast cancer treatments can also be developed using ANNs [11]. For example, ANNs may be used to predict how cancers will respond to various therapies. Clinicians may find it simpler to select the optimal course of action for each patient as a result. ANNs might generally have a significant impact on the early detection and management of breast cancer. As ANNs develop, they will likely become more important in the fight against this disease [14].

7 Conclusion

Using the image processing application, artificial neural networks (ANNs) have been effectively utilized to identify breast cancer [3]. Convolutional neural networks (CNNs) have demonstrated the ability to attain extraordinary accuracy rates, especially when trained on tiny datasets. In order to diagnose breast cancer, CNNs must be able to extract spatial information from pictures [15]. Additionally, breast tumors and mammograms may be distinguished using ANNs, helping radiologists diagnose and classify cancer more precisely. Because early identification of breast cancer improves patient outcomes, this application is crucial [8].

References

1. Nasien, D., Enjeslina, V., Hasmil Adiya, M., Baharum, Z.: Breast cancer prediction using artificial neural networks back propagation method. J. Phys.: Conf. Ser. **2319**(1), 012025 (2022). IOP Publishing
2. Bhattacharjee, N., Roy, S., Paul, S., Roy, P., Kausar, N., Dey, N.: Classification approach for breast cancer detection using back propagation neural network: a study. In: Deep Learning and Neural Networks: Concepts, Methodologies, Tools, and Applications, pp. 1410–1421. IGI Global (2020)
3. Li, M.: Research on the detection method of breast cancer deep convolutional neural network based on computer aid. In: 2021 IEEE Asia-Pacific Conference on Image Processing, Electronics and Computers (IPEC), pp. 536–540. IEEE (2021)
4. Desai, M., Shah, M.: An anatomization on breast cancer detection and diagnosis employing multi-layer perceptron neural network (MLP) and Convolutional neural network (CNN). Clinical eHealth **4**, 1–11 (2021)
5. Tomas, R.C., et al.: Detection of breast cancer by ATR-FTIR spectroscopy using artificial neural networks. PLoS ONE **17**(1), e0262489 (2022)
6. Mehdy, M.M., Ng, P.Y., Shair, E.F., Saleh, N.I., Gomes, C.: Artificial neural networks in image processing for early detection of breast cancer. Comput. Math. Methods Med. **2017**, 1–15 (2017)
7. Kaymak, S., Helwan, A., Uzun, D.: Breast cancer image classification using artificial neural networks. Proc. Comput. Sci. **120**, 126–131 (2017)
8. Gautam, A., Bhateja, V., Tiwari, A., Satapathy, S.C.: An improved mammogram classification approach using back propagation neural network. In: Data Engineering and Intelligent Computing: Proceedings of IC3T 2016, pp. 369–376. Springer, Singapore (2018)
9. Saini, S., Vijay, R.: Mammogram analysis using feed-forward back propagation and cascade-forward back propagation artificial neural network. In: 2019 Fifth International Conference on Communication Systems and Network Technologies, pp. 1177–1180. IEEE (2019)
10. Giri, P., Saravanakumar, K.: Breast cancer detection using image processing techniques. Orient. J. Comput. Sci. Technol. **10**(2), 391–399 (2017)
11. Sultana, Z., Rahman Khan, M.A., Jahan, N.: Early breast cancer detection utilizing artificial neural network. WSEAS Trans. Biol. Biomed. **18**, 32–42 (2021)
12. Bellary, M.Z., Deepthi, T.H.S., Mustafa, B.A., Ahmed, S.J., Sarkar, R.: Medical image analysis of knee osteoarthritis using modified deep CNN. J. Surv. Fisher. Sci. **10**(2S), 133–144 (2023)
13. Menendez, L., et al.: Artificial neural networks applied to cancer detection in a breast screening programme. Math. Comput. Model. **52**(7–8), 983–991 (2010)
14. Ayer, J., et al.: Breast cancer risk estimation with artificial neural networks revisited. Cancer **116**(14), 3310–3321 (2010)
15. Sharif, M.H.U.: Breast Cancer Detection using Artificial Neural Networks. ResearchGate [Preprint] (2021)
16. Singh, G.: Breast cancer prediction using machine learning. Int. J. Sci. Res. Comput. Sci. Eng. Inf. Technol. **8**(4), 278–284 (2020)
17. Sharma, S., Aggarwal, A., Choudhury, T.: Breast cancer detection using machine learning algorithms. In: 2018 International Conference on Computational Techniques, Electronics and Mechanical Systems (CTEMS), pp. 114–118. IEEE (2019)
18. Yixuan, L., Zixuan, C.: Performance evaluation of machine learning methods for breast cancer prediction. Appl. Comput. Math. **7**(4), 212–216 (2018)
19. Islam, M.M., Haque, M.R., Iqbal, H., Hasan, M.M., Hasan, M., Kabir, N.: Breast cancer prediction: a comparative study using machine learning techniques. SN Comput. Sci. **1**(5), 1–14 (2020)

20. Lulu, M.W., Sarifuddin, M., Eri, P.W., Arman, A.A.: Benign and malignant breast tumors classification based on texture analysis and backpropagation neural network. Компьютерная оптика **45**(2), 227–234 (2021)
21. Samah, A.A., et al.: Application of deep learning method in facilitating the detection of breast cancer. IOP Conf. Ser.: Mater. Sci. Eng. **864**(1), 012079 (2020). IOP Publishing
22. Ganggayah, M.D., Taib, N.A., Har, Y.C., Lio, P., Dhillon, S.K.: Predicting factors for survival of breast cancer patients using machine learning techniques. BMC Med. Inform. Decis. Mak. **19**(1), 1–17 (2020)

Supervised and Unsupervised Learning Techniques for Malware Classification Based on Opcode Frequency Features

K. A. Asmitha[1], P. Vinod[1,2], K. A. Rafidha Rehiman[1(✉)], Renil M. Cherian[1], Rohith Balachandran[1], Bins Titus[1], R. Rohith Raj[1], K. N. Fathima Afrah[1], and A. R. Archana[1]

[1] Department of Computer Applications, CUSAT, Kochi, Kerala 682022, India
asmitha@pg.cusat.ac.in, {vinod.p,rafidharehimanka}@cusat.ac.in
[2] Department of Mathematics, University of Padua, Padova, Italy

Abstract. Over the years, the complexity of malware has increased, presenting a significant threat to existing security systems. Malware, or malicious software, is designed to disrupt computer systems intentionally. The sophistication of malware has heightened, rendering it more challenging to detect. Notably, malware and benign files utilize opcodes differently, leading to variations in opcode usage frequencies. Malware files more frequently invoke some opcodes than benign files, while others see more significant usage in benign files than malware files. Leveraging these discrepancies in opcode usage frequencies enables the differentiation between malware and benign files. This paper investigates the use of supervised and unsupervised algorithms, emphasizing opcode frequency analysis. Here, we have utilized machine learning algorithms such as random forest, AdaBoost, decision tree, support vector machine (SVM), stacking classifier, and autoencoder. All these algorithms have demonstrated commendable performance. The stacking classifier had the highest performance, achieving an F1 score exceeding 99.6% for Dataset I and 99.3% for Dataset II. The autoencoder yielded an F1 score of approximately 96%. Furthermore, we employ SHAP to generate model explanations, providing insight into the interpretability of our models.

Keywords: malware · opcode · autoencoder · machine learning

1 Introduction

In our interconnected digital era, the looming threat of malware significantly risks the security of our data. The digitization of data, while advantageous, also comes with corresponding disadvantages in the form of potentially malicious attacks on the system. Criminals can orchestrate malicious attacks for various reasons, such as stealing information, gaining control of remote systems, committing identity theft and extorting money. This situation underscores the constant vulnerability of a system to external influences that may compromise the information it holds. Consequently, a robust security system with high-alert capabilities must be in place. The term "malware" refers to

T. Shreekumar et al. (Eds.): ISCComm 2023, CCIS 2231, pp. 221–231, 2025.
https://doi.org/10.1007/978-3-031-75605-4_15

any malicious piece of code or software designed to cause harm to the system. Any code or software deliberately designed to harm a computer or pilfer information falls under the malware umbrella. Malware encompasses various types, including viruses, worms, ransomware, trojan horses, and spyware. The recent ransomware attack[1] on the Colonial pipeline, a crucial component of the U.S. gas supply chain, vividly illustrates the severity of this menace. The attack compelled the pipeline to cease operations, triggering widespread panic, fuel shortages, and a subsequent surge in fuel prices. Another victim, the chemical distribution company Brenntag, fell prey to a ransomware attack. Ransomware, a specific type of malware, encrypts computer access and demands a ransom from the victim to restore it. In Brenntag's case, hackers encrypted files, forcing the company to pay a hefty 4.4 million dollars in Bitcoin to obtain the decryptor and prevent the unauthorized release of sensitive data. These incidents starkly exemplify the havoc that malware attacks can wreak. Cybercriminals continually devise new methods to evade malware detection, highlighting the necessity of analyzing malware for developing effective countermeasures. Malware analysis and detection encompass three overarching methods: static, dynamic, and hybrid malware analysis. Static analysis involves examining malware devoid of execution and checking the file for any signs of malicious intent. In dynamic methods, the malware runs in a sandbox—a secure environment for studying its behavior without compromising the computer system. The sandbox functions as a closed system where malware can be executed safely. Hybrid analysis integrates techniques from both static and dynamic analyses. It includes scrutinizing the malware signature and leveraging other behavioral patterns to gain a more comprehensive understanding of the malware. This integrated approach enhances the efficacy of malware analysis by actively considering both static and dynamic aspects.

In this work, we statically extract features from the assembly code of various executable files. We undertook a study to evaluate the efficacy of both supervised and unsupervised methods in creating a model proficient at identifying malware. Several supervised algorithms were employed, including SVM, Adaboost, Stacking Classifier, Decision Tree, and Random Forest. In the unsupervised method, we constructed an auto-encoder within a deep neural network framework to reconstruct features corresponding to each malware. Subsequently, we empirically analyzed the effectiveness of the auto-encoder in reconstructing malware features by adding a final dense layer within the auto-encoder structure to perform classification and detection tasks for malware. The principal contributions of the paper include:

- We performed an in-depth analysis of the effectiveness of various supervised and unsupervised algorithms in detecting malware.
- We present an unsupervised feature-learning approach using an auto-encoder based on a deep neural network designed to detect PE malware. We employ a single model for malware reconstruction and classification. Our approach utilizes nearly the minimum number of features.
- Comprehensive analysis on the effectiveness of features (opcode frequency) in detecting PE malware using supervised learning methods using two different datasets.

[1] https://www.cisa.gov/news-events/news/attack-colonial-pipeline-what-weve-learned-what-weve-done-over-past-two-years.

- We Provide model explanations through the utilization of the interpretability tool SHAP (SHapley Additive exPlanations).

The remainder of the paper is structured as follows: Section 2 reviews the literature from several research papers. Section 3 provides a detailed description of the methods employed. Section 4 presents the results obtained. Section 5 offers the conclusion and discusses the future scope.

2 Related Works

Static and dynamic analyses represent two fundamental approaches for identifying malware among benign programs. In static analysis, the program does not execute; instead, it extracts features such as opcode frequency [1], strings and byte sequences [2], function length [3], and API calls [4] from the file. The detection algorithm then utilizes these features for classification. On the other hand, dynamic analysis requires the execution of the program, capturing essential behavioral aspects and other parameters as features for the detection algorithm. In [5], the authors discuss a model to detect and classify malware. They employ machine learning algorithms for this objective, utilizing N-gram and API call features. Parameter variation and its effect were also studied in this paper. With experiment analysis, it was found that the proposed model was much superior to other models with an accuracy of 98%. The authors in [6] propose a new method for detecting unknown malware files. It leverages the frequency of opcode sequence occurrences. A vector representation of executables is generated using these opcode sequences, and machine learning algorithms are employed for the detection of unknown malware. [7] introduces an approach for classifying and detecting malware behavior. The process involves taking a malware executive history XML, converting it into a matrix, and using it as input in WEKA after translation. They measure performance by employing various machine learning algorithms in WEKA, and the regression classification method demonstrates the most effective results. The outlined method holds the potential for developing a behavioral antivirus. [8] proposes a bilayer behavioral abstraction method based on API sequences. The lower layer abstracts the low layer behavior of raw API calls and is combined with the high layer to produce high-layer behavior, which is interpretable. The layers are further combined into a higher dimensional vector space, which can then be used by machine learning algorithms. The paper also presents an improved one class SVM that makes use of negative samples to overcome class imbalance problems. Moreover, hybrid approaches exist, combining features from both static and dynamic analyses [9]. In [10], opcode frequency distribution was directly utilized for malware detection, whereas Karim et al. [11] employed opcode sequences and permutations for the same purpose.

A typical approach for the supervised machine learning classification of malware involves developing a binary model to distinguish between malware and benign programs. Occasionally, researchers construct multinomial models to differentiate samples across various malware families. Commonly employed supervised machine learning algorithms for malware classification encompass Support Vector Machine [6, 12], Naive Bayes [12], Decision Tree [12], J48 [1], AdaBoost, Random Forest [1], and TFIDF [6, 12]. Moreover, supervised learning can benefit from the application of feature extraction, selection, and dimension reduction algorithms to improve performance. In a traditional feature extraction method, an unsupervised learning algorithm generates more representative features during the supervised learning stage. Examples of feature selection methods comprise hierarchical methods, Weighted Term Frequency, unary variable removal, and Goodness evaluator. However, improper feature extraction may lead to lower accuracy, higher false favorable rates, and other issues. Deep Learning techniques are gaining prominence for feature extraction and beyond. Recent endeavors have found success with diverse Deep Learning techniques, including Deep Neural Networks [13], Recurrent Neural Networks (RNN), and Convolutional Neural Networks (CNN) [14], employed for supervised learning and classification. Likewise, researchers have employed Stacked Auto-Encoders, RNN-based Auto-Encoders, and other deep networks for optimal feature extraction.

[15] discusses the use of Windows audit log events for malware detection using neural networks. Features are extracted from audit log events and uses LSTM for detecting malware. They make use of the textual and sequential data from audit logs. By using different representations, the performance of the model is measured. Fast text provided the best representation, was able to generalize unseen data, and provided a good detection rate using LSTM. [16] recommends using Windows API calls extracted from PE files as a technique for detecting malware. They use a stacked autoencoder that includes a multi-layer Boltzmann machine and a layer of associative memory. The system comprises two stages: unsupervised pre-training and supervised parameter tuning. This system constitutes a heterogeneous deep learning architecture, surpassing the performance of other machine learning algorithms such as ANN, SVM, Decision Tree, and Naive Bayes.

3 Proposed Framework

Figure 1 illustrates our approach to classifying malware using supervised and unsupervised algorithms. This approach extracts opcodes from assembly files and creates datasets, documenting the opcode counts for malware and benign samples. Subsequently, various supervised and unsupervised learning models employ these datasets to distinguish a sample as malware.

3.1 Dataset

In this paper, we employed two different datasets. We prepare Dataset I using 5848 samples from the BIG 2015 dataset, and Dataset II consists of 10065 samples from the same BIG 2015 dataset. Additionally, we incorporate a benign dataset comprising 359 samples. We obtained the actual Dataset from the.asm files from the Microsoft Malware Classification Challenge dataset (BIG 2015). For each sample, they provide an asm file and a byte file, each uniquely identified by a hash value. The malware samples belong to nine malware families: Obfuscator.ACY, Lollipop, Ramnit, Kelihos ver3, Simda, Vundo, Kelihos ver1, Tracur, and Gatak. We create the benign Dataset from benign files.

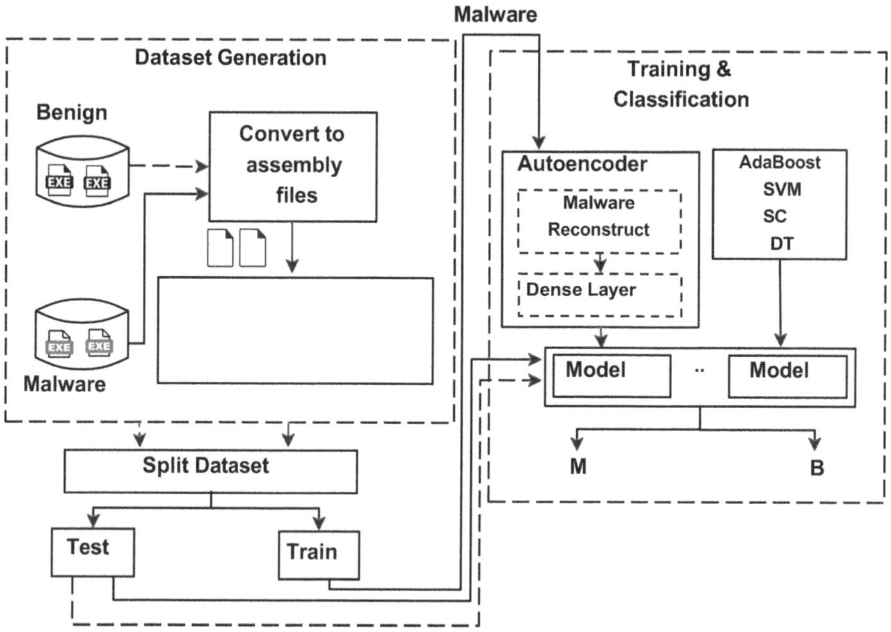

Fig. 1. Proposed architecture which contains the Dataset generation phase and a classification phase

3.2 Preprocessing and Feature Extraction

At the start, we extracted opcodes from all the malware and Benign executables, forming a comprehensive list of unique opcodes (18 opcodes). Following that, we calculated the number of occurances of each opcode in every executable and organized them according to the list of unique opcodes to compare the classifiers under investigation. The initial 18 opcodes include: 'and,' 'lea,' 'xor,' 'sub,' 'jmp,' 'mov,' 'pop,' 'test,' 'add,' 'call,' 'ret,' 'jne,' 'push,' 'je,' 'inc,' 'shl,' 'or,' 'cmp.' Among these opcodes, 'ret,' 'jne,' and 'je' were excluded since most values in these columns are zero for the malware dataset. Consequently, the final dataset for both malware and benign comprises 15 opcodes. We assigned a class label to malware and benign files for supervised classification. Before classifying the samples, we normalized them and checked for missing values. Additionally, we ensured that the dataset did not contain any duplicate samples.

We now randomly divide the dataset, including the opcode frequency for each executable, into distinct training and testing sets, maintaining an 80:20 ratio for malicious and benign datasets.

3.3 Models

Here, we discuss the supervised and unsupervised algorithms used for our experiment.

Autoencoder: Neural networks known as autoencoders consist of an encoder and decoder with identical input and output. The encoder compresses the input to produce the code, and the decoder reconstructs the input solely from the code. This unsupervised, data-specific, and lossy compression algorithm employs binary cross entropy or mean squared error as its loss function. These Autoencoders also function as generative models that can randomly generate new data resembling the input. They operate on data similar to their training set and typically have matching input and output dimensions, with the encoder and decoder as mirror images.

In this context, we emphasize the data (disassembled binaries) without labels, constituting unsupervised learning. The objective is to train a model for optimal malware reconstruction. Upon completion of the reconstruction model, an extra dense layer computes the error between the input (binary taken as input) and the output (reconstructed binary). We use only malicious files to train the auto-encoder and test using both malware and benign files. The autoencoder is exclusively trained on malicious binaries, resulting in elevated error rates for benign files and reduced rates for malware. Establishing an error threshold enables effective binary classification.

The autoencoder employed in this study comprises 18 layers, encompassing both input and output layers, with dimensions set at 15. The initial layer features 15 nodes, and subsequently, the number of nodes progressively decreases by two for the encoder layers and increases by two for the decoder layers. The activation function utilized throughout is tanh.

Decision Tree: A supervised machine-learning algorithm called a decision tree utilizes a tree-like structure designed for classification and regression tasks. Each node represents a specific attribute in the decision tree (DT). The algorithm classifies a sample by recursively traversing the tree from the root to a leaf node, testing attributes at internal nodes. The leaf nodes indicate the class label, and predictions are made based on learned decision rules from training.

Random Forest: It is an ensemble of decision trees where the majority vote determines the predicted class. It outperforms individual trees due to their uncorrelated nature. Even if some trees make errors, the random forest prediction is accurate as long as most are correct. It is capable of handling both continuous and categorical variables and serves for both classification and regression.

AdaBoost: AdaBoost, an ensemble method, combines weak classifiers to form a strong classifier. It initially assigns equal weights to all data points and subsequently prioritizes misclassified points by assigning higher weights in each iteration. Continue training until reaching a low error.

Support Vector Machine: SVM is a supervised algorithm for classification and regression that maps data points from a low-dimensional to a high-dimensional space, making them separable. The aim is to identify a hyperplane that effectively classifies the data, with support vectors influencing the hyperplane's position and orientation to maximize the classifier's margin.

Stacking Classifier: A stacking classifier is a meta-classifier and an ensemble technique to enhance performance by using predictions from different classifiers as features for training a model. Initially, train base classifiers on the data and use their predictions as new meta-classifiers' features. Split the dataset into two halves, using one for training base models and the other for obtaining predictions to train the metaclassifier. Figure 2 represents the architecture of the stacking classifier.

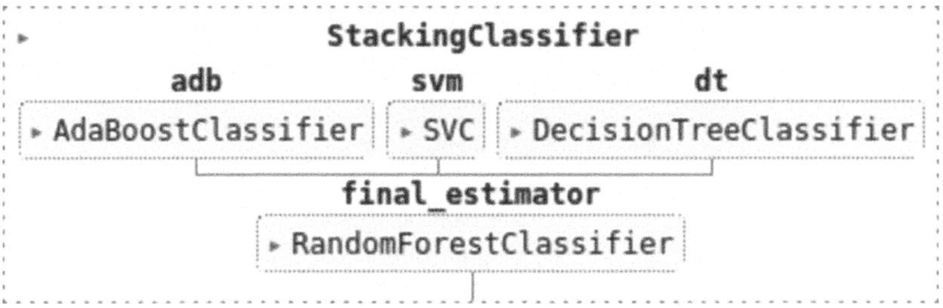

Fig. 2. The architecture of stacking classifier

4 Experiment Results

To assess the effectiveness of commonly used supervised machine learning classifiers against an unsupervised approach, we chose an Autoencoder with 15 layers. For the supervised algorithms, we made use of the Sci-Kit Learn Python library [17], while for the Autoencoder (AE), we employed the Keras Python library with TensorFlow as the backend. We set the activation function to tanh for all layers in the Deep Neural Network (DNN). We computed the F1 score to compare the performance to Defines as:

$$Precision(P) = \frac{TP}{(TP + FP)} \tag{1}$$

$$Recall(R) = \frac{TP}{TP + FN} \tag{2}$$

$$F1score(F1) = 2 * \frac{Precision * Recall}{Precision + Recall} \tag{3}$$

True Positive (TP): The system predicts the executable as malware, and it is indeed malware.

False Positive (FP): The system falsely predicts a benign executable as malware. False Negative (FN): The system incorrectly predicts a malicious executable as benign. True Negative (TN): The system correctly predicts a benign executable as benign.

We selected the F1 score as the evaluation metric for this imbalanced dataset since accuracy is unsuitable in such cases. The F1 score is the weighted average of Precision and Recall. Because we did not train the autoencoder using the benign dataset, it tends to yield a significantly larger mean squared error for benign data and a minor error for malware data. Based on this observation, we defined a threshold value, the average mean squared error of the malware data. We categorized samples with a mean squared error greater than this threshold as benign and those below the threshold as malware. The autoencoder used ten epochs and a batch size of 15. The autoencoder achieved an F1 score of 88% for Dataset I and 96.05% for Dataset II, as shown in Table 1. Notably, we observed a performance improvement when using more samples in the autoencoder. Additionally, we tested the autoencoder using normalized wannacry data. As shown in Fig. 3, the mean squared error, calculated as 0.008333, is much smaller than the threshold of 0.43257, enabling the autoencoder to classify wannacry as malware correctly. For dataset II, the threshold obtained is 0.70851.

Table 1. The parameters for each classifiers

Classifier	Parameters
Autoencoder	Epochs = 10 Batch size = 15 Activation = tanh Optimization Algorithm = Adam Loss Function = Mean Squared Error
SVC	kernel ='rbf'
Adaboost	max_depth = None n_estimators = 100 min_samples split = 2
Random Forest	criterion = gini n_estimators = 100 max_depth = None min_samples_split = 2 min_samples_leaf = 1
Decision Tree	criterion = gini max_depth = None min_samples_split = 2 min_samples_leaf = 1 splitter = best max_features = None

All classifiers achieved an F1 score exceeding 98% and performed exceptionally well on the test dataset. Notably, they successfully classified wannacry as malware. Examining Table 1, the Stacking Classifier outperformed others with F1 scores of 99.62% and 99.31% for Dataset I and Dataset II, respectively. Random Forest and AdaBoost also exhibited better performance, achieving F1 scores of 99.49% and 99.16% for Dataset I and 98.95% and 99.04% for Dataset II, respectively. SVM lagged, achieving the lowest F1 scores of 98.22% and 98.76% for Dataset I and Dataset II, respectively, among all the classifiers. Table 1 describes the parameters for each model (Table 2).

4.1 Model Interpretability

As machine learning models become more complex, comprehending their accurate predictions becomes challenging due to their inherently opaque nature. This lack of transparency underscores the importance of interpretability and explainability in machine learning. It is essential to enhance model transparency, ensuring that security analysts can discern the factors influencing decisions. We use the SHAP [18] to explain our models. SHAP values play a pivotal role in achieving interpretability by quantifying the contribution of each feature, such as "mov" and "call", etc., to the model's predictions.

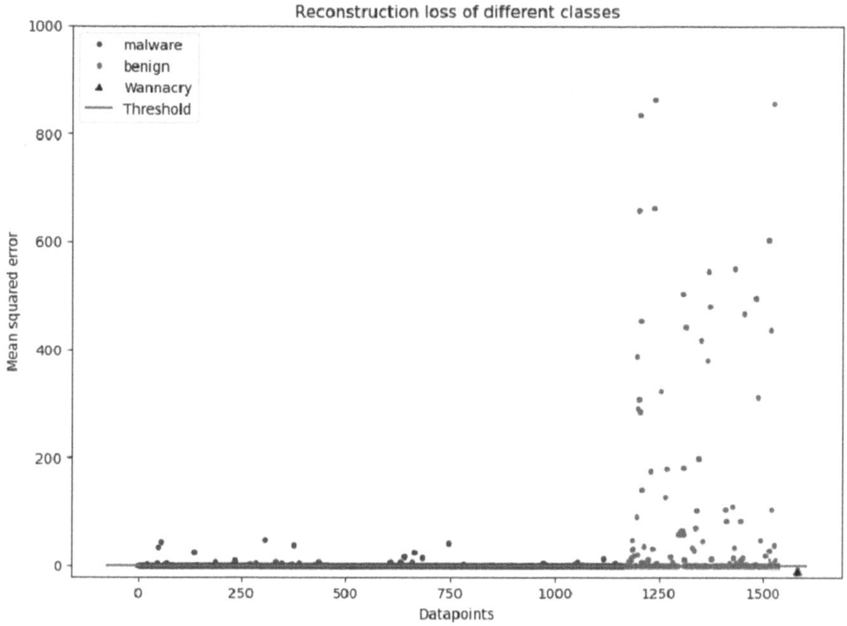

Fig. 3. The results for Testing Wannacry data in autoencoder. The blue dots represent the malware, and the green dots represent benign. The threshold is 0.43257

Table 2. The Results obtained for all the classification models in terms of F1-score

Classifier	Dataset I (F1)	Dataset II (F1)
AdaBoost	99.16	99.04
SVM	98.22	98.51
RF	99.49	98.95
DT	99.03	98.76
Autoencoder	88.04	96.05
Stacking classifier	**99.62**	**99.31**

Utilizing SHAP values allows us to pinpoint the most influential features and comprehend their impact on the model's outcomes. Figure 4 showcases the top five features that contribute the most to the predictions of our best model. The force plot in Fig. 5 illustrates the Stacking model's prediction. The model achieves a score of 0.56. Higher scores result in the model predicting Malware, while lower scores lead to predictions of Benign. Red and blue highlight essential features for predicting this observation. Red indicates features that elevate the model score, while blue represents features that reduce the score. Features exerting a more substantial impact on the score appear closer to the boundary between red and blue. The size of each bar corresponds to the magnitude of the impact. In this case, the sample is classified as Malware because the features shown in red (jmp, lea, call, pop, cmp, test, mov, xor) collectively elevated the score.

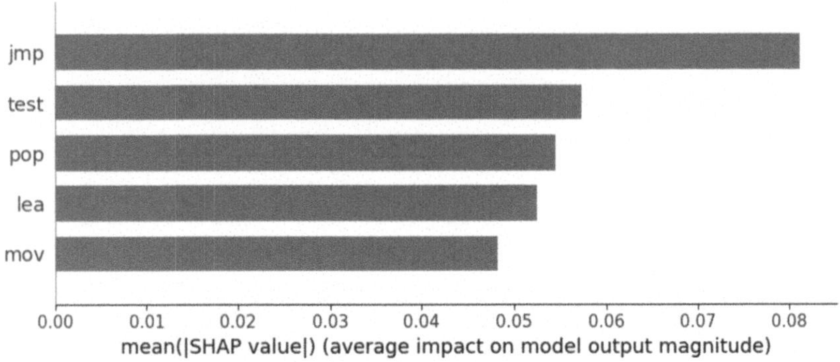

Fig. 4. The top contributing features of the best classifier

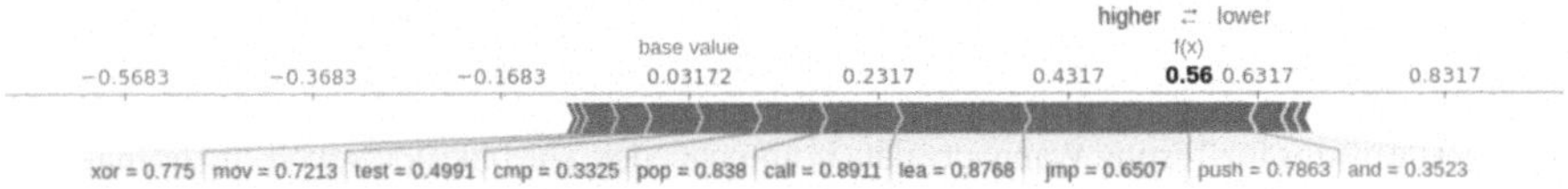

Fig. 5. The force plot for one particular sample

5 Conclusion

This paper evaluates the performance of both supervised and unsupervised learning techniques in classifying malware using opcode frequency features. Our findings reveal that the stacking classifier outperforms the other methods. The highest F1-score achieved by the supervised and unsupervised methods is 99.31% and 96.05%, respectively, using 15 features. This implies that using autoencoders based on deep neural networks for classification could be appropriate for detecting malware using opcode frequency as a feature. Consequently, classical machine learning, such as Random Forest, exhibits the highest F1 score. The classifiers were also tested using wannacry data, and all the classifiers were able to classify it as malware correctly. However, further investigation is necessary to comprehend the effectiveness of autoencoders when applied to extensive datasets employing sophisticated data processing and feature extraction methods.

References

1. Sharma, A., Sahay, S.K.: An effective approach for classification of advanced malware with high accuracy. arXiv preprint arXiv:1606.06897 (2016)
2. Schultz, M.G., et al.: Data mining methods for detection of new malicious executables. In: Proceedings 2001 IEEE Symposium on Security and Privacy (S&P 2001). IEEE (2000)

3. Tian, R., Batten, L.M., Versteeg, S.C.: Function length as a tool for malware classification. In: 2008 3rd International Conference on Malicious and Unwanted Software (MALWARE). IEEE (2008)
4. Wang, X., Yiu, S.M.: A multi-task learning model for malware classification with useful file access pattern from API call sequence. arXiv preprint arXiv:1610.05945 (2016)
5. Chowdhury, M., Rahman, A., Islam, R.: Malware analysis and detection using data mining and machine learning classification. In: International Conference on Applications and Techniques in Cyber Security and Intelligence: Applications and Techniques in Cyber Security and Intelligence. Springer (2018)
6. Santos, I., et al.: Opcode sequences as representation of executables for datamining-based unknown malware detection. Inf. Sci. **231**, 64–82 (2013)
7. Norouzi, M., Souri, A., Zamini, M.S.: A data mining classification approach for behavioral malware detection. J. Comput. Netw. Commun. **2016**, 1–9 (2016)
8. Miao, Q., et al.: Malware detection using bilayer behavior abstraction and improved one-class support vector machines. Int. J. Inf. Secur. **15**, 361–379 (2016)
9. Ahmadi, M., et al.: Novel feature extraction, selection and fusion for effective malware family classification. In: Proceedings of the Sixth ACM Conference on Data and Application Security and Privacy (2016)
10. Bilar, D.: Opcodes as predictor for malware. Int. J. Electron. Secur. Digit. Forens. **1**(2), 156–168 (2007)
11. Karim, M.E., et al.: Malware phylogeny generation using permutations of code. J. Comput. Virol. **1**(1–2), 13–23 (2005)
12. Kolter, J.Z., Maloof, M.A.: Learning to detect malicious executables in the wild. In: Proceedings of the Tenth ACM SIGKDD International Conference on Knowledge Discovery and Data Mining (2004)
13. Saxe, J., Berlin, K.: Deep neural network based malware detection using two-dimensional binary program features. In: 2015 10th International Conference on Malicious and Unwanted Software (MALWARE). IEEE (2015)
14. Kolosnjaji, B., et al.: Deep learning for classification of malware system call sequences. In: AI 2016: Advances in Artificial Intelligence: 29th Australasian Joint Conference, Hobart, 5–8 December 2016, Proceedings 29. Springer (2016)
15. Ring, M., Schlor, D., Wunderlich, S., Landes, D., Hotho, A.: Malware detection on windows audit logs using LSTMs. Comput. Secur. **109**, 102389 (2021)
16. Ye, Y., et al.: DeepAM: a heterogeneous deep learning framework for intelligent malware detection. Knowl. Inf. Syst. **54**, 265–285 (2017)
17. Pedregosa, F., et al.: Scikit-learn: machine learning in python. J. Mach. Learn. Res. **12**, 2825–2830 (2011)
18. Lundberg, S.M., Lee, S.-I.: A unified approach to interpreting model predictions. In: International Conference on Neural Information Processing Systems (2017)

Predictive Analytics for Diagnosing Alzheimer's Disease Using Artificial Intelligence and Machine Learning Algorithms

Hemanth Kumar Nichenametla[1], Suresh Kumar Peddoju[2(✉)], and Sudheer Shetty[3]

[1] Engineering Data Science, Cullen College of Engineering, University of Houston (UH), Houston, Texas, USA

[2] Department of Information Science Technology, Cullen College of Engineering, University of Houston (UH), Houston, Texas, USA
peddojusuresh@gmail.com

[3] Department of ISE, Alvas Institute of Engineering and Technology, Mangaluru, Karnataka, India

Abstract. Predictive analytics plays a vital role in the healthcare industry for providing timely insights and decisions to all the stakeholders of the healthcare industry. Machine Learning has become predominant in the healthcare industry for predicting various diseases that cause severe health issues in humans. One such disease is Alzheimer's, a common neurological condition that affects a sizable number of the elderly population. Although the symptoms of this disease are very minimal in the beginning, they become critical over time and can lead to dementia. Early prediction of this disease can improve a patient's health status and facilitate the improvement of targeted interventions. This study uses various machine learning algorithms such as Logistic Regression, KNN, Decision Tree, SVC, Random Forest, Hard Voting Classifier, Soft Voting Classifier, Gradient Boosting Classifier, Extreme Gradient Boosting Classifier, and Ensemble Learning Model to predict Alzheimer's disease. This research uses the Open Access Series of Imaging Studies (OASIS) longitudinal dataset to build different machine learning models by using multiple machine learning algorithms to predict Alzheimer's disease. The obtained results clearly indicate that ensemble model achieved a higher validation accuracy of 87.3% and test accuracy of 85.3%, in comparison with the other machine learning algorithms used in this work.

Keywords: Alzheimer's disease · Machine Learning · Ensemble Learning · Artificial Intelligence · Healthcare

1 Introduction

Alzheimer's disease, an ongoing neurological condition predominantly affecting the elderly population, is characterized by the progressive deterioration of cognitive abilities, including loss of memory, impaired thinking, and behavioral changes. As the disease progresses, symptoms escalate, eventually leading to severe functional deficits and

T. Shreekumar et al. (Eds.): ISCComm 2023, CCIS 2231, pp. 232–253, 2025.
https://doi.org/10.1007/978-3-031-75605-4_16

dementia [1]. Given the increasing commonness of Alzheimer's disease in an aging population, there is an immediate requirement to develop early detection and prediction methods to enhance patient outcomes and enable targeted interventions [2, 3].

In contemporary years, machine learning has evolved as a powerful computational tool in various disciplines, including healthcare. By analyzing large datasets encompassing clinical, neuroimaging, and demographic information, machine learning algorithms have determined promising results in predicting and classifying diseases, including Alzheimer's. These algorithms can identify patterns and markers indicative of the presence or progression of the disease [4].

This research study aims to harness the capabilities of machine learning techniques to construct a classification model for predicting Alzheimer's disease. The study utilizes the Open Access Series of Imaging Studies (OASIS) longitudinal dataset, which comprises comprehensive measurements of neuroimaging, clinical, and demographic data collected over a predetermined period from individuals with and without Alzheimer's disease. Leveraging longitudinal data enables the model to capture the disease's progression over time, enhancing the accuracy of predictions [4].

A variety of machine learning classification models are included in this study, including Logistic Regression, K-Nearest Neighbors (KNN), Decision Tree, Support Vector Classifier (SVC), Random Forest, Hard Voting Classifier, Soft Voting Classifier, Gradient Boosting Classifier, Extreme Gradient Boosting Classifier, and Ensemble Learning Model. By exploring a diverse range of models, the research aims to identify the best algorithm with the highest accuracy and robustness in predicting Alzheimer's disease. Ensuring the consistency and generalizability of the models, several data preprocessing techniques are applied. These techniques involve handling missing values, addressing categorical data, and scaling or normalizing features. However, outliers present in the data are not addressed due to the medical nature of the dataset and the potential clinical significance of extreme values.

To evaluate the performance of various models we applied a 5-fold cross-validation method to validate model accuracy, which provides insights into the models' consistency and stability. In addition, typical evaluation metrics, including Accuracy, Precision, Recall, and F1-score, are used to assess the models' prediction potential and their ability to reliably categorize individuals with and without Alzheimer's disease.

The following sections describe the remaining portion of the paper. Section 2 details about related literature and Sect. 3 provides detailed explanations of the data preprocessing steps. The methodology adopted in this research is elaborated in Sect. 4 along with detailing about different machine learning algorithms used in this paper for the prediction, offering a comprehensive overview. The achieved results are discussed in Sect. 5 which exhibits clear and concise plots, also including the outcomes associated with the various machine-learning algorithms used. Section 6 includes the conclusion part of the paper.

2 Related Work

During the current period, there observed a tremendous improvement in Research field in the usage of Machine Learning techniques to predict Alzheimer's disease (AD) in the early stages. Countless number of studies are going on into different ML models and various approaches aiming at the improvement of accuracy in terms of early AD prediction. This section of the paper gives an overview of different research approaches present in the literature with respect to the early detection of AD with the help of ML techniques.

In a series of research studies on Alzheimer's Disease classification, various machine learning models have been employed to predict and classify the disease. One study by Muhammad et al. utilized KNN, Decision Tree, Naïve Bayes, and Deep Learning models, with the highest accuracy of 78.32% achieved by a Deep Learning model. Another study by Zhao Fan et al. focused on Support Vector Machines (SVM) using different brain structure features, with the highest accuracy of 65% obtained for the Hippocampal Structure (HS) feature [5]. In another study by J. Neelaveni et al. compared SVM and Decision Tree models, with SVM outperforming Decision Tree, achieving an accuracy of approximately 85% [6].

Several other researchers also investigated Alzheimer's Disease classification using various machine learning models. Janani Venugopalan et al. achieved accuracies of 80.05% for K-Nearest Neighbors (KNN), 81.01% for Support Vector Machines (SVM), and 80.02% for Decision Trees [7]. Morshedul et al., explored SVM and Logistic Regression, achieving accuracies of 74.6% and 74.7%, [8] respectively. Additionally, Ning An et al. utilized deep ensemble learning models, with Multilayer Perceptron achieving 80.5% accuracy [9]. Lin Liu et al. employed Logistic Regression with Cross-Validation achieved accuracy of 83.3% and Linear Support Vector Classification with 80.6% accuracy [10]. C. V. Dolph et al. used LASSO-SAE to achieve 56.6% and LASSO-TEXTURE SAE 58% accuracy [11]. Finally, C. Kavitha et al. employed Decision Tree Classifier with accuracy value of 80.46%, and XG Boost with 85% to classify Alzheimer's Disease [12].

Taeho Jo et al. conducted research using Random Forest and XGBoost, achieving accuracies of approximately 68.7% and 73.08%, respectively [13]. Shojaie Mehdi et al. focused-on Alzheimer's classification using the Transfer Gradient Boost algorithm, achieving an accuracy of about 78.7% [14]. J. Sheng et al. employed Logistic Regression and attained an accuracy of 83.3% [15]. Xiao, R. et al. explored Sparse Logistic Regression, reaching an accuracy of around 84.2% [16]. Rohini, M. et al. utilized Logistic Regression and achieved an accuracy of approximately 89% [17]. Wang J et al. employed the Random Forest Classification model for Alzheimer's classification, obtaining an accuracy of 68% [18]. Kumari, R. et al. reported an accuracy of 73.6% using the Random Forest Classification model [19]. Shastry, K.A. et al. utilized an Ensemble Model in their research, obtaining an accuracy of 55.2% [20].These diverse results underscore the importance of selecting the right machine learning model for the task.

The presented studies collectively emphasize the significance of using machine learning techniques for Alzheimer's Disease classification. These models, including Deep Learning, SVM, Decision Trees, and ensemble learning, exhibit varying levels of success in predicting the disease, with accuracies ranging from 65% to 85%. Such research highlights the potential of data preprocessing and feature engineering to enhance accuracy in Alzheimer's Disease classification and underscores the importance of selecting appropriate machine learning algorithms to achieve robust results.

Table 1. Comparison of related work.

Sr. No	ML model	Approach	Accuracy
1	SVM	The Authors utilized Support Vector Machines (SVM) using different brain structure features,	65
2	SVM	This paper predicts Alzheimer's using machine learning with age, visits, MMSE, and education	85
3	KNN	This study uses single data modality to make predictions such as AD stages	80.05
4	SVM	This study uses single data modality to make predictions such as AD stages	81.01
5	DT	This study uses single data modality to make predictions such as AD stages	80.02
6	SVM	The approach involves utilizing the OASIS MRI dataset, preprocessing with feature selection and standardization	74.6
7	LR	The approach involves utilizing the OASIS MRI dataset, preprocessing with feature selection and standardization	74.7
8	Multi layered Perceptron	The authors utilized deep ensemble learning models, with Multilayer Perceptron	80.5
9	LR	The authors utilized the Logistic Regression model using extracted spectrogram features from speech data to identify AD is feasible	83.3
10	SVC	The authors collect a new speech dataset, which includes Alzheimer's disease patients and healthy control subjects and feed them to the models	80.6
11	Lasso Texture SAE	The study enhances various modalities and disease-specified datasets that detect and records images, signals, and speeches, etc	58
12	DT	The models were likely designed to provide insights into the stages of Alzheimer's based on patient responses	80.46
13	XGBoost	The models were likely designed to provide insights into the stages of Alzheimer's based on patient responses	85

(*continued*)

Table 1. *(continued)*

Sr. No	ML model	Approach	Accuracy
14	RF	The three-step approach involves fragmenting the genome	68.7
15	XGBoost	Applying a Sliding Window Association Test (SWAT) approach to identify SNPs	73.08
16	Transfer GB	The Transfer Gradient Boosting (TrGB) framework, applying an instance-based transfer learning approach for MRI modality	78.7
17	LR	The study introduces a Subclass Weighted Logistic Regression (SWLR) method for classifying Alzheimer's disease stages(LMCI vs AD)	83.3
18	Sparse LR	This study explores sparse logistic regression (SLR) for early AD and mild cognitive impairment (MCI) diagnosis (Fra_Heart)	84.2
20	RF	The simplified neuropsychological test battery NTB, featuring six key subtests, effectively distinguished MCI and dementia	68
21	RF	The study introduces a potent multimodal AD classification approach, employing the Adaptive HPT-RFE Classifier that outperforms existing techniques(NC vs MCI)	73.6
22	Ensemble model	The research introduces an ensemble model, LRFB, utilizing machine learning on the ADNI dataset,	55.2

Through a comprehensive analysis of the research conducted by the researchers, we have collectively developed numerous machine learning models aimed at advancing our understanding of Alzheimer's Disease (AD) classification. Our objective is to provide a comprehensive overview of the accuracies achieved by each of these models, thus contributing to the growing body of knowledge in this field.

3 Data Preparation

In this research work, we have used the Open Access Series of Imaging Studies (OASIS) longitudinal dataset consists of details about Alzimer's diseased (Demented Group) patient and a normal patient (Non-Demented Group) feature. This data set comprises of approximately 15 features, comprising of both numerical and categorical data. The Open Access Series of Imaging Studies (OASIS) dataset, with its 150 rows and 15 columns, is a pivotal resource in neuroimaging and cognitive research. Its longitudinal design allows researchers to study changes over time in individuals, including those with conditions like Alzheimer's disease. Despite its smaller size, the dataset offers a rich set of features and clinical data, supporting the development of machine learning models, diagnostic tools, and novel analytical approaches for cognitive disorder detection and monitoring. The OASIS dataset's columns provide essential information about subjects, their demographics, and neuroimaging and cognitive assessment measures, fostering advancements in brain health and cognitive disorder research. A snapshot of the dataset is provided in Table 1 and a comprehensive description of its features is provided in Table 2 [21, 22].

Table 2. Snapshot of the Dataset.

Subject ID	MRI ID	Group	Visit	MR Delay	M/F	Hand	Age	EDUC	SES	MMSE	CDR	eTIV	nWBV	ASF
OAS2_0001	OAS2_0001_MR1	Nondemented	1	0	M	R	87	14	2	27	0	1987	0.696	0.883
OAS2_0001	OAS2_0001_MR2	Nondemented	2	457	M	R	88	14	2	30	0	2004	0.681	0.876
OAS2_0002	OAS2_0002_MR1	Demented	1	0	M	R	75	12		23	0.5	1678	0.736	1.046
OAS2_0002	OAS2_0002_MR2	Demented	2	560	M	R	76	12		28	0.5	1738	0.713	1.010

Few features contain the missing values, and everything can be dealt with in data preprocessing phase and is explained in the below phase. Neat data preparation is crucial for accurate Early prediction of Alzheimer's disease using machine learning models. In this case study, we have implemented key preprocessing techniques on the Open Access Series of Imaging Studies (OASIS) longitudinal dataset. These techniques include handling missing values, addressing categorical data, and scaling or normalizing features.

There are few columns such as "SES" and "MMSE" consisting of null values, we employed imputation techniques such as median and mean imputation to eliminate null values from the data set. Categorical data was encoded using one-hot encoding or label encoding, depending on the cardinality of the variables. In this case the target variable that is "Group" column contains categorical data with three unique values "Demented", "Non-Demented", "Converted". Converted values are replaced with Demented as it is the status of showing dementia in a person. Now there are two unique values present in the target variable and one-hot encoding is implemented on this column. Finally, demented is converted into '1' and non-Demented is converted to '0'. These Data Preprocessing techniques ensured a complete and suitable dataset for training machine learning models, enhancing the accuracy and reliability of Alzheimer's disease predictions in early stages.

The following Table 3 shows the minimum, maximum, mean and median values of each column of the data set to see how the data is distributed and getting to know important data columns in the given data set.

Further, we implemented all the machine learning classification models and evaluated their performance using validation accuracies and test accuracies. We applied feature scaling to the model that achieves the highest accuracy. To ensure robustness of our

Table 3. Dataset description.

Sr. No	Feature Name	Feature Description
1	Subject ID	Identification number of the Subject
2	MRI ID	Identification number of the MRI
3	Group	To which group the patient belongs to either Demented or Non-Demented
4	Visit	Number of times the patient visited for check-up
5	MR Delay	Delay
6	M/F	Gender
7	Hand	Which hand(Right/Left)
8	Age	Age in years
9	EDUC	Number of Years of Education
10	SES	Social Economic Status
11	MMSE	Mini Mental State Examination
12	CDR	Clinical Dementia Rating
13	eTIV	Estimated Total Intracranial Volume
14	nWBV	Normalize Whole Brain Volume
15	ASF	Atlas Scaling Factor

Table 4. Minimum, Maximum, Median and Mean Values of each column.

Serial Number	Column Name	Min	Max	Median	Mean
1	EDUC	6	23	15	14.6
2	SES	1	5	2	2.5
3	MMSE	4	30	29	27.3
4	CDR	0	2	0	0.3
5	eTIV	1106	2004	1470	1488.13
6	nWBV	0.64	0.84	0.73	0.73
7	ASF	0.88	1.59	1.19	1.2

evaluation, we will employ the 5-fold cross-validation technique and obtain validation accuracies.

4 Methodology

The following list of machine learning algorithms are used in our research to classify Alzheimer's disease and evaluated the performance of these algorithms to diagnose the disease with greatest accuracy. The achieved results of these algorithms are discussed in the results section of this paper. The following figure (Fig. 1) shows the sequence of steps followed in the methodology in perform this research.

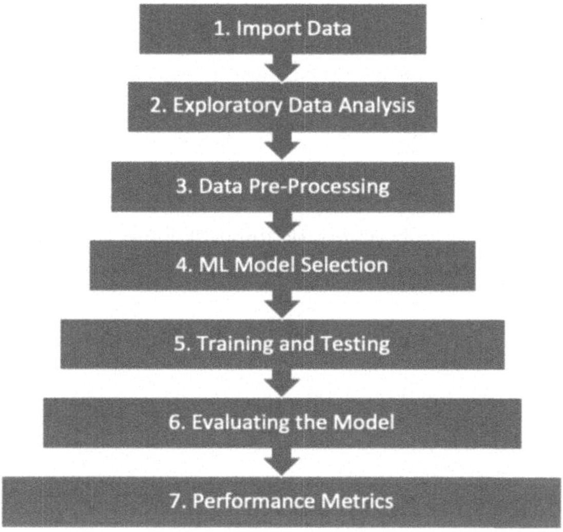

Fig. 1. Sequence of steps used in our Methodology.

4.1 Logistic Regression

The Logistic Regression is a Classification Machine Learning model used to classify the two-class categorical (Yes or No, Y or N, True or False) or numerical (0 or 1) data. It can be implemented in all the cases where the values of the target variable are in binary choices like as mentioned above True or False, Yes or No, 0 or 1 etc. This logistic Regression model mainly gets the probability of the input with which it belongs to the specific class. Based on these probability values the class name can be determined. The activation function to determine probabilities used is a sigmoid function. To be specific this function has the range of values between 0 and 1 so the probabilities. And the output from sigmoid function is nothing but the required probability values. So, if the aim is to predict the probabilities of the output in that case Logistic regression model can be used undoubtedly [23, 24].

$$L(\theta) = -\sum_{i=1}^{m} \left[y_i \log(h(x_i)) + (1 - y_i) \log(1 - h(x_i)) \right]$$

where:

- m is the number of training examples.
- y_i is the true label for the i_{th} example (0 or 1).
- $h(x_i)$ is the predicted probability for the i_{th} example.
- θ is the vector of model parameters.

4.2 Decision Tree

A decision tree is a type of classification technique that has a flowchart-like visual representation. Internal nodes in this structure act as key nodes for running tests on features. The results of these tests decide the direction to travel, leading us to the suitable leaf node that corresponds to a specific class label. The mix of features that go into the classification process is represented by the branches that join the nodes. As a result, by exploring the routes from the root to the leaf nodes, we reveal the fundamental principles behind efficient categorization [25].

Entropy is a measure of uncertainty or impurity in a set of data. The higher the entropy, the more uncertain the data is. The formula for entropy is:

$$E(S) = -\sum\nolimits_{i=1}^{m} p_i \log_2(p_i)$$

where:

- $E(S)$ is the entropy of the set S
- m is the number of possible classes.
- p_i is the proportion of examples in class i

Information gain measures the amount of information that is gained by splitting a set of data into two subsets. The formula for information gain is:

$$IG(S, A) = E(S) - \sum\nolimits_{v=1}^{V} |Sv|/|S|E(Sv)$$

where:

- $IG(S, A)$ is the information gain for splitting set S on attribute A
- v is the number of possible values for attribute A
- Sv is the subset of S containing examples with value V for attribute A

4.3 Support Vector Machine

SVM is a machine learning model which is used for both regression and classification tasks. They are especially used for classification of complex small and medium size data. Through non-linear mapping, it transforms the original training data into elevated higher dimensions. It looks for a linear optimal separating hyperplane through those new dimensions. With the help of nonlinear mapping, the data from both the classes can be separated by using a hyperplane. The hyperplane can be found using the support vectors and margins. By maximizing the margin and minimizing the classification errors, SVM implements the classification task [26].

The decision boundary is represented as:

$$f(x) = sign(w \cdot x + b)$$

where:

- w is the weight vector.
- x is the input vector.
- b is the bias term.

4.4 Random Forest

Random Forest is one of the easy-to-use algorithms in machine learning which provides accurate results even when hyper-tuning is not used. Random Forest is a kind of ensemble learning which overcomes the limitation of decision trees of overfitting. Ensemble learning uses multiple algorithms or same algorithms multiple times. Random forest is a collection of the decision trees. By harnessing the collective wisdom of these trees, random forest surpasses the performance of individual trees by mitigating errors and reducing prediction correlations. To achieve accurate class predictions, it requires informative features and strives to maintain low correlations among the trees by randomly sampling the training data and selecting a subset of features for each tree [27, 28].

Majority voting formula, which is used for classification problems. This formula simply takes the majority class predicted by the individual decision trees as the final prediction for the new data point.

$$\mathrm{Pr}edictedclass = \mathrm{argmax}(P(class = c|x))$$

where:

- $P(class = c|x)$ is the probability that the new data point x belongs to class c.
- arg max() is the function that returns the argument with the maximum value.

4.5 Soft Voting Classifier

Soft voting is an ensemble learning approach used to combine the probabilities predicted by different ML algorithms, resulting in a more refined and probabilistic final prediction. Unlike hard voting, which relies on majority voting, soft voting calculates the average of the class probabilities assigned by each model. By considering the weighted average of these probability values, soft voting provides a more nuanced decision-making process. For instance, if one model predicts a stone with a 40% probability and another predicts it as a oil with an 80% probability, the ensemble prediction would assign a probability of 60% for the object being an oil. This technique is applicable in fields such as machine learning classification tasks, accounts for the confidence levels of each algorithm and enhances accuracy and robustness by leveraging the collective knowledge of multiple algorithms. In simple words it is the combination of multiple classifiers to make collective predictions. By aggregating the predicted probabilities from each classifier, the Soft Voting Classifier achieves accurate and robust predictions [29].

$$y_{pred} = \underset{c}{\text{argmax}} P(c|x)$$

where:

- y_{pred} is the predicted class.
- c is the class label.
- x is the input data.
- $P(c|x)$ is the probability that the input data belongs to class c, as predicted by the ensemble.

4.6 Hard Voting Classifier

Hard voting is a widely employed technique. It falls under the family of ensemble learning, Here the final prediction is determined by selecting the class with the highest number of votes among individual ML algorithms. This approach involves independent predictions from each algorithm, and the ensemble combines these predictions to arrive at a collective decision in terms of voting. For instance, in the case of predicting the color of a specific drink, if three algorithms individually predict "Pink", "blue" and "blue" the ensemble will ultimately predict "blue" due to its majority number of votes. Hard voting simplifies the decision-making process by considering the majority consensus among the algorithms. This technique finds extensive application across various fields, particularly in machine learning classification tasks, where the combination of predictions from multiple models enhances accuracy and robustness. Leveraging the collective wisdom of the ensemble, hard voting offers a straightforward and effective approach to making class predictions [30].

$$y_{pred} = np.\text{argmax}\left(np.bincount\left(y_{pred_{individual}}\right)\right)$$

where:

- y_{pred} is the predicted class label for the current sample.
- $np.bincount\left(y_{pred_{individual}}\right)$ is a vector that counts the number of times each class label was predicted by the individual classifiers.
- $np.\text{argmax}()$ is a function that returns the index of the largest element in a vector.

4.7 Gradient Boosting Classifier

Gradient Boosting is a powerful classification algorithm that iteratively improves predictions by fitting new predictors to the residual errors made by previous predictors. It starts by calculating the log odds and converting them to probabilities using a logistic function. Residuals are then computed and used to build a decision tree. To make predictions, the algorithm combines the base log odds with scaled output values from each tree, adjusting for bias and variance using a learning rate. This process continues until a stopping criterion is met. The resulting algorithm provides accurate predictions for unseen instances and is a valuable tool in various classification tasks [31].

$$F_m(x) = F_{m-1}(x) + \alpha_m h_m(x)$$

where:

- $F_m(x)$ is the prediction of the ensemble after m iterations.
- $F_{m-1}(x)$ is the prediction of the ensemble after $m - 1$ iterations.
- α_m is the learning rate, which controls the impact of each new decision tree.
- $h_m(x)$ is the prediction of the mth decision tree.

4.8 Extreme Gradient Classifier

Extreme Gradient Boosting (XGBoost) is a highly optimized and efficient algorithm that offers a unique approach to tree construction and pruning, resulting in accelerated training speed and improved performance, particularly when dealing with large-scale datasets. XGBoost employs the Similarity Score and Gain metrics to determine the most effective node splits by considering the residuals (the disparity between actual and predicted values) and the previous probability calculated at each iteration. By incorporating advanced techniques such as the Approximate Greedy Algorithm, Parallel Learning, Sparsity-Aware Split Finding, and Cash-Aware Access, XGBoost achieves faster processing and higher accuracy compared to conventional Gradient Boosting methods. This algorithm has emerged as a crucial component in enhancing model performance and has gained substantial popularity across diverse machine learning applications [32].

$$F_m(x) = F_{m-1}(x) + f_m(x)$$

where:

- $F_m(x)$ is the prediction for the m_{th} tree.
- $F_{m-1}(x)$ is the prediction for the $(m - 1)_{th}$ tree.
- $f_m(x)$ is the prediction for the m_{th} tree.

The $f_m(x)$ term is calculated using the following formula:

$$f_m(x) = \gamma I_m(x)$$

where:

- γ is the learning rate.
- $I_m(x)$ is the indicator function for the m_{th} tree.

4.9 Ensemble Classifier Model

Ensemble learning, a prominent machine learning technique, has gained widespread recognition for its ability to significantly improve model accuracy. It accomplishes this by amalgamating multiple machine learning models. This approach encompasses four key features: Bagging, which employs majority voting and selects random subsets of training data to train diverse base learners, ensuring their dissimilarity; AdaBoost, a boosting technique that sequentially creates base classifiers using bootstrap samples with instance-weight adjustments based on misclassification rates; and Stacking, a strategy that combines various machine learning models to enhance predictive accuracy by leveraging the strengths of diverse algorithms. These ensemble methods collectively contribute to achieving superior model accuracy and performance [29].

4.10 KNN Classifier Model

K-Nearest Neighbors (KNN) is a versatile and non-parametric supervised learning method widely applied in areas like pattern recognition, data mining, and intrusion detection. It groups unclassified data points based on their proximity to training set neighbors, utilizing distance metrics like Euclidean, Manhattan, and Minkowski distances. By analyzing clusters of data points, KNN intuitively determines a point's classification based on its nearest neighbors' group. Selecting the right 'k' value (number of neighbors) is vital and can be determined through cross-validation. KNN is used in data preprocessing, pattern recognition, and recommendation engines, offering benefits like ease of implementation, adaptability to new data, and minimal hyperparameters. Nonetheless, it faces scalability, dimensionality, and overfitting challenges, which can be addressed using techniques like feature selection and dimensionality reduction to enhance its performance [33].

4.11 Naïve Bayes Classifier

It is a supervised machine learning model used for a classification problem designed to differentiate objects based on specific features. It utilizes algorithms to assign objects to predefined categories, playing a crucial role in various applications. Among the classifiers, the Naive Bayes classifier operates on the principle of Bayes theorem. This probabilistic model assumes feature independence and calculates the probability of a hypothesis given the evidence. Naive Bayes classifiers find extensive application in sentiment analysis, spam filtering, and recommendation systems, offering fast implementation and ease of use. However, their performance may be impacted when predictors are not truly independent, highlighting a limitation in real-life scenarios [26].

$$P(C|X) = (P(X|C) * P(C))/P(X)$$

where:

- $P(C|X)$ is the posterior probability of class C given instance X
- $P(X|C)$ is the likelihood of instance X given class C
- $P(C)$ is the prior probability of class C
- $P(X)$ is the evidence, which is the probability of instance X occurring, regardless of the class.

4.12 Multilayer Perceptron Classifier

The Multilayer Perceptron (MLP) classifier is a powerful feed-forward artificial neural network that is widely used for supervised machine learning tasks. It utilizes the back propagation learning technique, making it capable of handling complex non-linear problems. The MLP demonstrates excellent performance, even with large input datasets, and maintains a high accuracy ratio even with smaller data. These qualities make the MLP highly valuable in various classification tasks within the field of machine learning [34].

The dataset is split into an 80:20 ratio, with 80% allocated to the training data and the remaining 20% designated as the testing data set. Following this, all the classification models are fitted using the training data, and their performance is evaluated using the test data. To ensure robustness in the evaluation process, a 5-fold cross-validation technique is employed on the training data.

$$y_{hat} = W_{out} * h(W_{in} * x + b_{in}) + b_{out}$$

where:

- x is the input vector.
- W_{in} is the weight matrix for the input layer.
- b_{in} is the bias vector for the input layer.
- h is the activation function for the hidden layer(s)
- W_{out} is the weight matrix for the output layer.
- b_{out} is the bias vector for the output layer.
- y_{hat} is the predicted output.

5 Results

This section provides a more in-depth analysis of the machine learning algorithms utilized in our study for diagnosing Alzheimer's disease. The primary goal is to offer insights into the reasons behind the superior performance of a particular algorithm and to discuss potential drawbacks and limitations.

In this research paper, we have evaluated the performance of all the models using different metrics such as Precision, Recall, F-1 Score, Validation Accuracy, Accuracy, Area under the Curve (AUC) for the Receiver Operator Characteristics (ROC) curve. Our goal is to find out the most accurate model out of all the models developed. We have analyzed the confusion matrix for all the models to gain deeper insights into each model's performance. These matrices provide valuable information on accurate instance classification and identification of true positives, true negatives, false positives, and false negatives.

All the plots, bar graphs and snapshots related to ROC curves, Confusion matrix, precision, recall, F-1 score, Validation Accuracy, Accuracy and Area under the curve (AUC) comparisons for all the models are show cased clearly in Table 5, Table 6 and Table 7. Among the diverse machine learning algorithms assessed in our study, the ensemble model emerged as the top performer, exhibiting the highest accuracy, validation accuracy, and area under the curve (AUC). Understanding the factors contributing to its superior performance is crucial.

The ensemble model harnesses the collective knowledge of multiple algorithms through techniques like bagging, boosting, and stacking. By amalgamating diverse models, each with its unique strengths and weaknesses, the ensemble approach mitigates individual model limitations and enhances overall robustness. The synergy among different algorithms contributes to improved accuracy and generalization, making the ensemble model well-suited for complex tasks like Alzheimer's disease prediction.

Acknowledging the limitation in our study, which lacks a detailed comparison with State-of-the-Art (SOTA) deep learning algorithms, future research should extend the comparison to include deep learning approaches. SOTA deep learning models, such as convolutional neural networks (CNNs) and recurrent neural networks (RNNs), have demonstrated remarkable success in various medical imaging tasks, including neurological disorder detection. Integrating these models into our analysis could provide a more comprehensive understanding of the landscape and potentially uncover novel insights.

Table 5. Confusion Matrices are plotted below for all the Model.

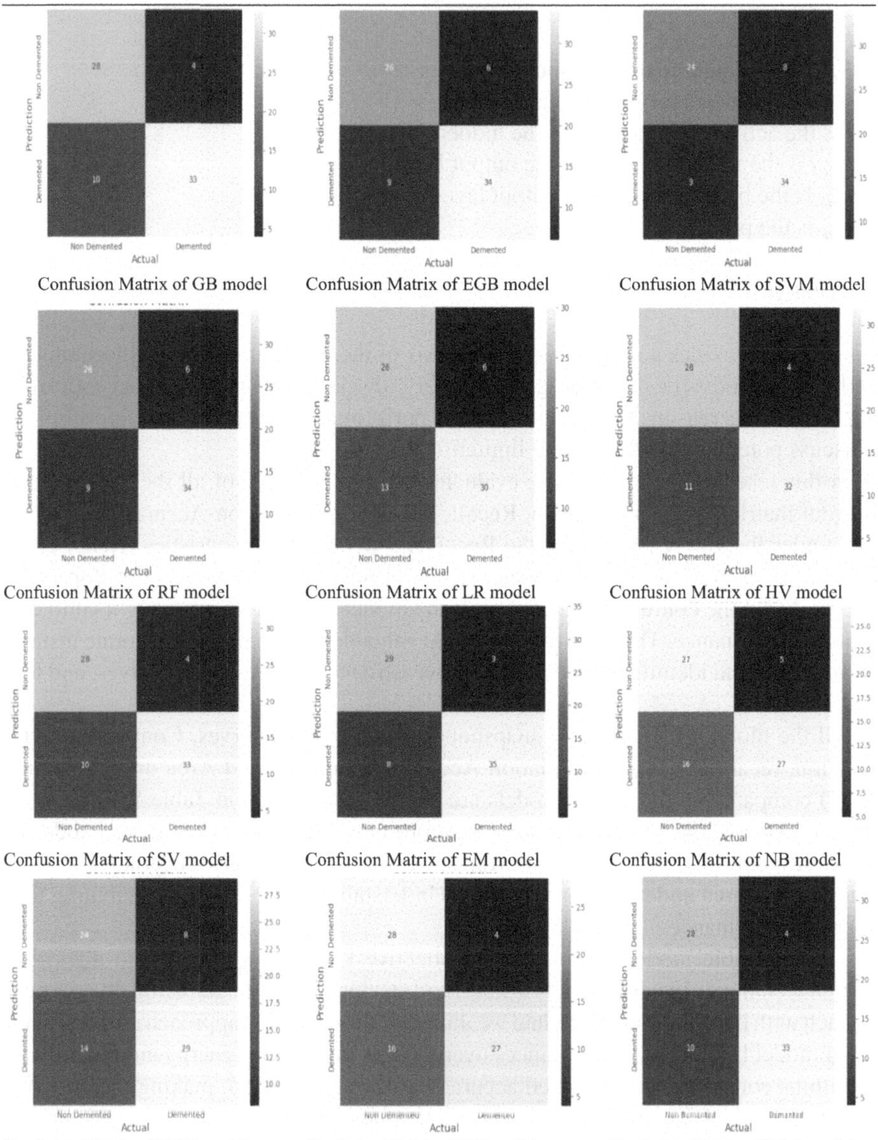

Confusion Matrix of GB model — Confusion Matrix of EGB model — Confusion Matrix of SVM model

Confusion Matrix of RF model — Confusion Matrix of LR model — Confusion Matrix of HV model

Confusion Matrix of SV model — Confusion Matrix of EM model — Confusion Matrix of NB model

Confusion Matrix of KNN model — Confusion Matrix of DT model — Confusion Matrix of MLP model

Table 6. ROC Curves are plotted below for all the Models.

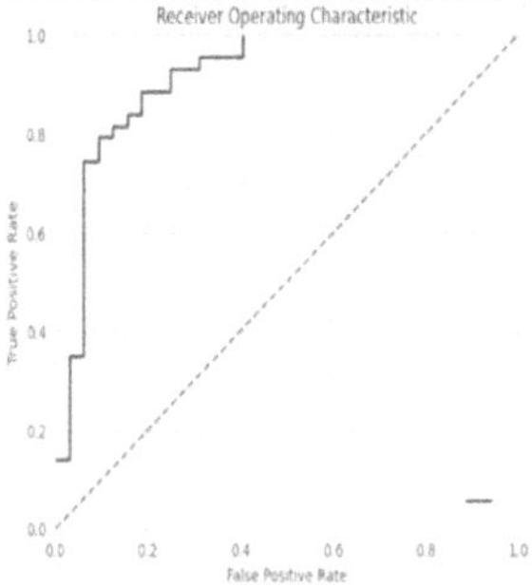

ROC Curve of GB model

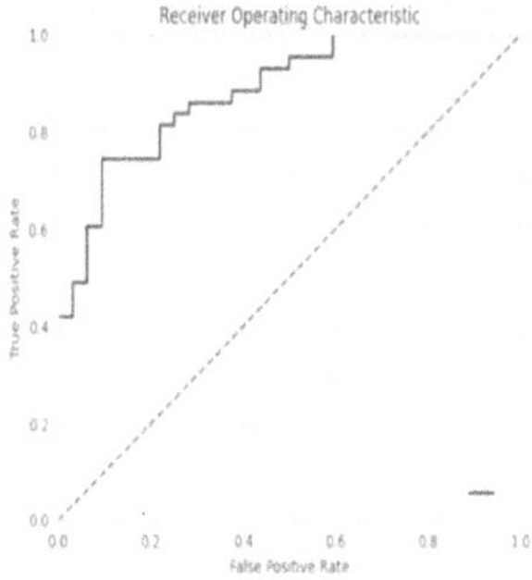

ROC curve of EGB model

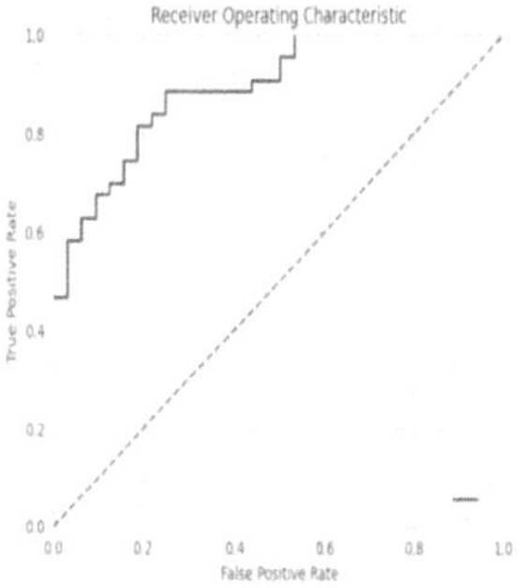

ROC Curve of RF model

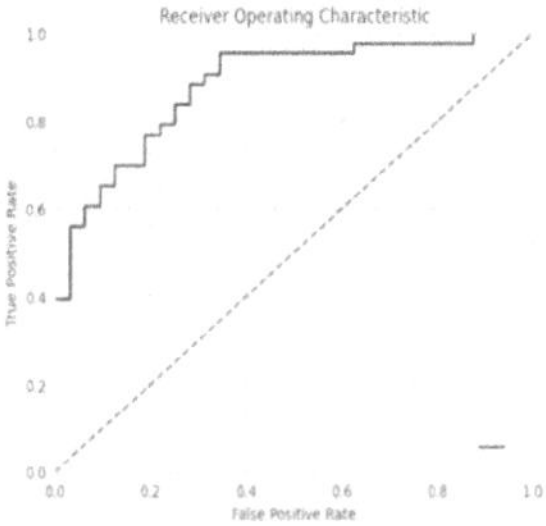

ROC Curve of LR model

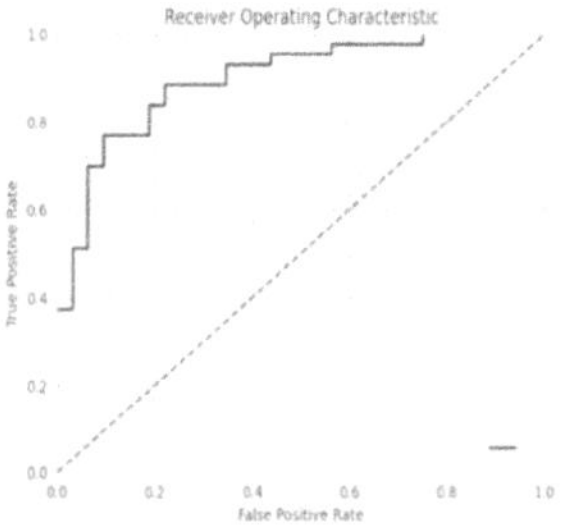

ROC Curve of SV model

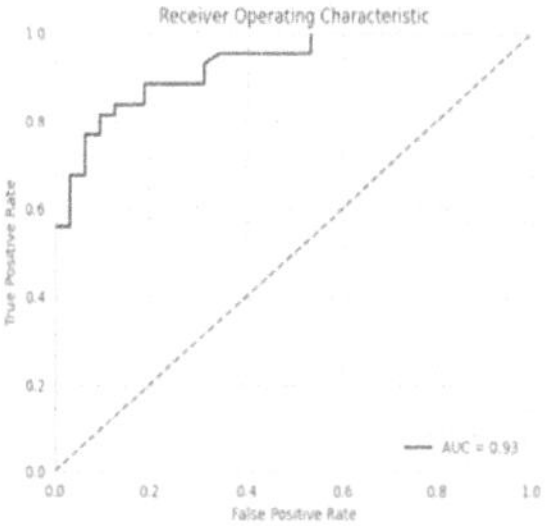

ROC Curve of EM model

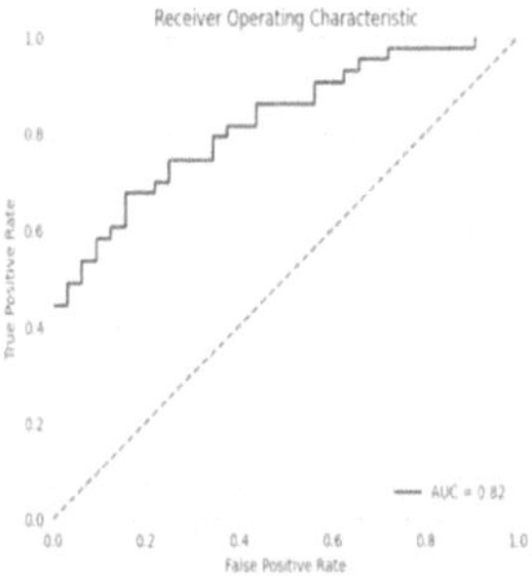

ROC Curve of NB model

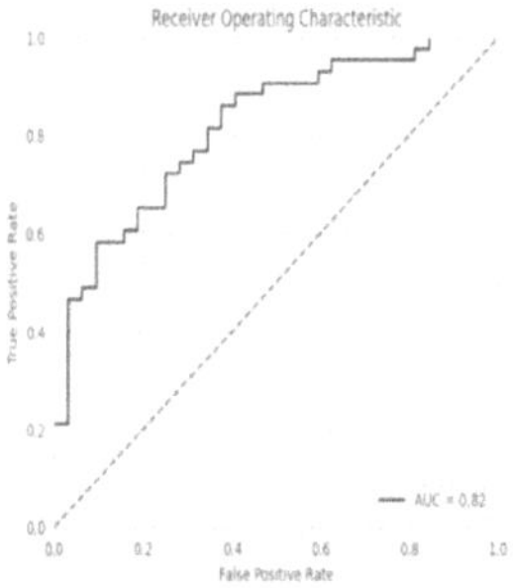

ROC Curve of KNN model

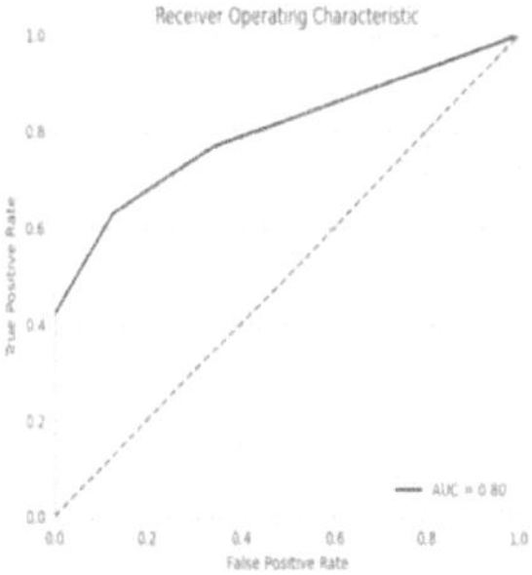

ROC Curve of DT model

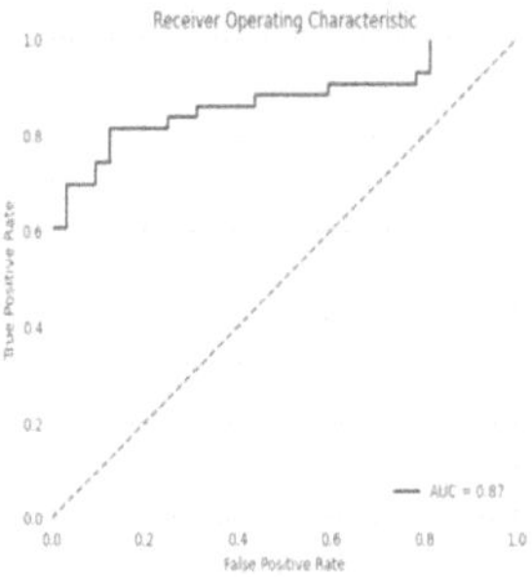

ROC Curve of MLP model

Table 7. Metrics details of all the Machine Learning Models.

Machine Learning Model	Validation Accuracy	Accuracy	Precision	Recall	F1 score	AUC
Logistic Regression	79.5	74.7	83.3	69.8	75.9	75.5
Extreme Gradient Boosting Classifier	80.6	76	82.1	74.4	78	80.2
Support Vector Machine Classifier	79.9	77.3	81	79.1	80	77
Random Forest Classifier	82.2	80	85	79.1	81.9	80.2
Hard Voting Classifier	81.2	80	88.9	74.4	81	80.1
Soft Voting Classier	83.6	81.3	89.2	76.7	82.5	82.1
Gradient Boosting Classifier	81.9	82.7	87.5	81.4	84.3	82.9
Ensemble Model	87.3	85.3	92.1	81.4	86.4	86
KNN Classifier	64.1	70.7	78.4	67.4	72.5	71.2
Naïve Bayes Classifier	64.1	72	84.4	62.8	72	73.6
Multi-Layer Perceptron	76.5	81.3	93.9	72.1	81.6	82.9
Decision Tree Classifier	77.6	73.3	87.1	62.8	73	75.1

5.1 Performance Metrics

The following performance metrics [35, 36] were used in this work to evaluate the performance of all the algorithms. The results are depicted in Table 8.

$$\text{Precision} = \frac{TP}{(TP + FP)}$$

$$\text{Recall} = \frac{TP}{(TP + FN)}$$

$$\text{Accuracy} = \frac{(TP + TN)}{(TP + TN + FP + FN)}$$

$$F \quad 1\ \text{Score} = \frac{2 + Precision + Recall}{Precision + Recall}$$

AUC-ROC is a metric to evaluate a classification machine learning problem. Various threshold values are utilized while evaluating this metric.

Validation Accuracy - Accuracy of a model on the validation data.

Table 8. Bar graph comparison of all the metrics for all the Machine Learning Models.

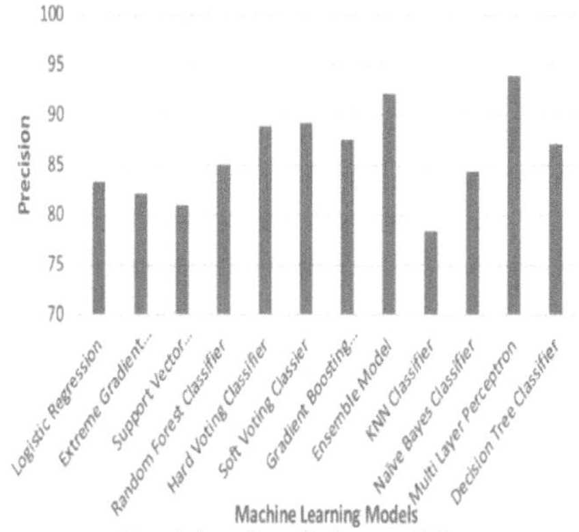

Precision for all the models

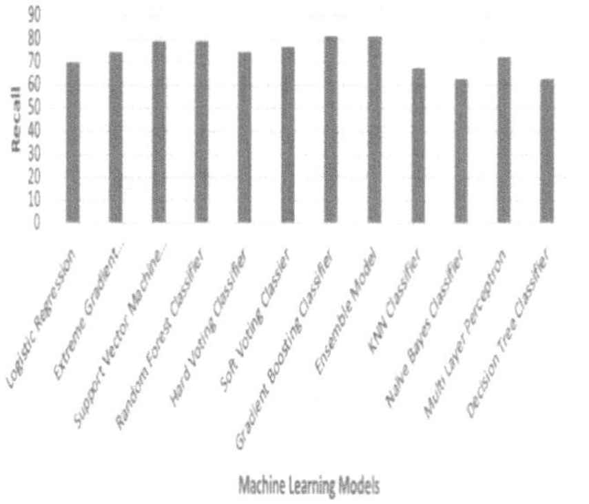

Recall for all the models

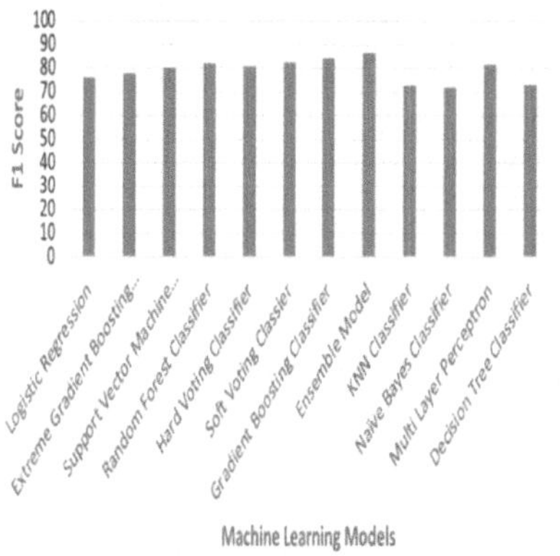

F1 Score for all the models

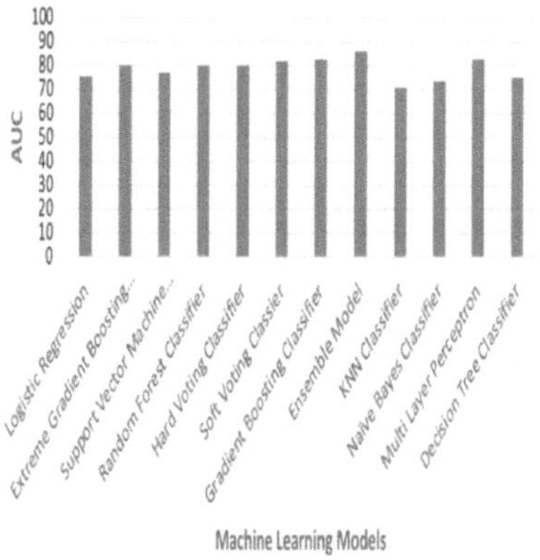

AUC Scores for all the models

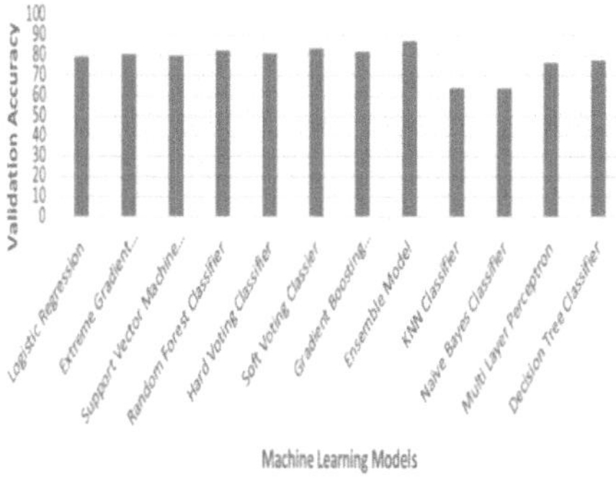

Validation Accuracy for all the models

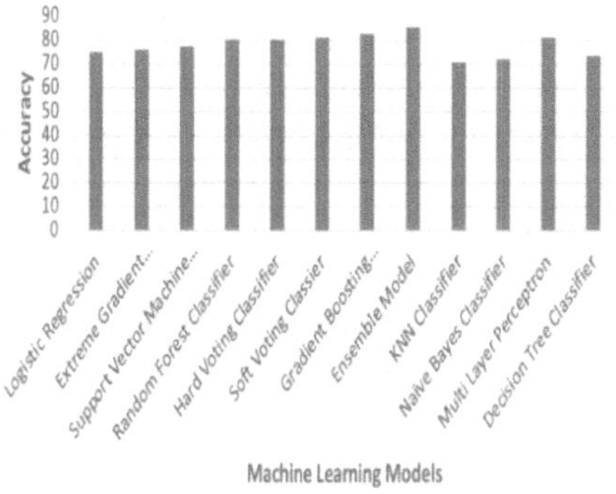

Accuracy for all the models

6 Conclusion

Alzheimer's Disease is a severe neurodegenerative disorder that greatly impacts individuals and their families. Detecting the disease in its early stages is of paramount importance, as it allows for timely medical interventions and support services. In this research, we aimed to develop multiple machine learning models to diagnose Alzheimer's Disease (AD) very early and evaluated the performance of each algorithm with different performance metrics. Among the models evaluated, the ensemble model emerged as the top performer, exhibiting the highest accuracy, validation accuracy, and area under the curve (AUC). The ensemble model's superior performance provides a valuable tool for identifying AD cases at an early stage, potentially leading to improved patient outcomes and quality of life. Furthermore, the ensemble model's excellent AUC highlights its capacity to discriminate between AD and non-AD individuals accurately. This discrimination ability is crucial for accurate diagnosis and the development of targeted treatment strategies. In conclusion, our research demonstrates that the ensemble model surpasses other models in terms of highest accuracy of 85.3%, validation accuracy of 87.3%, and AUC of 86%. The ensemble model offers a promising technique for reliable and accurate AD prediction. Further research and validation studies are encouraged to validate its effectiveness and explore its integration into clinical practice.

7 Future Work

Acknowledging the limitation in our study, which lacks a detailed comparison with State-of-the-Art (SOTA) deep learning algorithms, future research should extend the comparison to include deep learning approaches. SOTA deep learning models, such as convolutional neural networks (CNNs) and recurrent neural networks (RNNs), have demonstrated remarkable success in various medical imaging tasks, including neurological disorder detection. Integrating these models into our analysis could provide a more comprehensive understanding of the landscape and potentially uncover novel insights.

In extension to our current research, there is an opportunity to improve the model accuracy by identifying the most relevant features with feature engineering techniques and construct refined models based on these essential features. Such an approach holds the potential to further enhance the accuracy of our predictive models. We will explore this avenue of research in subsequent phases to improve the overall effectiveness of our Alzheimer's Disease prediction model.

To address the identified limitations and further enhance diagnostic capabilities, future research directions include:

Interpretable Ensemble Models: Developing techniques to enhance the interpretability of ensemble models, ensuring that healthcare professionals can trust and understand the model's decisions.

Integration of Deep Learning: Extending the comparison to include state-of-the-art deep learning algorithms, exploring their potential for improved accuracy in Alzheimer's disease prediction, especially in the context of neuroimaging data.

These future directions aim to advance the understanding of machine learning models in Alzheimer's disease prediction and contribute to ongoing efforts to improve early detection and intervention strategies.

References

1. Alzheimer's Association. 2015 Alzheimer's disease facts and figures. Alzheimer's Dementia **11**(3), 332–384 (2015). https://doi.org/10.1016/j.jalz.2015.02.003
2. Scheltens, P., De Strooper, B., Kivipelto, M., et al.: Alzheimer's disease. The Lancet **397**(10284), 1577–1590 (2021). https://doi.org/10.1016/S0140-6736(20)32205-4
3. Finder, V.H.: Alzheimer's disease: a general introduction and pathomechanism. J. Alzheimer's Disease **22**(s3), S5–S19 (2010). https://doi.org/10.3233/JAD-2010-100975
4. Tanveer, M., Richhariya, B., Khan, R.U., et al.: Machine learning techniques for the diagnosis of Alzheimer's Disease. ACM Trans. Multimed. Comput. Commun. Appl. **16**(1s), 1–35 (2020). https://doi.org/10.1145/3344998
5. Fan, Z., Xu, F., Qi, X., Li, C., Yao, L.: Classification of Alzheimer's Disease Based on Brain MRI and Machine Learning. Springer, Heidelberg (2019)
6. Neelaveni, J., Devasana M.S.G.: Alzheimer disease prediction using machine learning algorithms. In: 2020 6th International Conference on Advanced Computing and Communication Systems (ICACCS), pp. 101–104. IEEE (2020). https://doi.org/10.1109/ICACCS48705.2020.9074248
7. Venugopalan, J., Tong, L., Hassanzadeh, H.R., et al.: Multimodal deep learning models for early detection of Alzheimer's disease stage. Sci. Rep. **11**, 3254 (2021). https://doi.org/10.1038/s41598-020-74399-w
8. Morshedul Bari Antor, A.H.M., et al.: A Comparative Analysis of Machine Learning Algorithms to Predict Alzheimer's Disease
9. An, N., Ding, H., Yang, J., Au, R., Ang, T.F.A.: Deep ensemble learning for Alzheimer's disease classification. J. Biomed. Informat. **105**, 103411 (2020). https://doi.org/10.1016/j.jbi.2020.103411. ISSN 1532-0464
10. Liu, L., Zhao, S., Chen, H., Wang, A.: A new machine learning method for identifying Alzheimer's disease. Simulat. Model. Pract. Theory **99**, 102023 (2020). https://doi.org/10.1016/j.simpat.2019.102023. ISSN 1569-190X
11. Dolph, C.V., Alam, M., Shboul, Z., Samad, M.D., Iftekharuddin, K.M.: Deep learning of texture and structural features for multiclass Alzheimer's disease classification. In: 2017 International Joint Conference on Neural Networks (IJCNN), Anchorage, pp. 2259–2266 (2017). https://doi.org/10.1109/IJCNN.2017.7966129
12. Kavitha, C., Mani, V., Srividhya, S.R., Khalaf, O.I., Romero, C.A.T.: Early-Stage Alzheimer's Disease Prediction Using Machine Learning Models
13. Jo, T., Nho, K., Bice, P., Saykin, A.J.: For the Alzheimer's Disease Neuroimaging Initiative, Deep learning-based identification of genetic variants: application to Alzheimer's disease classification. Brief. Bioinform. **23**(2), bbac022 (2022). https://doi.org/10.1093/bib/bbac022
14. Mehdi, S., et al.: A transfer learning approach based on gradient boosting machine for diagnosis of Alzheimer's disease. Front. Aging Neurosci. **14**, 1663–4365 (2022). https://doi.org/10.3389/fnagi.2022.966883
15. Sheng, J., Wu, S., Zhang, Q., Li, Z., Huang, H.: A binary classification study of Alzheimer's disease based on a novel subclass weighted logistic regression method. IEEE Access **10**, 68846–68856 (2022). https://doi.org/10.1109/ACCESS.2022.3186888
16. Xiao, R., Cui, X., Qiao, H., et al.: Early diagnosis model of Alzheimer's disease based on sparse logistic regression. Multimed. Tools Appl. **80**, 3969–3980 (2021). https://doi.org/10.1007/s11042-020-09738-0
17. Rohini, M., Surendran, D.: Toward Alzheimer's disease classification through machine learning. Soft. Comput. **25**, 2589–2597 (2021). https://doi.org/10.1007/s00500-020-05292-x
18. Wang, J., et al.: Random forest model in the diagnosis of dementia patients with normal mini-mental state examination scores. J. Personaliz. Med. **12**(1), 37 (2022). https://doi.org/10.3390/jpm12010037

19. Kumari, R., Nigam, A., Pushkar, S.: An efficient combination of quadruple biomarkers in binary classification using ensemble machine learning technique for early onset of Alzheimer disease. Neural Comput. Applic. **34**, 11865–11884 (2022). https://doi.org/10.1007/s00521-022-07076-w
20. Shastry, K.A., Sattar, S.A.: Logistic random forest boosting technique for Alzheimer's diagnosis. Int. J. Inf. Tecnol. **15**, 1719–1731 (2023). https://doi.org/10.1007/s41870-023-01187-w
21. Shahbaz, M., Ali, S., Guergachi, A., Niazi, A., Umer, A.: Classification of Alzheimer's disease using machine learning techniques. In: Proceedings of the 8th International Conference on Data Science, Technology and Applications, pp. 296–303. SCITEPRESS - Science and Technology Publications (2019). https://doi.org/10.5220/0007949902960303
22. Baglat, P., Salehi, A.W., Gupta, A., Gupta, G.: Multiple machine learning models for detection of Alzheimer's disease using OASIS Dataset. 614–622 (2020). https://doi.org/10.1007/978-3-030-64849-7_54
23. Nick, T.G., Campbell, K.M.: Logistic Regression, pp. 273–301 (2007). https://doi.org/10.1007/978-1-59745-530-5_14
24. Xiao, R., Cui, X., Qiao, H., Zheng, X., Zhang, Y.: Early diagnosis model of Alzheimer's Disease based on sparse logistic regression. Multimed. Tools Appl. **80**(3), 3969–3980 (2021). https://doi.org/10.1007/s11042-020-09738-0
25. Song, Y.Y., Lu, Y.: Decision tree methods: applications for classification and prediction. Shanghai Arch Psychiatry **27**(2), 130–135 (2015). https://doi.org/10.11919/j.issn.1002-0829.215044
26. Mohan, V., Dhayanand, S.: Liver Disease Prediction Using SVM and Naïve Bayes Algorithms, vol. 4 (2015). https://www.researchgate.net/publication/339551659
27. Paul, S., Ranjan, P., Kumar, S., Kumar, A.: Disease predictor using random forest classifier. In: 2022 International Conference for Advancement in Technology (ICONAT), pp. 1–4. IEEE (2022). https://doi.org/10.1109/ICONAT53423.2022.9726023
28. Dimitriadis, S., Liparas, D.: Alzheimer's DNI. How random is the random forest? Random forest algorithm on the service of structural imaging biomarkers for Alzheimer's disease: from Alzheimer's disease neuroimaging initiative (ADNI) database. Neural Regen. Res. **13**(6), 962 (2018). https://doi.org/10.4103/1673-5374.233433
29. Kumari, S., Kumar, D., Mittal, M.: An ensemble approach for classification and prediction of diabetes mellitus using soft voting classifier. Int. J. Cognit. Comput. Eng. **2**, 40–46 (2021). https://doi.org/10.1016/j.ijcce.2021.01.001
30. Rojarath, A., Songpan, W., Pong-inwong, C.: Improved ensemble learning for classification techniques based on majority voting. In: 2016 7th IEEE International Conference on Software Engineering and Service Science (ICSESS), pp. 107–110. IEEE (2016). https://doi.org/10.1109/ICSESS.2016.7883026
31. Blagus, R., Lusa, L.: Gradient boosting for high-dimensional prediction of rare events. Comput. Stat. Data Anal. **113**, 19–37 (2017). https://doi.org/10.1016/j.csda.2016.07.016
32. Wu, Z., Wang, X., Jiang, B.: Fault diagnosis for wind turbines based on ReliefF and eXtreme gradient boosting. Appl. Sci. **10**(9), 3258 (2020). https://doi.org/10.3390/app10093258
33. Sinha, P., Sinha, P.: Comparative Study of Chronic Kidney Disease Prediction Using KNN and SVM (2015). www.ijert.org
34. Surendiran, J., Yuvaraj, N., Ramkumar, M., Ravi, C.N., Vidhya, R.G.: Classification of diabetes using multilayer perceptron. In: 2022 IEEE International Conference on Distributed Computing and Electrical Circuits and Electronics (ICDCECE), pp. 1–5. IEEE (2022). https://doi.org/10.1109/ICDCECE53908.2022.9793085

35. Fan, J., Upadhye, S., Worster, A.: Understanding receiver operating characteristic (ROC) curves. CJEM **8**(01), 19–20 (2006). https://doi.org/10.1017/S1481803500013336
36. Fan, Z., Xu, F., Qi, X., et al.: Classification of Alzheimer's disease based on brain MRI and machine learning. Neural Comput. Appl. **32**, 1927–1936 (2020). https://doi.org/10.1007/s00521-019-04495-0

A Comprehensive Study on Artificial Intelligence (AI) Driven Internet of Healthcare Things (IOHT)

Astha Sah[1,2], Bharat Bhushan[1,2(✉)], and Chinmai Shetty[1,2]

[1] Department of Computer Science and Engineering, School of Engineering and Technology, Sharda University, Greater Noida 201310, India
bharat_bhushan1989@yahoo.com, chinmai@nitte.edu.in
[2] Department of Information Science and Technology, NMAM Institute of Technology, Affiliated to Nitte (Deemed to be University), Nitte, India

Abstract. In a wide range of industries including healthcare, smart cities, agriculture and industrial automation IoT systems are utilized to carry out a number of crucial functions. The Internet of Healthcare Things (IoHT) has drawn a lot of attention as an opportunity for remote patient tracking and accuracy treatment. However, IoHT is used so extensively privacy violations like IoHT leakage of data have sparked significant concern among the public. At the contrary hand AI technologies has shown considerable promise to strengthen IoHT reliability and privacy protection. Using Artificial Intelligence (AI) techniques, the majority of these systems contain intelligence for making intelligent choices. The edge devices must be methodically built with an emphasis on the relevant AI capabilities from exploration through deployment in order to assure good performance. Technologies for healthcare has risen to the level of healthcare 5.0, which replaces earlier technologies. The objective of healthcare 5.0 is to build an entirely independent healthcare system that analyses how a patient's many health issues interact with one another. This paper conducts a survey to determine the viability of merging AI and IoHT in protecting privacy, a comprehensive analysis of current AI-based IoHT systems is discussed. Additionally, many privacy issues in IoHT are highlighted. By addressing numerous research fields in AI, IoT and IoHT health data and security fusion, this paper review at existing research. This study also highlights some authors works to minimize the security issues by using AI technology. It additionally addresses current studies obstacles and provide some recommendations for future study. This paper concludes that there is lack of some security issues in healthcare sector which is needed to be enhanced.

Keywords: Internet of Things · Healthcare · IoHT · Artificial Intelligence · Security · privacy

1 Introduction

In this modern technological age, Artificial intelligence (AI) has gained popularity and has been developing throughout the period. AI is the term for intelligence displayed by a machine or robots that is comparable to human intelligence. The applications based

T. Shreekumar et al. (Eds.): ISCComm 2023, CCIS 2231, pp. 254–269, 2025.
https://doi.org/10.1007/978-3-031-75605-4_17

on AI can assist in the healthcare industry in tackling challenges with accessibility to facilities for healthcare. Among other things, the applications are mostly employed in emergencies, early disease detection, and personalized therapy [1]. Although AI is now being used in most of the sectors, it exceeded expectations from humans and raised service quality. As a consequence, AI has grown deployed to the scheduling and administrative aspects of medicine. Nursing assistants, robotically aided surgery and numerous other operational helpers are the most popular application of AI in healthcare. The major clinical requirements can be met by AI [2].

Everything will be dependent on Information Technology (IT) in the modern era. Everything in the global community of the future will be outfitted with the latest innovations that the Internet of Things (IoT) can use to illuminate the globe. Information technology, or IoT, is utilized extensively in smart cities, numerous businesses and other applications, to create clever, effective, and advantageous to human existence [3]. An aspect of the IoT used in the healthcare industry or Internet of Healthcare Things (IoHT) for short, are growing in popularity because of their capabilities to enhance the care of patients in a number of ways, such as dependability, ease of use, accuracy and ongoing connectivity. The IoHT is characterized as the connecting of the digital and physical worlds via devices with internet in the medical field [4]. IoHT has been referred to as "an expanded framework for linking virtual and physical things for improved healthcare services" [5].

The healthcare sector is essential for people's economic as well as the physical health. Even though the healthcare industry is continually evolving, a nation like India, with its expanding population always lacks sufficient facilities. For a secure and efficient management, it is necessary to take into account a few distinct properties associated with the medical data which is gathered from numerous sources and regions, is unstructured or structured, time-sensitive, abundant, and has a variety of access permissions because it is sensitive secret information. Considering data are vital to the overall healthcare sector, efforts are concentrated on developing methods for gathering, storing, aggregating, processing, and analyzing a huge quantity of medical records with the objective to gain beneficial knowledge about it.

This paper provides a thorough analysis of IoT and IoHT-based smart health care systems. The aims to examine a variety of associated research fields, including the most advanced IoT-based intelligent healthcare, the data integration of IoTs and AI in intelligent healthcare based smart medical care, and security of IoT-based innovative healthcare challenges. Taking advantage of illustrative case instances, this paper examines the drawbacks of conventional IoT design flows and also highlighted the transformation of healthcare sector using AI.

The entire paper presents as Sect. 2 introduce the concept of IoT and its architecture which consist of applications area and security issues. Section 3 discusses the concept of traditional healthcare system and evolution of healthcare sector as technology emerges and challenges faced by healthcare due to security issues. Section 4 represent the concept of AI and application of AI in healthcare which deals by resolving the security issues of IoHT system. Section 5 explains recent advances and Sect. 6 conclude the paper.

2 IoT Background

The terms "Internet" and "things" are combined to form the phrase "Internet of Things" (IoT). Every item that can connect to the internet is referred to as a "Thing," which includes context-aware objects, sensors and smart devices that are widespread and informative. IoT devices require to have three particular fundamental characteristics enumerated as below.

Ability to Sense: IoT gadgets need to be enabled to sense and analyse the data they generate. In the medical field, for instance, many dedicated sensor technologies are used to sense and gather information on cholesterol level, body temperature, heart rate and blood glucose. The data synthesis is self-contained.

Communicable: IoT devices need to be able to pass on data to the multiple intended information distribution centres after aggregating and sensing using a wide range of networking methods such as Mobile Adhoc Networks (MANET), Wireless Sensor Networks (WSNs), Wireless Local Area Networks (WLAN), Mobile networks (3G, 4G, 5G) and Wireless Technologies etc.

Actionable: Gathered information by itself is useless, so IoT devices that are responsive to user input require being capable to interpret the information with the goal to execute a kind of action. If a patient's pulse rate or blood pressure level is measured and transmitted, for instance, this information is meaningless until technology that processes is applied to identify whether it is outside or within a level that is considered normal. Additionally, IoT devices must be able to autonomously warn medical professionals when further actions are necessary.

2.1 IOT Architecture and Component

As urbanization rapidly increasing the emergence of technologies like IoT has also evolved greatly and enhanced the life of citizens. IoT has huge applications in smart cities, homes, industries and healthcare. IoT Architecture consists of three layers which includes Perception layers, Network layers, and Application Layers and the components use in each layer are demonstrated in Fig. 1 and discussed in the subsections below.

The Perception Layer: The Perception layers, which primarily identifies items and collect details, is comparable to the five senses and the skin on the face of the Internet of Things. The 2-D bar code tags and viewers, tags with RFID and reader-writers, cameras, the aid of GPS, detectors, devices, and networked sensors constitute every part of the perception layers. Its primary responsibility is to determine the object via obtaining data.

The Network Layer: Conveying and analysing data is the primary role of the network layer, and it's similar to the brain and the neural network of the IoT. An integrated network for Web and interaction, a system management converging, a data converging, a smart analysis converging, and other components constitute the network layers. The data gathered from the layer of perception will be transmitted and processed by the network layer. Components such as Wi-Fi, Router, Access point, Bluetooth and many more are part of this layer.

The Application Layer: In order to achieve the comprehensive intellectual development, the Application Layer combines the societal segregation of IoT with requirements from the industry. In order to construct an intellectualized industry that is associated to people's socialization of work and ultimately forms humanity, the layer of application is a consequence of the deeper confluence of commercial and web technologies with the requirements of the industry. At the outset of expansion, this framework layer characterization of IoT as a technological standpoint seems sense. Smart home, Smart cities, Smart hospitals moreover come under the domain of Application layer.

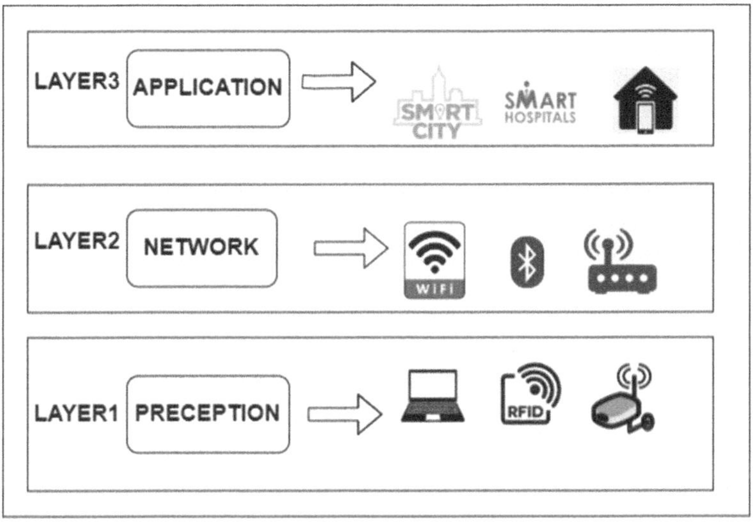

Fig. 1. Architecture of IOT

2.2 Application Area

A large number of modern-day apps have grown intelligent, but are currently incapable to interact with one another. By allowing them to do so and exchange helpful information, a variety of novel applications will be developed. The value of our lives would undoubtedly increase with these new applications that have some autonomy, all because of the idea of the IoT.

- **Smart Farming:** Smart agricultural systems can assist in monitoring many aspects of crop fields such as sunlight, humidity, temperature, rain forecast, and moisture in the soil using linked sensors. Watering system automation is another benefit of IoT. Smart agriculture offers several advantages including improved company productivity through the automation of processes, greater autonomy over manufacture, improved item quality and volume, climate condition recording, and crop management.
- **Smart Cities:** To fully adopt IoT technologies, cities that are smart must have official and public cooperation and thorough preparation at every stage. Towns are able

to enhanced through the use of IoT in a number of ways, including by bolstering transport systems, maintaining the health and safety of the population, and upgrading infrastructure.

Healthcare: A concept known as IoT sees everything as being an element of the web and having distinct online identities. IoT has significantly improved the quantity, quality and precision of medical data as well as the accuracy of medical care delivery. IoT is an architecture allowing the utilization of technologies like cloud computing, AI, network communication, sensors and big data as well as computation to deliver efficient alternatives. It created health unit more productive by providing facilities such as improved interaction with patients, adherence to sanitation standards, monitoring from far and based on location services that are real-time. There are many applications of IoT in healthcare as shown in Fig. 2. Some of applications are highlighted in the subsections below.

- *Telemedicine and Telehealth:* Remote Patient Monitoring (RPM), telemedicine and medical connected devices are more important now than ever before since they enable clinicians to screen patients without physically interacting with them. RPM technology is practically identical to improvements in telemedicine since it naturally collects data on patients with ongoing illnesses, so that medical personnel can watch patients from distant [6].

  5G Connected Ambulances Integrated with IoT and IoT Based Smart Ventilators: Ambulance can be combined with IoT to give health care solutions by providing another layer of movement touch and feel together with voice via LTE (Long Term Evolution), which is projected to have higher bandwidth and low latency capabilities. IoT and its linked tool Blockchain technology have made it feasible to remotely monitor many ventilators without directly contacting doctors, lowering the risk of infection and reducing the number of bed-site visits.

  Remote monitoring and Real-Time Location-Based Service: One of the most important uses of IoT is to track health remotely. Continual observation aids in providing individuals with quality medical care. Numerous peoples worldwide pass away as they lack the opportunity prompt medical intervention. IoT has a solution for it. IoT assists in giving them greater healthcare services and concern at remote locations. Health care providers may monitor the locations of their patients using IoT. It can be especially helpful when someone who has been hurt is seeking emergency medical attention. IoT gadgets supports monitoring the environment in real-time.

2.3 Security of IoT Devices and Applications

The prevalence of recently developed IoT devices and the rapid rise in their use in healthcare facilities has raised the possibility of vulnerabilities and other securities like issues such as data leakage and data stolen related problems. According to [7], the amount of IoT devices per sec will increase to 152K by 2025. The introduction of 5G and Wi-Fi 6 networks boosts bandwidth and offers higher rates of data for enabling novel applications. Following are the primary causes of security concerns:

- The type and degree of severity of the information transmitted by medical equipment.

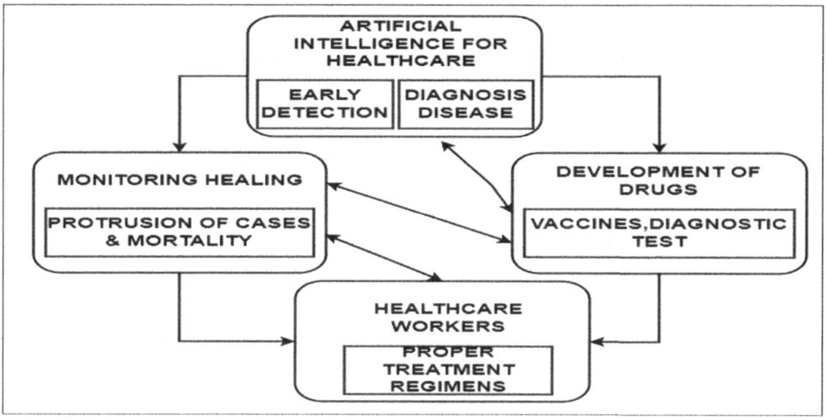

Fig. 2. Application of AI in healthcare

- Data compatibility issues and complexity.
- The disregard for security concerns by healthcare equipment vendors.

The factors mentioned above are causing a surge in privacy-related worries for connected devices in regard to confidentiality, integrity, and availability (CIA) [8]. Application layer is not absolutely safe and are subject to the following set of attacks. The following include a few examples of these assaults:

- **Software Attacks:** Applications for the IoT have problems with security, such as Denial of Service, Malicious script, Spyware, Trojan horse and Phishing attacks.
- **Encryption Attacks:** In addition to software vulnerabilities, authentication attacks can be used to bypass system authentication. Here are a few instances: Man in the Middle attack, Crypt-analysis attacks and Side channel attacks.

3 Healthcare

The WHO describes health care as the preservation as well as enhancement of an entire level for mental, social, and physical health with the primary goals of improving patients' standard of life and prolonging the average life span. In accordance to existing procedures, customized medical care is a practice which customizes health services for each individual by identifying prevalent traits such as inheritance, transmission and lifestyle [9]. The following account, yet lacks the ability to customize the general well-being of an individual via a number of medical conditions, as this requires knowledge of the biological connections between multiple illnesses. Instead, it assists in customize an individual medical issue with particular factors (e.g., genes, inheritance).

3.1 Tradition Healthcare System

The study of recent deployments of the IoT in medical fields allows for a classification of the existing technology research into three categories as Applications for professional help, selfcare monitoring and patient monitoring. The majority of systems are focused on monitoring and medical assistance but neglect to pay attention to self-care alternatives. The SHARP (Sustainable, Holistic, Adaption, Real-time, and Precision) framework has been used to evaluate the benefits and drawbacks of the research reported. This approach is primarily used to healthcare-related systems. While some want to adapt their program to numerous channels for better engagement and visibility, mobile gadget developers typically choose to focus their emphasis on a single channel to connect with the audience they are targeting. The Health Prognosis Integrated Systems (HPIS) was developed to address a number of these issues and has the capacity to serve an extensive variety of individuals with cutting-edge technology a user-friendly interface, an interactive chatbot and an open for further medical help. The possibility for developing a lightweight instrument for Real-time tracking and wellness management has increased with the rise of the internet in response to the growing demand for proactive treatment of persistent illnesses. Assistance from family and friends is necessary for regularly maintaining an eye on the ill and old in order to prevent accidents.

3.2 Transformation of Healthcare Technology Using IOT

The structure of medical care system was deeply interrelated to medicines in the former time. On well trained doctors and healthcare service providers especially on biological herbs the history of medicine1.0 was based on the past time. Healthcare 1.0 was in effect at the time and important health issues were being addressed through smart public health strategies [10]. Healthcare 2.0, the name was provided when massive hospital was developed as antibiotics originated and advanced diagnostic systems were introduced such as X-ray. Mass manufacturing was the main feature of healthcare 2.0 during the industrialisation era. With the expansion in micro-technology and electronics later surgery greatly benefited. Medicine 3.0 was created as a result of evidence-based therapeutics, the use of image recognition, navigation surgery and Surgical robots. The term "Medicine 4.0" (also known as "smart-health") refers to the integration of intelligent devices with network technologies, fast transmission and smart microelectronics into healthcare systems and their significant contribution to better lifestyles and improved healthcare services that ultimately lead to the healthier life. Intelligent occlusal splint, intelligent implants for cancer therapy, personalized telematic therapy and personalized chemotherapy these are few examples of medicine 4.0 [11]. The main motive of healthcare 4.0 is the creation of commercial models using ML techniques and AI smart wearables with built-in sensors assist in the collection, monitoring and diagnosis of diseases from patient data. The vision of healthcare 5.0 has gained a lot of popularity recently because of the progress made in technological advances. Digital wellness is giving rise to a new concept termed "healthcare 5.0" which integrates physical (such as biological functions) and cyber (such as monitoring and control) elements to offer customised healthcare. A completely autonomous system of healthcare that considers the interconnections between a patient's various health issues is the objective of healthcare 5.0. It achieves this by making use of cutting-edge information technologies including cloud computing, blockchain, AI, big data analysis and IoT. Transformation of healthcare technology also led to the introduction of IoHT technology. By supporting the collection of a vast amount of diverse medical information the IoT is establishing itself as a crucial component in transforming the healthcare system globally. The IoHT is viewed as "an expanded framework for integrating virtual and physical things for improved healthcare service" [12].The following are some of the key benefits of IoHT for remote monitoring: easy patient location; support for the patient's independent life; access to a larger amount of real-time data about the patients; safer surroundings; a significant reduction in healthcare errors and collisions and ongoing surveillance of the patients' daily activities for better outcomes of healthcare management.

3.3 Major Challenges in IOT

Despite the fact that the Internet of Things has many advantages for the medical industry, we cannot ignore the challenges it faces and regarding the challenges some solutions are shown in Fig. 3. The privacy and security of information pose the biggest threat to IoT-based medical services. The vast amount of data that is constantly being kept and exchanged can be easily exploited and used against the patient and the healthcare provider. Hackers frequently use fake identification to buy pharmaceuticals and prescription medications with the intention of misusing them. Nowadays, everyone owns a distinctive gadget, and this variation in gadgets is hindering the display of innovation. The speed of data transfer, analysis, and retrieval has decreased since IoHT is still incapable to resolve the issue of different mobile device' compatibility with the IoT network.

By installing the idea of portable encryption algorithms and creating a feasible security framework that is sufficiently intelligent to handle any obstacles faced by IoT healthcare sectors, it can be accomplished to defeat countless threats and attacks on IoT equipment such as sinkhole attacks, grayhole, wormhole sybil, injection and message replication, side channel attack, denial of service and sniffing.

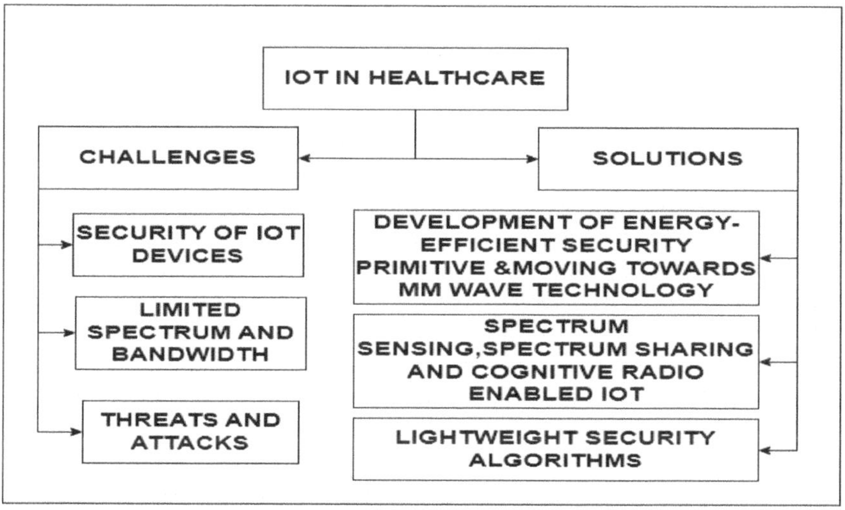

Fig. 3. Challenges and respective solution of IOT in healthcare

4 Artificial Intelligence

With the tremendous growth of technology AI rooted system are being used in many sectors such as in healthcare, retail and marketing, media, entertainment, aerospace, defence, insurance, financial services and industrial etc. AI provides benefits like reduced cost estimation, high performance, scalability, high design interpretability, better inventory management and higher customer retention rate. The cutting-edge trends in AI remarks in notable advantages as flaw detection, enhanced management and increased customer retention.

4.1 AI in Healthcare

The healthcare sector acquires impact and AI is giving resolution. Data analysis is a key component of healthcare 5.0 because it helps to match commercial solutions to patient requirements. By contributing with healthcare picture diagnosis and analysis AI for healthcare offers the ability to enhance both staff productivity and patient care. Inside IBM's smart cloud where Watson resides, AI is used frequently in the medical field. Innovative medical treatments have always highlighted customized care, which is simple to obtain with machine learning. Machine learning helps significant medical sector stakeholders acquire an edge and accelerate the treatment delivery line particularly in maintaining efficient digital medical records [13]. AI can help with the creation of better drugs and antigens proactive strategies as well as effective treatment programs [14]. Advantages of AI in sector of healthcare such as shorter wait times, easy appointment scheduling, prompt departure, prompt payment of bills, diagnosis earlier, quicker treatment, saving money, better encounter Identification of disease outbreaks and some are discussed below in details.

- **Diagnosis/Disease Identification:** One of the numerous healthcare difficulties that machine learning is being applied to is the diagnosis and identification of illnesses and other health issues. The field of precision healthcare continues to be enhanced by the Watson Genomics, an agreement between IBM Watson Quest Diagnostics and Health, which intends to combine artificial intelligence with hereditary tumours analysis.
- **Personalized Treatment/Behavioural Modification:** Better illness assessment is strongly associated with customised healthcare, or improved counselling based on personal health data augmented by machine learning. Currently, guided instruction is in charge of this field. It enables doctors to choose from a fewer number of treatments or quantify risk for patients based on genetic data and symptoms, for instance.
- **Drug Discovery and Development:** AI has several potential uses in the discovery and development of drugs, both today and in the years to come, ranging from next-generation sequencing to applications in precise healthcare. Machine learning algorithms might be used for early drug evaluation and initial evaluation as well as to forecast a drug's effectiveness after considering a wide range of biological parameters.
- **Radiation Treatments:** Initial findings from an AI system that might be useful in planning radiation treatments for individuals with neck and head cancer are encouraging.

- **Automation:** One method mechanisation might be included into medical facilities in the future is through machine learning algorithms in robotic surgeon. Surgery performed by robots may one day be totally computerised with no need of human intervention, owing to methods like deep learning and machine learning.

In advanced health care systems, the privacy and security of healthcare data are of the outmost significance. Data about a patient should be treated with confidence. Patients who have their privacy violated may face bullying in public, which can leave them sad and traumatized. If healthcare data about athletes is disclosed, participants on competing sports teams might utilise it to obtain unfair advantages. Healthcare information should be treated confidentially and securely sent via channels of communication as a result [14]. Traditional system shows some missing links of AI specifications in software development as shown in Fig. 4. The use of AI has a number of opportunities for healthcare industry. Highlighting the major improvement in healthcare from traditional healthcare system to Transformational healthcare system are discussed in Table 1.

4.2 Security Approaches Based on AI

- The authors of [15] provided evidence of the communication vulnerabilities of individual healthcare equipment to various cyberattacks. In particular, whether a third-party attacker can use man-in-the-middle, rewind, counterfeits data injections including denial-of-service assaults to intercept confidential health information transmitted by an individual's healthcare equipment. The authors suggested using HEKA, an intrusion detection system (IDS), to continually keep track on gadget connectivity and security breaches. The suggested system uses various machine learning approaches for detecting attacks along with an n-gram-based technique to identify unusual traffic-flow pattern.
- The author of [16] covered four key aspects that are transparency, informed authorization, confidentiality and accountability which are the important factors in preventing security issues in healthcare sector. In these areas, optimal practises and norms have significance in preventing unexpected or undesirable outcomes in healthcare sector.
- The authors of [17] addressed the drawbacks and aspects to be taken into account while implementing AI-based safety measures such as Intrusion Detection and Prevention Systems (IDPS), within a power utility system for creating a safe and assaults free environment for healthcare sector.
- The authors of [18] focused at an attack strategy for IoT systems as well as IoT safety measures utilising supervised learning alongside other machine learning (ML) approaches, reinforcement learning (RL) along with unsupervised learning. The primary focus was on IoT authentication, access regulation, secure privacy as well as identification of malware for information protection techniques. Machine learning also has the potential to address security risks like denial-of-service attacks, counterfeiting, eavesdropping, biometric fingerprints and data leaks.

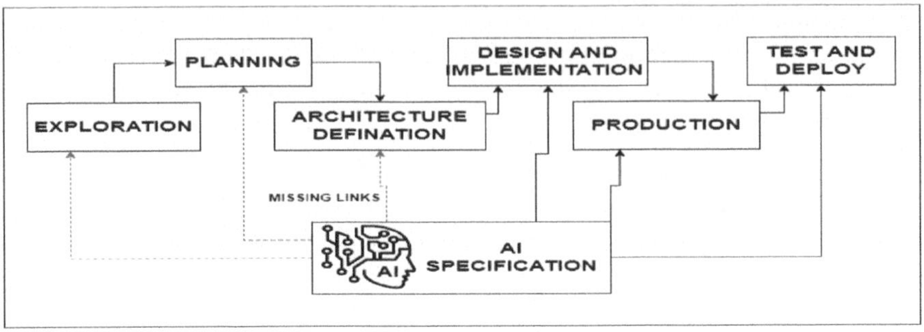

Fig. 4. System design without AI specification

Table 1. Comparative table of Traditional and transformational healthcare system

Characteristic	Traditional healthcare system	Transformation Healthcare system
Data Holding	PACS networks (Picture Archiving and Communication System) are used for storing data	In centralized networks like IPFS, which stands for Interplanetary file system data is kept
Mechanism for Database Distribution	It depends entirely on based on the cloud processes in alongside a third-party-managed HER	It relies on blockchain-enabled deployment processes and healthcare managers' management of the EHR database
Scalability of Management	There are much fewer transactions that are handled in a given amount of period	The more transactions completed, the greater the degree of scalability attained
Data Transparency	A significantly greater chance of data breaches	Data tracing and monitoring ability are made simple by Blockchain, commencing with the original state
Cost of Implementation	Very simple to put into practice	Excessive regards of execution costs

5 Recent Advances

Anjum et al. [19] represented an idea for an earphone of the future dubbed "eNext." In contrast to currently available traditional earphones, eNext uses AI and IoT-driven technologies to address the risks to hearing health, ensure pedestrian safety, and provide health monitoring capabilities. Chamola et al. [20] presented an extensive overview of the current state-of-the-art in creating a reliable and understandable a AI, keeping in mind that AI is a mystery with limited knowledge of its fundamental principles. The study also covers the different TAI elements, the biases associated with them, and the tendencies that lead to system unreliability. Baker et al. [21] proposed the synergy in detail by looking at the AIoT for medical domain. In this paper, a brief overview of the integrated framework of IoT in the medical field is provided, along with information on new communication technologies, cross-layer AI, devices and sensors. Saraswat et al. [22] proposed architecture for the digital 3tomography (CT) image segmentation and classification procedure is presented in this study together with an EXAI group. There is a proposal for an EXAI system taxonomy in Healthcare 5.0, along with a presentation of operational issues. Firouzi et al. [23] represented all significant facets of IoT data monetization will be covered in this essay, with an emphasis on the health care sector. The associated difficulties, including handling data, scalability, norms, compatibility, privacy, and confidentiality, will also be covered. TAIMOOR et al. [24] presented research which conducts an extensive survey on personalized healthcare services and also provide measures in order to overcome the noted limitations of current systems and also suggest a methodology for creating personalized, resilient and reliable healthcare services. Abdel-Basset et al. [25] proposed a guided dual-channel network that combines a neural residual network for the space and time fusion of sensor data from the inertial devices, a focus mechanism and long short-term memory (LSTM). M.M. Kamruzzaman et al. [26] proposed a number of studies published in the past six years, from 2016 to 2021. The Prisma flow chart serves as a visual representation of the selecting process. This paper recognized the growing healthcare problems can be avoided by applying AI, ML, DL, Edge IoT, 6G, cloud computing, and AI. Alshehri et al. [27] presented a thorough analysis of edge-intelligent advanced medical devices based on IoT and IoMT, primarily concentrating on journal publications released during 2014 and 2020. It addressed a number of research topics in this literature survey, including edge and internet of things security, medical signal fusion, AI, and IoT and IoMT. Alladi et al. [28] proposed a two-stage, two-way identification mechanism utilizing Physical Unclonable Functions (PUFs), a type of hardware security primitive, is given in this work. Bharadwaj et al. [29] purpose of the work is to review and compile the numerous cutting-edge ML algorithm applications that have recently been combined with IOHT. A brief overview of some of the most popular machine learning algorithms has been provided, along with an analysis of their benefits, limitations, and potential for development across a range of IOHT application. AIRTON SILVA et al. [30] represented an M/M/c/K queuing network model for evaluating the performance of IOHT network along with three layers fog/edge/cloud computing continuum. LI et al. [31] proposed a reliable structure for Edge computing in an IoT enabled medical facility using SDN. The Edge servers in the suggested architecture use a simple way of authentication to confirm the identity of the IoT devices. Following authentication, these gadgets gather medical data and transmit it to the Edge server systems for processing, storing, and analysis. Table 2 contains an overview of the various papers discussed above concerning IoT and AI in the field of healthcare.

Table 2. Recent advances table

Author	Year	Description
Anjum et al. [19]	2023	Represented an idea for an earphone of the future dubbed "eNext"
Chamola et al. [20]	2023	Proposed a reliable and understandable AI
Baker et al. [21]	2023	Proposed an investigate the synergy in detail by looking at the AIoT for medical domain
Saraswat et al. [22]	2022	Proposed architecture for the digital tomography (CT) image segmentation and classification Procedure
Firouzi et al. [23]	2022	Represented all significant facets of IoT data in healthcare
TAIMOOR et al. [24]	2022	Presented research which conducts an extensive survey on personalized healthcare services
Abdel-Basset et al. [25]	2021	Proposed a guided dual-channel network that combine a neural residual network
M. M. Kamruzzaman et al. [26]	2021	Proposed the Prisma flow chart serves as a visual representation of the selecting process
Alshehri et al. [27]	2021	Presented a thorough analysis of edge-intelligent advanced medical devices based on IoT and IoMT
Alladi et al. [28]	2021	Proposed a two stage, two-way identification mechanism utilizing Physical Unclonable Functions (PUFs)
Bharadwaj et al. [29]	2021	Reviewed and compiled the numerous cutting-edge ML algorithm applications
AIRTON SILVA et al. [30]	2021	Represented an M/M/c/K queuing network model for evaluating the performance of IOHT network along with three layers fog/edge/cloud computing continuum
LI et al. [31]	2020	Proposed a reliable structure for Edge computing in an IoT enabled medical facility using SDN

6 Conclusion

The state of healthcare has evolved and is still evolving. The term "Internet of Health Thing" refers to the essential function which includes the emergence of automated 5G ambulances, intelligent ventilator, telemedicine and remote monitoring which is playing a crucial role in providing an intervention to the outdated healthcare and medical system. Technological advances in the healthcare field assist medical personnel in remotely monitoring patients as well as responding rapidly. When IoT and AI technology are coupled they're working wonders particularly in the healthcare industry by preventing the transmission of false data and identifying data forgeries. However, it is also impossible to overlook the challenges in deploying IoHT networks such as scalability concern, spectrum scarcity and mainly assaults and data confidentiality. IoT and AI integration has the ability to transform healthcare by cutting costs while enhancing performance, precision and security concerns. Based on earlier survey papers and journal articles, this survey gives a review of tremendous growth in healthcare sector. Transformation of healthcare sector over the years has also been highlighted in this paper. It is observed that if the AI requirements are established and taken into consideration from the very beginning of the design process enhanced system design is achieved. This paper concludes different applications of IOT and AI in healthcare sector and the security challenges raised by IOT which later solved by AI till some extend. Future research might look at the integration of various explainable AI models into wellness and healthcare management offerings.

References

1. Agrebi, S., Larbi, A.: Use of artificial intelligence in infectious diseases. Artif. Intell. Precis. Heal. 415–438 (2020). https://doi.org/10.1016/B978-0-12-817133-2.00018.5
2. Accenture LLP US. Artifical Intelligence (AI) in Healthcarel Accenture (2017)
3. Bahalul Haque, A.K.M., Bhushan, B., Nawar, A., Talha, K.R., Ayesha, S.J.: Attacks and countermeasures in IoT based smart healthcare applications. In: Balas, V.E., Solanki, V.K., Kumar, R. (eds.) Recent Advances in Internet of Things and Machine Learning. Intelligent Systems Reference Library, vol. 215. Springer, Cham (2022). https://doi.org/10.1007/978-3-030-90119-6_6
4. Sethi, R., Bhushan, B., Sharma, N., Kumar, R., Kaushik, I.: Applicability of industrial IoT in diversified sectors: evolution, applications and challenges. Stud. Big Data Multim. Technol. Internet Things Environ. **79**, 45–67 (2020). https://doi.org/10.1007/978-981-15-7965-3_4
5. Kalis, B., Wisdom, R.: Accenture 2017 – Internet of Health Things Survey (2017). https://www.accenture.com/t20180423T065104Z__w__/us-en/_acnmedia/PDF-42/AccentureHealth-2017-Internet-of-Health-Things-Survey.pdfla=en#zoom=50
6. Dang, L.M., Piran, M.J., Han, D., Min, K., Moon, H.: A survey on Internet of Things and cloud computing for healthcare. Electronics **8**(7), 768 (2019). [39] Chen, C., Loh, E.-W., Kuo, K., Tam, K.-W.: The times they are a-Changin'–healthcare 4.0 is coming! J. Med. Syst. **44**, 40 (2019)
7. Zhan, K.: Sports and health big data system based on 5G network and Internet of Things system. Microprocess. Microsyst. **80**, 103363 (2021)
8. Nasiri, S., Sadoughi, F., Tadayon, M.H., Dehnad, A.: Security requirements of Internet of Things-based healthcare system: a survey study. Acta Inform. Medica **27**(4), 253258 (2019). https://www.ncbi.nlm.nih.gov/pmc/articles/PMC7004290/pdf/AIM-27-253.pdf
9. Khamparia, A., Singh, P.K., Rani, P., Samanta, D., Khanna, A., Bhushan, B.: An internet of health things-driven deep learning framework for detection and classification of skin cancer using transfer learning. Trans. Emerg. Telecommun. Technol. (2020). https://doi.org/10.1002/ett.3963
10. Chen, C., Loh, E.-W., Kuo, K., Tam, K.-W.: The times they are a-Changin healthcare 4.0 is coming! J. Med. Syst. **44**, 40 (2019)
11. Wolf, B., Scholze, C.: Omedicine 4.0. Curr. Direct. Biomed. Eng. **3**, 183186 (2017)
12. Azimi, I., Rahmani, A.M., Liljeberg, P.: Internet of things for remote elderly monitoring: a study from user-centered perspective. J. Ambient Intell. Human Comput. **8**(2) (2017). https://doi.org/10.1007/s12652-016-0387-y

13. Hu, Z., Ge, Q., Li, S., Jin, L., Xiong, M.: Artificial intelligence forecasting of covid-19 in china. arXiv preprint arXiv:2002.07112 (2020)
14. Abou-Nassar, E.M., Iliyasu, A.M., El-Kafrawy, P.M., Song, O.-Y., Bashir, A.K., El-Latif, A.A.A.: DITrust chain: towards blockchain based trust models for sustainable healthcare IoT systems. IEEE Access **8**, 111223111238 (2020)
15. Newaz, A.I., Sikder, A.K., Babun, L., Uluagac, A.S.: 'HEKA: a novel intrusion detection system for attacks to personal medical devices. In: Proceedings of the IEEE Conference on Communication Network Security (CNS), pp. 1–9 (2020)
16. Lamanna, C., Byrne, L.: 'Should artificial intelligence augment medical decision making? The case for an autonomy algorithm.' AMA J. Ethics **20**(9), 902–910 (2018)
17. Hasan, K., Shetty, S., Ullah, S.: Artificial intelligence empowered cyber threat detection and protection for power utilities. In: Proceedings of the IEEE 5th International Conference on Collaboration Internet Computing (CIC), pp. 354–359 (2019)
18. Xiao, L., Wan, X., Lu, X., Zhang, Y., Wu, D.: 'IoT security techniques based on machine learning: how do IoT devices use AI to enhance security?' IEEE Signal Process. Mag. **35**(5), 41–49 (2018)
19. Anjum, M.N.: eNext: an IoT and AI driven solution to the plugged-ear pandemic. IEEE Internet Things J. **10**(13), 11940–11941 (2023). https://doi.org/10.1109/JIOT.2023.3244800
20. Chamola, V., Hassija, V., Sulthana, A.R., Ghosh, D., Dhingra, D., Sikdar, B.: A review of trustworthy and Explainable Artificial Intelligence (XAI). IEEE Access **11**, 78994–79015 (2023). https://doi.org/10.1109/ACCESS.2023.3294569
21. Baker, S., Xiang, W.: Artificial Intelligence of Things for smarter healthcare: a survey of advancements, challenges, and opportunities. In: IEEE Commun. Surv. Tutor. **25**(2), 1261–1293 (2023). https://doi.org/10.1109/COMST.2023.3256323. Second quarter
22. Saraswat, D., et al.: Explainable AI for healthcare 5.0: opportunities and challenges. IEEE Access **10**, 84486–84517 (2022). https://doi.org/10.1109/ACCESS.2022.31976711111
23. Firouzi, F., Farahani, B., Barzegari, M., Daneshmand, M.: AI-driven data monetization: the other face of data in IoT-based smart and connected health. IEEE Internet Things J. **9**(8), 5581–5599 (2022). https://doi.org/10.1109/JIOT.2020.3027971
24. Taimoor, N., Rehman, S.: Reliable and resilient AI and IoT-based personalised healthcare services: a survey. IEEE Access **10**, 535–563 (2022). https://doi.org/10.1109/ACCESS.2021.3137364
25. Abdel-Basset, M., Hawash, H., Chakrabortty, R.K., Ryan, M., Elhoseny, M., Song, H.: ST-DeepHAR: deep learning model for human activity recognition in IoHT applications. IEEE Internet Things J. **8**(6), 4969–4979 (2021). https://doi.org/10.1109/JIOT.2020.3033430
26. Kamruzzaman, M.M.: New opportunities, challenges, and applications of edge-AI for connected healthcare in smart cities. In: 2021 IEEE Globecom Workshops (GC Workshops), Madrid, pp. 1–6 (2021). https://doi.org/10.1109/GCWkshps52748.2021.9682055.
27. Alshehri, F., Muhammad, G.: A comprehensive survey of the Internet of Things (IoT) and AI-based smart healthcare. IEEE Access **9**, 3660–3678 (2021). https://doi.org/10.1109/ACCESS.2020.3047960
28. Alladi, T., Chamola, V., Naren: HARCI: a two-way authentication protocol for three entity healthcare IoT networks. IEEE J. Select. Areas Commun. **39**(2), 361–369 (2021). https://doi.org/10.1109/JSAC.2020.3020605
29. Bharadwaj, H.K., et al.: A review on the role of machine learning in enabling IoT based healthcare applications. IEEE Access **9**, 38859–38890 (2021). https://doi.org/10.1109/ACCESS.2021.3059858

30. Silva, F.A., Nguyen, T.A., Fé, I., Brito, C., Min, D., Lee, J.-W.: Performance evaluation of an internet of healthcare things for medical monitoring using M/M/c/K queuing models. IEEE Access **9**, 55271–55283 (2021). https://doi.org/10.1109/ACCESS.2021.3071508
31. Li, J., et al.: A secured framework for SDN-based edge computing in IoT-enabled healthcare system. IEEE Access **8**, 135479–135490 (2020). https://doi.org/10.1109/ACCESS.2020.3011503

Predictive Models for the Early Diagnosis and Prognosis of Knee Osteoarthritis Using Deep Learning Techniques

S. Y. Malathi[1(✉)], Geeta Bharamagoudar[1], Sharada K. Shiragudikar[2], and Shashikumar G. Totad[3]

[1] CSE Department, KLEIT, Hubballi, Karnataka, India
{malathisy,geetarb}@kleit.ac.in

[2] School of Computer Science and Engineering, KLE Technological University, Hubballi, Karnataka, India
Sharada.shiragudikar@kletech.ac.in

[3] Department of Automation and Robotics, KLE Technological University, Hubballi, Karnataka, India
totad@kletech.ac.in

Abstract. Uncovering hidden patterns and relationships in knee OA data using machine learning algorithms can be incredibly useful for predicting illness progression and developing more effective treatments. By analyzing large amounts of data, machine learning algorithms can identify patterns and relationships that might not be immediately apparent to human researchers. This can help us better understand the mechanisms behind knee OA and identify potential targets for treatment. By developing more potent treatments, we may be able to slow the progression of the disease and improve patient outcomes. The knee osteoarthritis dataset information was initially acquired from Kaggle and the information on knee osteoarthritis is processed before model training. The study aims to develop and apply advanced machine learning methods, such as deep learning techniques like CNN-SVM, Mobilenet v2, Resnet152v2, Resnet 50, and EfficientnetB5, to analyze knee OA data and develop accurate prediction models for early diagnosis and prognosis. The study evaluates the performance of these models on a dataset of individuals with knee OA to determine their efficacy and accuracy.

Keywords: CNN · EfficientNetB5 · Knee osteoarthritis · mobilenetV2 · ResNet50

1 Introduction

Knee osteoarthritis (OA) is one of the most frequent and disabling forms of the condition, which affects millions of people globally. For effective management and therapy, as well as for enhancing patient outcomes and quality of life, early diagnosis and prognosis are essential. Though it may happen in the early ages too, due to various knee injuries. However Middle Ages and old aged people are highly affective due to KOA due to

T. Shreekumar et al. (Eds.): ISCComm 2023, CCIS 2231, pp. 270–284, 2025.
https://doi.org/10.1007/978-3-031-75605-4_18

the changes into the translucent bluish-white cartilage called hyaline cartilage. The modern facilities of computing platforms, tools and methodologies provides enormous opportunities for both accurate diagnosis (**AD**) of the KOA and effective early prediction (EP) of the probability of the KOA. Both AD and EP helps medical professional to provide adequate procedure to minimize the ill effect due to the Knee osteoarthritis (KOA). The typical framework for both AD and ED using computer vision and medical imagining require three effective system that includes: i) pre-processing, ii) segmentation and classification with or without grading of the severity of the Knee osteoarthritis.

Deep learning (DL) techniques have recently demonstrated tremendous promise in the analysis and interpretation of large data sets, including clinical and medical imaging data. In order to improve early diagnosis and prognosis, this study seeks to develop and apply cutting-edge DL approaches for knee OA data analysis. A wide range of deep learning (ML) algorithms, including deep learning and ensemble learning, are investigated in this research in order to analyze a wide range of knee OA data, including clinical data, patient-reported outcomes, medical imaging data, and data collected in the field. The goal is to identify biomarkers that could be used to detect and forecast knee OA early and to develop a more specialized therapy plan based on data about individual patients. We aim to uncover hidden patterns and relationships in the data using the power of DL algorithms, which may be essential for predicting illness progression and developing potent treatments. A research study that aims to use deep learning (DL) techniques to improve the early diagnosis and prognosis of knee osteoarthritis (OA). The study will analyze a wide range of knee OA data, including clinical data, patient-reported outcomes, and medical imaging data, in order to identify biomarkers and hidden patterns that could help detect and forecast knee OA early, and develop a more specialized therapy plan based on data about individual patients. The study could significantly advance the diagnosis and treatment of knee OA, which affects millions of people globally and is currently frequently undiagnosed and untreated. Illustrations of the KL grade method for assessing the intensity of knee OA are shown in Fig. 1.

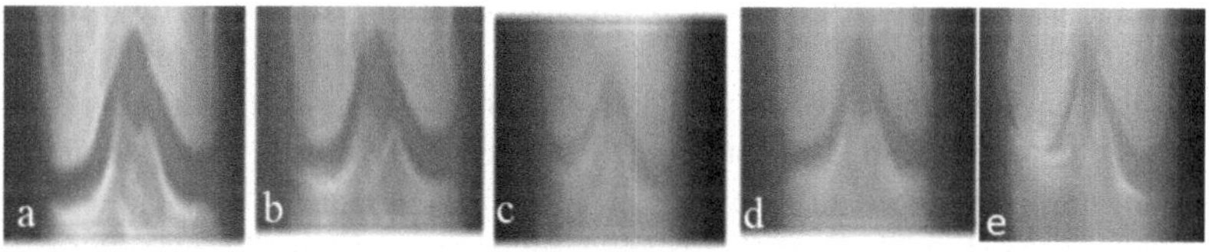

Fig. 1. Illustrations a, b, c, d, and e representing KL grades 0, 1, 2, 3, and 4, respectively, to assess knee OA intensity.

2 Related Works

Using demographic, clinical, and radiographic data, Liu et al. 2017 [1] suggested a machine learning-based method to predict the progression of knee osteoarthritis. They obtained a 76.7% accuracy using a random forest classifier. One of the early contributions to the use of machine learning algorithms for the analysis of data related to knee osteoarthritis was this research. Using MRI data, Wang et al. 2019 [2] suggested a deep learning-based method to reliably identify knee osteoarthritis. The authors obtained a precision of 91.8% by using a convolutional neural network (CNN). It demonstrated the potential of deep learning methods for the diagnosis and prediction of knee osteoarthritis. Kim et al. 2019 [3] suggested using demographic, clinical, and radiographic data to determine the severity of knee osteoarthritis. Support vector computers were utilized. With the aid of support vector machines, they attained 81.3% precision. This research showed the potential of machine learning algorithms for estimating the severity of knee osteoarthritis. Using MRI data, Li et al. 2020 [4] suggested a deep learning-based method to identify knee osteoarthritis. They used CNN and had a 96% success rate. This research showed that deep learning techniques for diagnosing knee osteoarthritis had a high degree of accuracy. The use of demographic, clinical, and radiographic data by Karimi et al. 2018 [5] to predict knee osteoarthritis was suggested as a machine learning-based method. Their success was 84% using a random forest classifier. This research showed how machine learning algorithms can be used to predict knee osteoarthritis. An artificial neural network-based strategy to forecast knee osteoarthritis was put forth by Yu et al. 2018 [6]. They obtained 83.7% accuracy using a feedforward neural network. This research showed how artificial neural networks could be used to predict knee osteoarthritis. Using demographic, clinical, and radiographic data, Li et al. 2019 [7] suggested a machine learning-based method to predict knee osteoarthritis. With a random forest classification, they were 86.3% accurate. This research showed how machine learning algorithms can be used to predict knee osteoarthritis. Hu et al. 2018 [8] examined the effectiveness of decision tree, logistic regression, and support vector machine learning algorithms for predicting knee osteoarthritis. (SVM). They came to the conclusion that SVM worked the best. This research illustrated how crucial it is to choose the right machine learning algorithm for predicting knee osteoarthritis. Kim et al. 2019 [9] suggested a deep learning-based method to categorize the degree of knee osteoarthritis using MRI data. They used CNN and had a 93.6% success rate. This research illustrated the high classification accuracy of deep learning methods for knee osteoarthritis. Using MRI data, Hu et al. 2020 [10] suggested a machine learning-based method to identify knee osteoarthritis. They used CNN and had a 94% success rate. This research showed that machine learning algorithms for diagnosing knee osteoarthritis had a high degree of accuracy. In order to analyze and forecast knee osteoarthritis using demographic, clinical, and radiographic data, Gupta et al. 2019 [11] suggested a machine learning-based approach. With the help of Bayesian learning, they obtained 80% accuracy. This research illustrated the potential of Bayesian learning for the analysis and prediction of knee osteoarthritis. In their survey of the literature on machine learning methods for diagnosing and predicting knee osteoarthritis, Chen et al. 2020 [12] looked at research on deep learning, SVM, logistic regression, and random forest. This research gave a thorough analysis of the most recent machine learning methods for analyzing knee osteoarthritis. Using MRI data, Yang et al. 2019 [13] suggested a deep learning-based method to identify knee osteoarthritis. They used CNN and had a 93.6% success rate. This research showed that deep learning techniques for diagnosing knee osteoarthritis had a high degree of accuracy.

A machine learning-based strategy was put forth by Zhang et al. 2020 [14] to forecast the development of knee osteoarthritis using radiomics and demographic information. Their success was 78% using a random forest classifier. This research showed how radiomics could be used to predict the progression of knee osteoarthritis. The authors proposed a comparative modeling approach based on patient data and plain x-ray images to predict the severity of knee osteoarthritis. They used several machine learning algorithms, including decision trees, random forests, and support vector regression, to predict the Knee Injury and Osteoarthritis Outcome Score (KOOS) of patients. The results showed that machine learning models based on patient data and plain x-ray images can accurately predict the severity of knee osteoarthritis [15]. This review article provides an overview of the current state of evidence on the epidemiology of osteoarthritis. The authors discuss the prevalence, incidence, and risk factors of osteoarthritis, as well as its impact on individuals and healthcare systems. They also highlight the need for further research to better understand the burden of osteoarthritis and develop effective prevention and management strategies. Knee osteoarthritis [16]. Identified other potential prognostic factors, such as gender, pain, and muscle strength, but noted that the evidence was less clear for these factors. The authors concluded that a better understanding of the prognostic factors for knee osteoarthritis could aid in the development of more effective treatment strategies for this common and debilitating condition [17].

3 Material and Methods

Initially the knee osteoarthritis dataset information was acquired from Kaggle and the information on knee osteoarthritis needs to be processed before model training. Which involve processes like resizing the pictures, pixel value normalization, and dividing the dataset into training, validation, and test sets.

Transfer Learning: The pre-trained model EfficientNet B5 has attained cutting-edge success on a number of image classification tasks. The model on our dataset for knee osteoarthritis can be improved by using transfer learning to take advantage of this. In order to achieve this, we must freeze the model's previously trained layers and train only the new classification layers that we put on top of the model. The suggested model can be made up of a pre-trained Efficient Net B5 design as a foundation, followed by a few dense classification layers. Experimentation can be used to optimize both the number of thick layers and the number of neurons in each layer. For training and assessing the model, it is crucial to select the loss function and assessment metrics. Here, the assessment metric is accuracy, and the loss function is categorical cross-entropy. Next we move To fine-tune the model's hyperparameters, such as the learning rate, batch size, and number of epochs, train the model on the training set and verify it on the validation set. Next create confusion matrix and calculate the performance metrics like accuracy, precision, recall and F1-score.To identify the model's performance on data that has not yet been seen, evaluate it on the test set. An overview of the proposed work adopted is shown in Fig. 2.

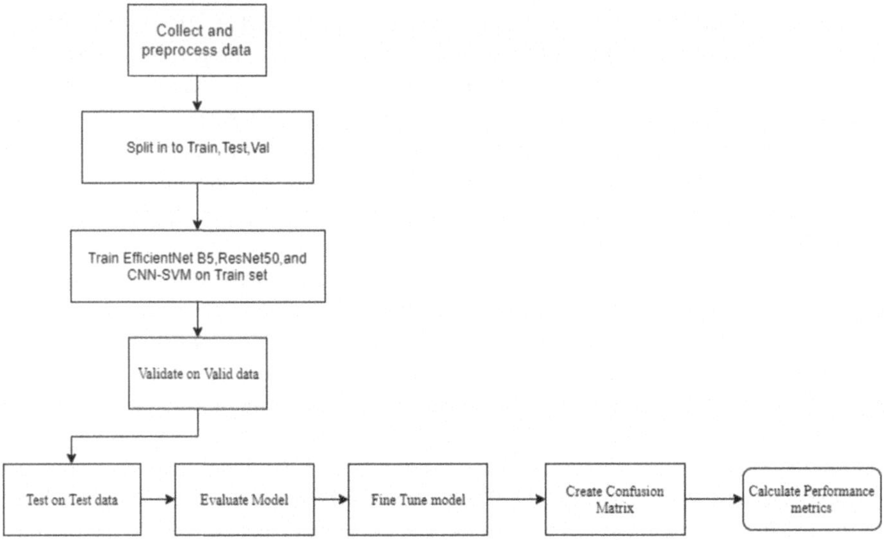

Fig. 2. Proposed method to assess knee OA intensity and classification using CNN

3.1 Data Description

This dataset contains knee X-ray data for both knee joint detection and knee KL grading. The dataset is organized from OAI [18]. Articular cartilage, a supple, slippery substance that usually protects bones from joint friction and impact, deteriorates in osteoarthritic knees. Additionally to having an effect on the soft tissues close by, the condition can alter the bone that rests beneath the cartilage. Knee osteoarthritis, which is commonly referred to as just knee arthritis, is by far the most common type of arthritis to cause knee discomfort. Knee pain is also a frequent symptom of reactive arthritis, rheumatoid arthritis, and numerous other uncommon forms of arthritis.

3.2 Data Visualization

Understanding the multiple grades or labels used to categorize knee images in a dataset and input image of 5 different classes has been rendered simpler by Table 1. By listing the number of images included in each collection of data, Fig. 3(a-b) gives details about the size of the dataset and the resources required for creating and testing a machine learning model.

Figure 3(a-b) shows the number of images in the train, test, and validation sets as well as the number of images in each class.

Figure 4 illustrates the distribution of data across different classes and sets to give a clear picture of how the data is split for training, testing, and validation purposes.

3.3 EfficientNetB5

The EfficientNet family of models is the basis for EfficientNetB5, a deep convolutional neural network architecture that is intended to achieve high accuracy while being more

Table 1. Data Visualization of Knee osteoarthritis dataset

Grades	Label	Description
Grade 0	Healthy	Healthy Knee image
Grade 1	Doubtful	Joint constriction is doubtful, and osteophytic lipping may be present
Grade 2	Minimal	Osteophytes are definitely present, and potential joint space narrowing
Grade 3	Moderate	Numerous osteophytes, obvious narrowing of the joint area, and mild sclerosis
Grade 4	Severe	Significant joint narrowing, large osteophytes, and serious sclerosis

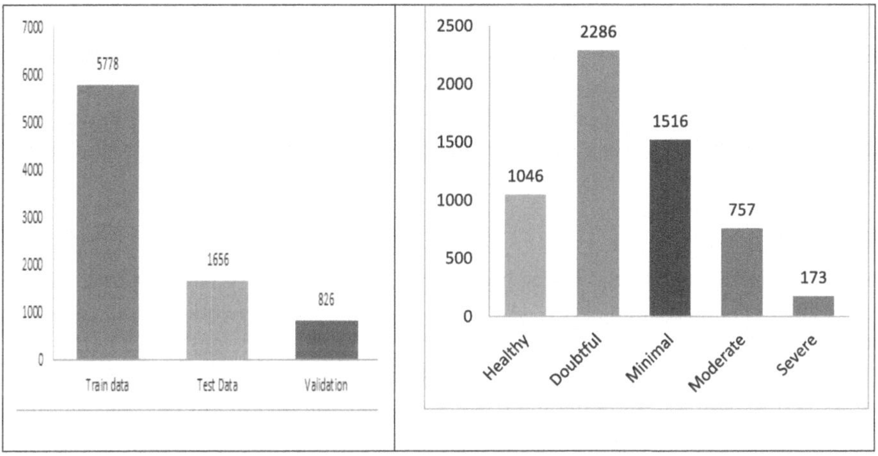

Fig. 3. a. No of images in the Train, Test and Validation set b. Image count in each Class.

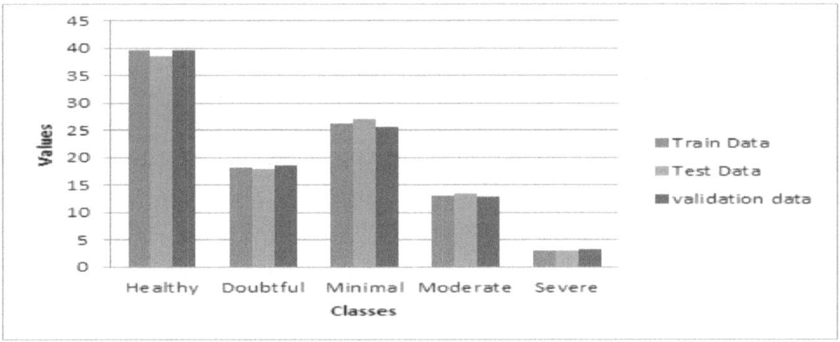

Fig. 4. Distribution of Data in 5 classes.

computationally efficient than other architectures of a comparable size. With a total of 30 million parameters, EfficientNetB5 is the fifth largest model in the EfficientNetseries.The Mobile Inverted Residual component with Bottleneck Connections (MBConv), a variant of the Inverted Residual Block used in MobileNetV2, serves as the fundamental building component of EfficientNetB5. Three major parts make up the MBConv block is shown in Table 2. The EfficientNetB5 architecture's primary building block, the MBConv block, can be 170 divided into three main components. The Depthwise Separable Convolution procedure is 171 the first step. To do this, first execute a pointwise convolution, then a depthwise convolution. 172 The pointwise convolution uses a 1x1 convolution to combine the output of the depthwise 173 convolution, which performs distinct convolution operations for each channel of the input. 174 This can be expressed mathematically as

$$Y = PW(DW(x)) \tag{1}$$

where x is the input, DW is the depthwise convolution operation, PW is the pointwise convolution operation, and y is the output.This method preserves incoming spatial information while requiring fewer parameters and computational resources than a typical convolution. The Linear Bottleneck is the second component. In this case, the number of channels is first compressed with a 1x1 convolution, then extended with a ReLU activation function, and finally compressed once more with another 1x1 convolution. The output of the linear bottleneck is combined with the input of the MBConv block using a skip connection.

$$y = F(x) + x \tag{2}$$

where x is the input, F is the linear bottleneck operation, and y is the output, can be used to mathematically depict this. As a result, even as the network's depth increases noticeably, it can continue to add helpful features. Finally, EfficientNetB5 is made up of more and bigger MBConv blocks as the network gets deeper. The network also uses global average pooling, dropout, and batch normalization in order to increase accuracy as well as speed. A comprehensive description of EfficientNetB5's architecture can be found in Table 2. With the number of classes denoted in parenthesis.

Table 2. The architecture of EfficientNetB5

Layer Type	Output Size	No of Filters	Contd		
Input	$456 \times 456 \times 3$	-	MBConv6	$57 \times 57 \times 576$	576
Convolution	$228 \times 228 \times 48$	48	MBConv6	$29 \times 29 \times 960$	960
Batch Norm	$228 \times 228 \times 48$	-	MBConv6	$29 \times 29 \times 960$	960
Swish	$228 \times 228 \times 48$	-	Convolution	$29 \times 29 \times 160$	160
MBConv6	$114 \times 114 \times 48$	48	Batch Norm	$29 \times 29 \times 160$	-
MBConv6	$114 \times 114 \times 88$	88	Swish	$29 \times 29 \times 160$	-
MBConv6	$114 \times 114 \times 144$	144	Global Pooling	$1 \times 1 \times 960$	-
MBConv6	$57 \times 57 \times 240$	240	Dropout	$1 \times 1 \times 960$	-
MBConv6	$57 \times 57 \times 384$	384	Dense	1000	-
MBConv6	$57 \times 57 \times 384$	384	Output	1000	-

3.3.1 Depthwise Separable Convolution

This operation performs a depthwise convolution followed by a pointwise convolution and is the first operation in an MBConv block. In comparison to a standard convolution, this process requires fewer parameters and computational resources while maintaining input spatial information.

3.3.2 Linear Bottleneck

After the result of the depthwise separable convolution, the number of channels is compressed using a 1x1 convolution, expanded using a ReLU activation function, and then compressed again using a 1x1 convolution. Skip connection is used to combine the input to the MBConv block with the output of the linear bottleneck, ensuring that the network can continue to acquire useful features even as its depth grows significantly.As the network grows deeper, EfficientNetB5 is composed of MBConv blocks that get bigger and more numerous. To further boost accuracy and speed, the network also has a number of additional features like global average pooling, dropout, and batch normalisation.

4 Results and Discussion

We compare the accuracy of each model by evaluating various performance parameters such as accuracy, precision, recall, and F1 score from the transfer learning with EfficientNetB5 with 5 classes and 3 classes. Figure 5 details the accuracy of various neural network designs on various classification tasks.

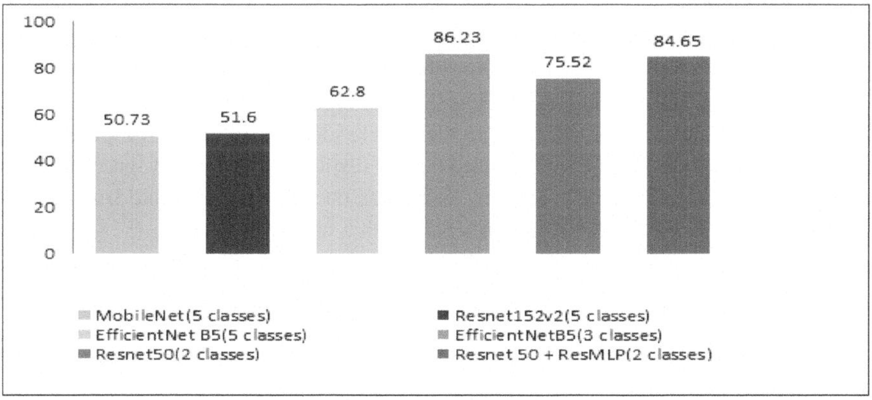

Fig. 5. Accuracy of different neural network architectures on various classification tasks.

4.1 Performance Evaluation for EfficientNetB5 with 5 Classes

It is crucial to keep the number of image examples for each class to a minimum during training to prevent class imbalance. To achieve this, the train data frame can be trimmed to have a limit of 500 images for each class. After trimming the data frame, we can balance the dataset using data augmentation methods to make sure that each class has exactly 500 images. After balancing the dataset, we can continue to create a model using transfer learning with EfficientNetB5, a powerful pre-trained model for image classification tasks. Here the base model trainable from the outset, rather than freezing its weights during the initial training. This approach can lead to faster convergence and lower validation loss for the same number of total epochs, as the base model can start adapting to the new task right away. By allowing the base model to update its weights during the initial training, the model can better incorporate information from the new task and avoid getting stuck in local minima. Therefore, making the base model trainable from the outset can be a useful strategy for improving the training of deep neural networks.

Better efficiency and shorter training periods may result from this. We can design a unique Keras callback to enhance the model's efficiency during training. This callback can be used to decide whether to keep training going or stop it depending on certain factors, such as the precision or validation loss. The callback can also be used to set other training parameters, such as the learning rate, to help us get improved results. Once the model has been trained, we can define a function that takes an integer amount of test steps and a test generator. This function can yield predictions on the test set, a confusion matrix, and a classification report, all of which can be used to assess the model's effectiveness and pinpoint areas for development. With an accuracy of 0.959 and an AUC score of 0.98, we can see that the algorithm classified the "healthy" class with good accuracy. Similar to this, the algorithm classified the "doubtful" class with an accuracy of 0.822 and an AUC score of 0.935, performing reasonably well. The "minimal" and "moderate" classes, however, presented difficulties for the algorithm, resulting in reduced accuracy and AUC scores. In particular, the algorithm had trouble distinguishing the "minimal" class with a low sensitivity of 0.66 and the "moderate" class with a low specificity of 0.333. Notably, there were no true positive or true negative predictions for the "severe" class, proving that the algorithm was unable to accurately classify any cases in this category. To improve the algorithm's overall performance in the classification job, it must continue to accurately identify all classes, especially the "minimal" and "moderate" classes. Figure 6:(a-c) displays the accuracy, precision, and recall for EfficientNetB5 with 5 classes, as well as the false positive and true positive rates and the TN,TP,FP,FN for EfficientNetB5 with 5 classes.

4.2 Performance Evaluation for EfficientNetB5 with 3classes

To further improve the accuracy of our classification algorithm, we can modify the class labels. In particular, while the "moderate" and "severe" classes stay distinct, the "healthy doubtful" and "minimal" classes can be combined into a single "healthy" class. This change may make the model's conclusions more precise and understandable. By reducing the number of groups and streamlining the classification process, we can make the model more approachable and straightforward to understand. TN, TP, FP, FN for EfficientNetB5 with 3 classes and Precision, Recall, and F1 Score for EfficientNetB5 with 3 classes are shown in Fig. 7(a-b).

Figure 8 shows a comparison of the datasets for the 5 and 3 groups in terms of accuracy.

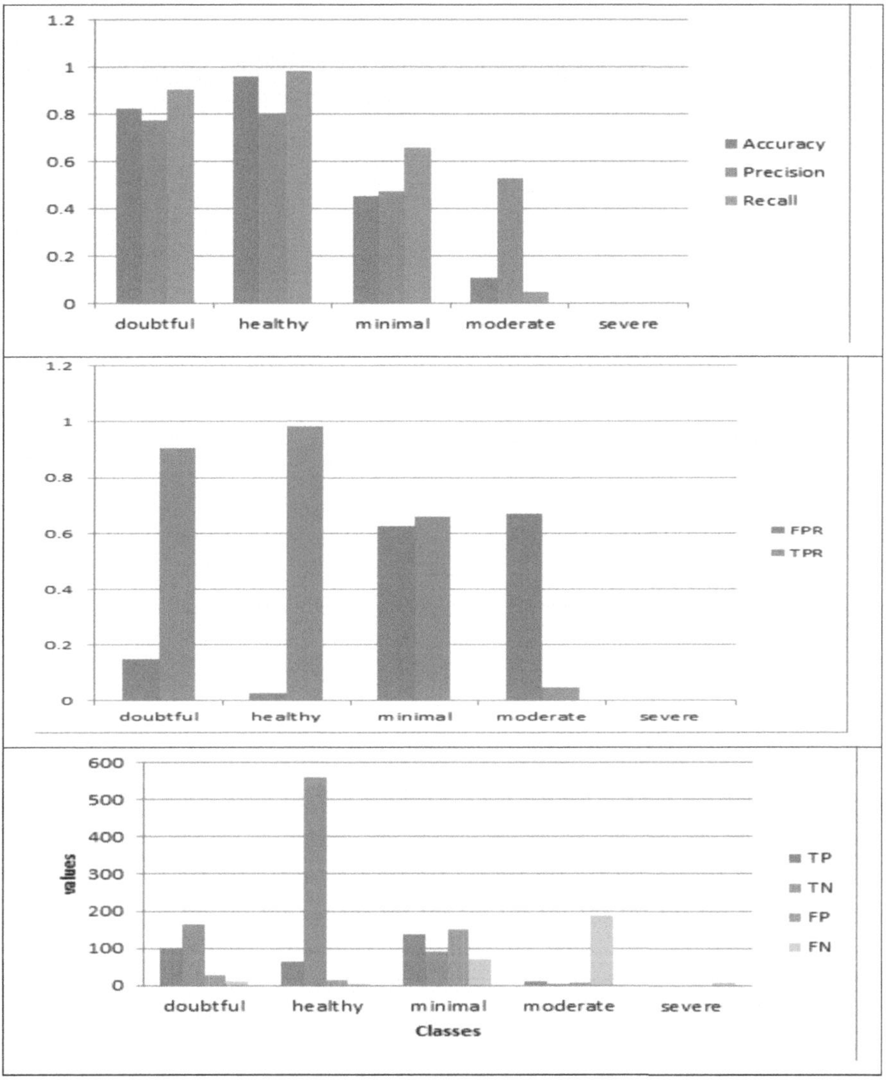

Fig. 6. (a)Accuracy,precision and Recall for EfficientNetB5 with 5 classes (b) False positive rate and True positive rate for EfficientNetB5 with 5 classes(c) TN,TP,FP,FN for EfficientNetB5 with 5 classes

Table 3 shows the performance comparison between multi class classification of knee osteoarthritis prediction using efficientNetB5. This classification report illustrates how a model performed on a collection of data. Five distinct classes are Doubtful, Healthy, Minimal, Moderate and Severe have been forecasted by the model. With a precision of 0.3168, recall of 0.3243, and F1-score of 0.3205, the model did badly on the Doubtful class. With an accuracy of 0.7475, a recall of 0.3378 and an F1-score of 0.4653, the Minimal class also performed poorly. The model's accuracy is 0.6280, meaning that

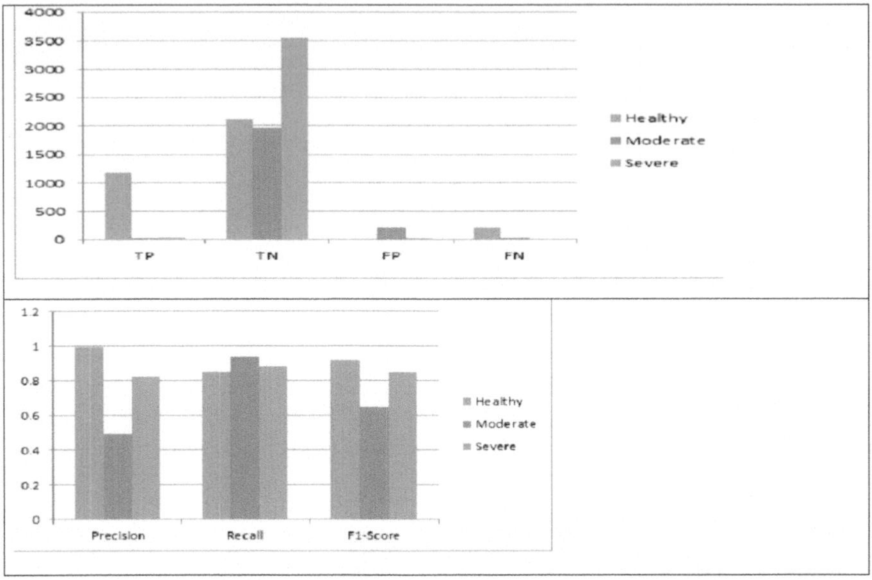

Fig. 7. a)TN,TP,FP,FN for EfficientNetB5 with 3 classes b) Precision,Recall and F1 Score for EfficientNetB5 with 3 classes.

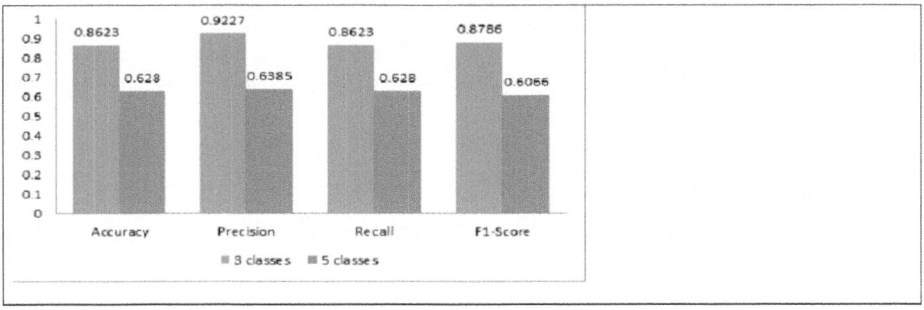

Fig. 8. Comparison of accuracy between the datasets for the 5 and 3 classes.

it accurately identified the class label in 62.8% of the instances. This appears to be a multi-class classification model's evaluation criteria. On a dataset of 1656 instances, broken down into three classes are Healthy, Moderate, and Severe. The percentage of instances with properly identified knee disease affected over all instances is known as Accuracy. Metrics such as accuracy, precision, recall, and f1-score are used to assess the success of the model. This dataset has fewer classes than the 5-class classification, but the algorithm still produced an accuracy score of 0.8623. The "Healthy" class had the greatest precision (0.9958), while the "Moderate" class had the highest recall (0.9327). The'Severe' class had the best f1-score (0.8491). An accuracy of 99.93% on the training set and an accuracy of 89.23% on the validation set were attained by the model after 75 training iterations using a dataset for a classification task. According to the dataset, the model's 260 total accuracy is 86.23%. This indicates that the model is operating effectively on the dataset 261 and may be applied to forecast new data.

Table 3. Performance Comparison between Multiclass classification.

Metrics	5 Classes		3 Classes	
Accuracy		0.628		0.862
Precision	Doubtful	0.31	Healthy	0.99
	Healthy	0.68	Moderate	0.49
	minimal	0.74	Severe	0.81
	Moderate	0.68		
	Severe	0.75		
Recall	Doubtful	0.32	Healthy	0.85
	Healthy	0.877	Moderate	0.93
	minimal	0.337	Severe	0.88
	Moderate	0.84		
	Severe	0.86		
F1-Score	Doubtful	0.32	Healthy	0.91
	Healthy	0.77	Moderate	0.64
	minimal	0.46	Severe	0.84
	Moderate	0.75		

Various Knee picture cuts are combined with the results of the fixed split, as seen in the images above, to produce the graphical elements. The district growth process, the suggested strategy, the Snake process, the Level Set computing process, and the traditional fix-based divisional tactic are all represented by these components, which represent the division outcomes for various techniques. Each technique consistently demonstrates a high degree of precision and efficiency over time when used frequently. In Fig. 9 the training and validation curve is shown for 5 classes. In Fig. 10 the training and validation curve is shown for 3 classes.

The sensitive rate, accuracy, and efficiency of the KL classifier are demonstrated by the confusion matrix in Fig. 11.

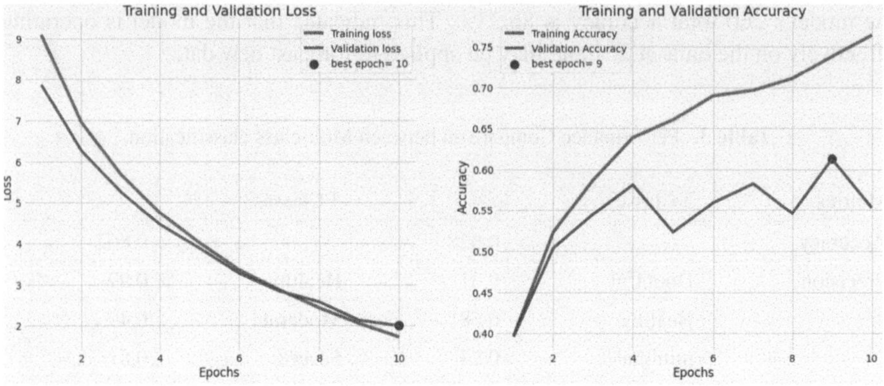

Fig. 9. Training and validation loss,Training and validation Accuracy for Efficient netB5(5 classes) for Knee osteoarthritis dataset.

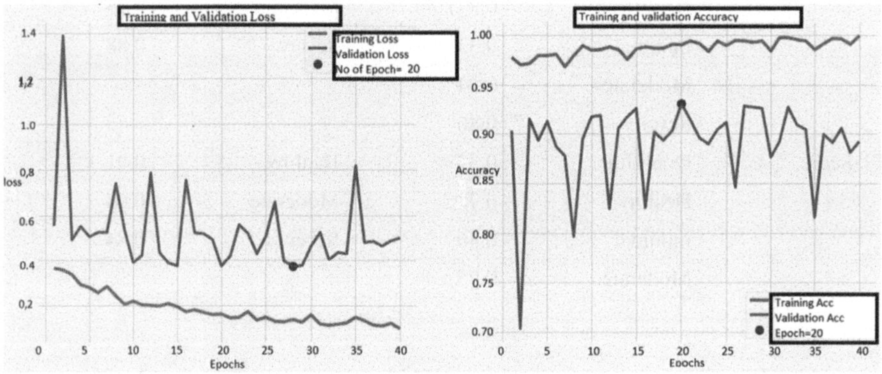

Fig. 10. Training and validation loss, Training and validation Accuracy for Efficient netB5(3 classes) for Knee osteoarthritis dataset.

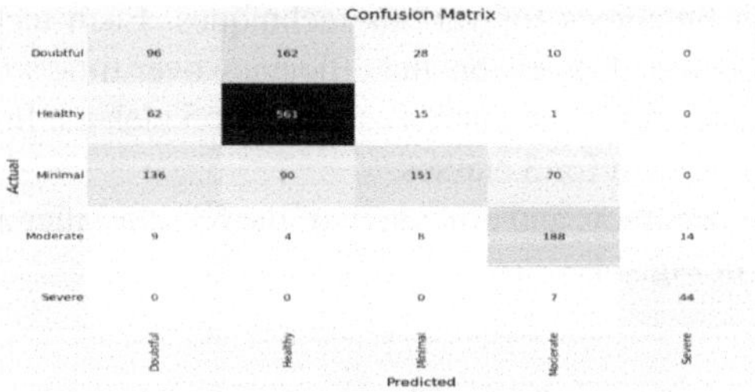

Fig. 11. Confusion matrix

5 Conclusion

In the study on the development and application of advanced machine learning approaches for the analysis of knee osteoarthritis data, the EfficientNet B5 architecture outperformed other architectures, achieving a high accuracy of 86.23% for 3 classes and 62% for 5 classes in the early diagnosis and prognosis of multiclass classification using the base model trainable from the outset, rather than freezing its weights during the initial training. For the same total number of epochs, this method can result in faster convergence and reduced validation loss. According to this research, EfficientNet B5 is a very promising tool for analysing data related to knee osteoarthritis and has the potential greatly increase the precision of disease detection and prognosis. However, more research and validation are required to validate the efficacy of this strategy in a larger and more diverse community. Nevertheless, this research represents a major step forward in the study of knee osteoarthritis and paves the way for further investigation and the creation of machine learning-based methods for the early detection and prognosis of this condition. However, as the researchers noted, more studies are needed to validate the efficacy of this approach in a more diverse population. Additionally, the use of data from multiple sources can provide more robustness to the model and improve its generalization performance.

Acknowledgment. We gratefully acknowledge the assistance and cooperation extended by Dr. Madhuchandra R, Associate professor, department of Orthopedics, Karnataka Institute of Medical Science(KIMS), Hubballi.

Declarations.

AuthorContribution

MSY: Data Creation, Visualization, Software, Validation, Writing-original draft preparation; GRB: Data collection, Methodology; SKS: Conceptualization, Writing-review and editing.

Ethical Responsibilities of Authors

All authors have read, understood, and have complied as applicable with the statement on "Ethical responsibilities of authors" as found in the Instructions for Authors.

Funding

There was no specific grant from a funding agency in the public, commercial, or nonprofit sectors for this research.

Conflict of Interest

The authors declare that they have no conflict of interest.

Data Availability

The dataset will be made available on request.

References

1. Liu, J., Zhang, C., Xu, Y., Wang, X.: Predicting knee osteoarthritis progression using machine learning techniques. J. Med. Imag. Health Inf. **7**, 592–598 (2017)
2. Wang, S., Zhao, Y., Zhou, W.: Automatic detection of knee osteoarthritis using deep learning techniques. J. Healthc. Eng. **1**, 1–10 (2019)
3. Kim, S., Lee, J., Lee, B.: Predicting knee osteoarthritis severity using machine learning models. J. Med. Syst. **43**, 1–9 (2019)
4. Li, Y., Li, M., Li, X.: Deep learning-based diagnosis of knee osteoarthritis using magnetic resonance ımaging. J. Magn. Reson. Imag. **52**, 1415–1421 (2020)
5. Karimi, A., Saeedi, S., Armin, S.: Towards predictive modeling of knee osteoarthritis using machine learning. J. Biomed. Inf. **77**, 63–76 (2018)
6. Yu, H., Guo, L., Li, T.: Knee osteoarthritis prediction using artificial neural networks. Chin. J. Med. Imag. Technol. **34**, 1189–1192 (2018)
7. Li, X., Liang, L., Li, Y.: Predicting knee osteoarthritis using random forest. J. Med. Syst. **43**, 1–8 (2019)
8. Hu, S., Wang, X., Liu, J.: A comparative study of machine learning algorithms for knee osteoarthritis prediction. Int. J. Comput. Sci. Netw. Sec. **18**(9), 66–70 (2018)
9. Kim, J., Lee, J., Lee, K.: Classification of knee osteoarthritis severity using convolutional neural networks. Int. J. Imag. Syst. Technol. **29**, 302–310 (2019)
10. Hu, Y., Wei, D., Zong, X.: Machine learning-based knee osteoarthritis diagnosis using MRI data. J. Healthc. Eng. **2020**, 1–8 (2020)
11. Gupta, S., Mathur, M., Singh, R.: Automated analysis and prediction of knee osteoarthritis using machine learning techniques. J. Med. Syst. **43**, 1–10 (2019)
12. Chen, L., Zhang, H., Wang, S.: A review on machine learning techniques for knee osteoarthritis diagnosis and prediction. J. Healthc. Eng. **2020**, 1–12 (2020)
13. Yang, Y., Chang, C., Lin, Y.: Deep learning for automated knee osteoarthritis diagnosis using MRI data. J. Magn. Reson. Imag. **49**, 1057–1064 (2019)
14. Zhang, Y., Wang, Z., Wang, H.: Predicting knee osteoarthritis progression using machine learning and radiomics. J. Med. Imag. Health Inf. **10**, 530–537 (2020)
15. Abedin, J., et al.: Predicting knee osteoarthritis severity: comparative modeling based on patient's data and plain X-ray images. Sci. Rep. **9**, 5761 (2019)
16. Allen, K.D., Golightly, Y.M.: Epidemiology of osteoarthritis: state of the evidence. Curr. Opin. Rheumatol. **27**, 276 (2015)
17. Belo, J., Berger, M., Reijman, M., Koes, B., Bierma-Zeinstra, S.: Prognostic factors of progression of osteoarthritis of the knee: a systematic review of observational studies. Arthritis Care Res. Official J. Am. Coll. Rheumatol. **57**(1), 13–26 (2007)
18. Chen, P.: Knee Osteoarthritis Severity Grading Dataset. In: Mendeley Data, vol. V1 (2018). https://doi.org/10.17632/56rmx5bjcr.1

Predicting Salinity Resistance of Rice at the Seedling Stage: An Evaluation of Transfer Learning Methods

Sharada K. Shiragudikar[1(✉)], Geeta Bharamagoudar[2], Manohara K. K.[3], Malathi S. Y.[2], and Shashikumar G.Totad[4]

[1] School of Computer Science and Engineering, KLE Technological University, Hubballi, Karnataka, India
sharada.shiragudikar@kletech.ac.in

[2] CSE Department, KLEIT, Hubballi, Karnataka, India
{geetarb,malathisy}@kleit.ac.in

[3] Central Coastal Agricultural Research Institute, Old Goa, India
manohar.gpb@gmail.com

[4] Department of Automation and Robotics, KLE Technological University, Hubballi, India
totad@kletech.ac.in

Abstract. Rice has become one of the largest and most widely cultivated food crops in the world. Nevertheless, there is a significant probability that high amounts of salt, especially at the stage of seedling development, would have a detrimental effect on rice output. As a result, it is critical to quickly discover and develop salinity-tolerant rice crop types, particularly at the stage of seedling development, to prevent a decrease in rice output. Expertise from humans is required for the categorization of visual signals and the conventional way of a standard assessment system for identifying rice crop salinity stress. It's time demanding and prone to mistakes, both of which may lead to inaccurate classification. Recent developments in deep learning and transfer training have opened the road for the automated categorization of a wide range of agricultural jobs, from the detection of plant diseases and pests to the identification of weeds and invasive species. The lack of plant-domain-specific pre-trained models and the limited number of available datasets remain significant obstacles to agricultural automation. We use the AgriNet dataset, a database of 850 image graphs of farms and agricultural settings. The work shows the need for a deep learning developed model over the conventional way of measuring rice crop sensitivity to salt stress throughout the seedling stage to detect and categorize stress caused by salinity in the seedling stage of rice using field images. Thus, we use one of the most advanced deep learning approaches, VGG 16, VGG 19, Resnet 50, Resnet 100, exception, Resnet 50V2, Inception V3, MobileNetV2, DenseNet121, and ResNet101V2 which was built in Python's Jupyter Notebook, to construct the classification model. To further demonstrate the need for a computer-based classification system for salinity estimation that could be used as a tool for automating the classification process for rice development to aid scientists and farmers in the rice crop management system, the model can classify rice seedling images in accordance to the scores as Grade1, Grade3, Grade5, Grade7, and Grade9.

T. Shreekumar et al. (Eds.): ISCComm 2023, CCIS 2231, pp. 285–308, 2025.
https://doi.org/10.1007/978-3-031-75605-4_19

Keywords: rice crop salinity stress · VGG 16 · VGG 19 · Resnet 50 · Resnet 100 · exception · Resnet 50V2 · Inception V3 · MobileNetV2 · DenseNet121 · and ResNet101V2 · classification

1 Introduction

Rice is a major crop that provides food for the population. To keep the world's population from starving and to ensure continued food security, rice output must be expanded. The production capacity of modern rice varieties is being hampered by a variety of abiotic stressors caused by the rapidly evolving environment. These challenges include lengthy dry periods, repetitive foods, exposure to ocean water, etc. After dry weather, saltiness is the subsequent most lethal abiotic hazard to rice production, impacting almost 1 billion acres of land throughout the globe [1]. This accounts for more than 6% of all farmable land [2] and over 20% of all irrigated land [3].

Rice is especially susceptible to salt stress throughout the seeding and early vegetative stages [4], as well as later during the reproductive stages. When root absorption capacity is reduced due to osmotic stress. Excessive soil salinity has a detrimental effect on rice plant development and output [5]. High salinity stress factors like those seen in Fig. 1 have a major effect on agricultural productivity worldwide, often reducing it by more than half.

Productivity is reduced throughout various phases of development because of salinity stress. Therefore, better genotype selection will aid in assuring the increased rice output. Since the region of saline-contaminated soil is so large, it is crucial to employ even simple methods of genotype selection that reliably raise crop yields for the sake of food safety. Phenotyping populations of breeding and germplasms precisely and accurately is a necessary step toward this goal.

Classifying stress levels by traditional phenotyping methods requires a lot of time, and effort, and may even be dangerous. However, rice seedling stress may be diagnosed and categorized in different periods utilizing.

With a projected global population of over 9 billion in 2050, food production would need to rise by 70%. Artificial intelligence (AI) in agriculture is necessary because of resource constraints and climate change [6]. The applications of artificial intelligence in agriculture are many, ranging from the diagnosis of plant illnesses to the robotic eradication of weeds and pests to the identification of new medicinal plants. In addition to posing a threat to food security, plant diseases have severe impacts on smallholder farmers, who may lose as much as half of their crop to pests and weeds [7]. Free diagnostic services through mobile apps may help smallholder farmers via automated detection of weeds, pests, and plant diseases. Robotic weed killers and sensor-based crop monitoring are two further examples of automation used in areas with a scarcity of trained farmers. Automatic plant species identification is a crucial detection problem that has applications in areas such as medicinal herb research and the protection of yet-to-be-discovered plant species.

Several simple discriminative models, including linear classifiers and support vector machines (SVM), have traditionally been used to accomplish the recognition task. Deep learning (DL) has been extensively employed in agricultural research for plant categorization tasks after dominating all computer vision challenges). Due to their nature as data-hungry neural networks, DL models need extremely huge datasets to provide accurate results. Since convolutional models go from low-level to high-level features, this presents a significant difficulty since it might result in a negative transfer. To better segment soybeans in vitro, for instance, they suggested a transfer learning system based on synthetic images. Despite the large amount of available data for soybean applications, the suggested framework improved accuracy by 8%. The data samples of Grade 1, 3, 5, 7, 9 leaf images are shown below Fig. 1.

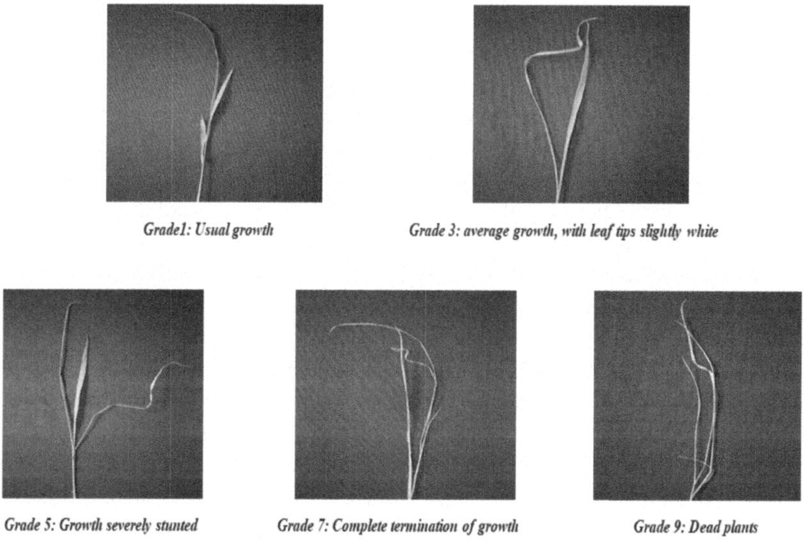

Fig. 1. Sample Grades 1,3,5,7,9 Dataset

2 Related Works

Soil salinity in coastal places is a significant hindrance to soil [2], even though there is a long history of coastal mudflat rehabilitation and usage [1] and coastal mudflats represent a crucial reserve resource for agricultural production. When compared to other harmful compounds [4, 5], salinity is by far the most significant abiotic issue that lowers the growth rate, production, and quality of all crops [3, 5]. Compared to when the soil is inundated, the salt concentration is lower when the surface soil is dry. Waterlogging is a need for rice cultivation [6], and rice is the first crop to be planted on coastal saline soils following salt leaching and enhanced tidal flat soil [7]. The effects of salinity on rice production may be mitigated by the development and testing of salt-tolerant rice cultivars, which also helps advance global food security.

The growth and production of rice, a glycophytic plant, may be negatively impacted by salt stress on the stage of the crop, the degree, and duration of the stress, and the rice variety's tolerance [8]. Various germplasms and various rice growing seasons within the same variety have variable salt tolerances. Rice is particularly vulnerable to salt during the seedling stage [9]. Growing salt-tolerant rice in salty environments is an effective strategy for reducing rice waste [10, 11]. Plants' salinity resistance can be evaluated in several ways [12]. It took more than a month [13] to test 500 rice cultivars and identify the 34 that could survive in a 0.3% salt chloride solution while they were still in the seedling stage. It took Sun et al. about six months to perform an experiment in which they irrigated 550 rice cultivars with salt water for both vegetative and reproductive development [14]. Salt tolerance during germination was measured in 549 rice germplasm resources [15] over seven days. Rice takes around half a year to produce; growing more than one kind of rice is time-consuming and labour-intensive regardless of how many different types of rice are being grown [16, 17]. However, screening trials for salt tolerance during rice germination only take seven to ten days [18, 19]. The salt content of the water is uniform throughout. Therefore, evaluating the plant's reaction to salt stress via measuring germination rates is a valid approach. Roots are unable to take in as much water during the germination process because of the strong osmotic pressure in solution caused by the comparatively high salt levels in the germination medium [20]. The germination rate of plants is slowed and decreased by high salt. Reduced germination, shorter roots, shorter coleoptiles, and shorter seedlings are all effects of high salinity [21]. The results of this screening for salt resistance during germination are encouraging for further research into salt-resistant rice varieties [22]. This study aimed to identify highly salt-tolerant and salt-sensitive varieties among 114 rice germplasms by evaluating their response to non-saline and saline treatments during the germination process.

The rice (Oryza spp.) crop is significant since it is mostly consumed by humans. Found that 100 grams of rice have 345.0 calories, 78.2 grams of carbs, 6.8 grams of protein, and a good deal of the zinc and niacin your body needs. Rice protein is the most bioavailable because it is so easily digested (88%). For the poorest members of society, rice accounts for between half and eighty percent of their daily caloric intake.

To this end, it is crucial to prioritize screening and assessing the potential of salt tolerance in rice genotypes during the seedling stage, when a significant amount of time may be saved. Therefore, one viable approach to enhancing the productivity of salt-affected soils is the discovery and cultivation of salinity-tolerant genotypes and cultivars. Traditional approaches to this problem are too laborious and clumsy to manage a huge population of genotypes, thus new approaches are needed. Screening by a computer might be one of the time-efficient, trustworthy options here. The purpose of this research was to provide non-destructive, high-throughput phenotyping strategies for monitoring rice salinity stress.

Rice is a major crop that provides food for the population. To keep the world's population from starving and to ensure continued food security, rice output must be expanded. The production capacity of modern rice varieties is being hampered by a variety of abiotic stressors caused by the rapidly evolving environment. These challenges include lengthy dry periods, repetitive foods, exposure to ocean water, etc. About 1 billion acres of land throughout the globe are affected by saltiness, making it the second most deadly demand in rice cultivation after the dry season [4]. This accounts for more than 6% of all farmable land [4] and over 20% of all irrigated land [10].

Rice is especially susceptible to salt stress during its early vegetative and reproductive stages [15]. When root absorption capacity is reduced due to osmotic stress. Excessive soil salinity hinders the development and production of rice plants [11]. Rice is a unique and widely-cultivated plant because it thrives in salty environments. Soil salinity over 4DeciSiemens per meter is considered moderate salinity by IRRI [7], while salinity beyond 8DeciSiemens per meter is considered excessive over coastal salinity [1] in India, particularly for rice. Too much salt causes oxidative stress and ionic toxic effects on rice plants. When rice plants are subjected to extreme salt stress, they may exhibit a wide variety of morphological, metabolic, and molecular abnormalities and symptoms, including death.

Productivity is reduced throughout various phases of development because of salinity stress. Enhanced genotype selection will undoubtedly play a role in assuring increased rice productivity. Since the region of saline-contaminated soil is so large, it is crucial to employ even simple methods of genotype selection that reliably raise crop yields for the sake of food safety. Phenotyping breeding populations and germplasms precisely and accurately is a necessary step toward this goal.

Computer-based phenotyping is non-invasive, quick, and easier on the workers than traditional phenotyping procedure discussed in one of the previous works [38] for stress classification. For this reason, it is crucial to employ a computer-based decision system for more precise phenotyping prediction, which requires the identification of symptoms on the leaves of the rice crop while it is still in the seedling stage. Some Agriculturists and scientists have used this method to make more accurate such as PCA based [39] and timely forecasts about paddy crop stress. One of the quick, recursive, and trustworthy alternatives to this work might be deep learning, which would assist us to save time and avoid the need for human intervention. Injured leave and healthy leave are used in this study as training and testing data for deep learning models that perform automated pre-diction and classification.

3 Materials and Methods

3.1 Proposed Method

Learning from raw data without intentionally extracting features is called implicit processing, whereas feature extraction is the process of picking relevant features to represent the raw data in a meaningful manner. Finding and categorizing rice crop pests and illnesses may be made much easier using this method. Pre-trained CNN models may often be reused, but not without some tweaking to make them better fit the new job at hand.

The classification layer of a deep learning model is updated once the layers above it have mastered the art of deriving significant features from the input data. In order to quickly adapt a model to a new task or dataset, the classification layer is frequently modified first during fine-tuning, even if all of the model's layers may be useful for feature extraction. After the classification layer has been updated, adjustments to the earlier layers may be necessary to maximize feature extraction for the new task. The model will be better able to predict the appropriate class for each input if this is done.

Subsequently, a learnable layer was replaced by a layer that aggregates information from multiple tiers beneath it. The process of replacing a learnable layer involves selecting the appropriate learnable layer, generating a new instance of it, inserting it into the network (perhaps after freezing its settings), training the network, and evaluating its performance. Depending on the CNN model being used, this layer could be a completely linked layer or a convolution2D layer. By include this extra layer, the input image may yield more insightful data that may be used to improve the accuracy of the model. Features such as the shape, size, and color of the lesions should be extracted by the new layer, for example, if the to increase training speed, some of the model's layers could be left unchecked. Trial and error as well as performance analysis of the task and dataset can be used to identify the ideal number of frozen layers. By using this technique, the model's reliance on the training set of data is decreased and its generalizability is improved. Owing to the progress of technology, every level involved in this project received comprehensive training.

The dataset cannot be used with the CNN model until the images have been scaled. Before supplying the image sizes to the model, typical procedure calls for standardizing them. Dividing the dataset into a training set and a validation set is also advised. It is imperative that the model be trained and evaluated efficiently if it is to generalize successfully to new, unknown inputs.

At last, we retrained the CNNs from scratch using the rice salinity disease dataset and evaluated their results. This will enable for an assessment of the model's precision and shed light on its advantages and disadvantages. For a more thorough knowledge of

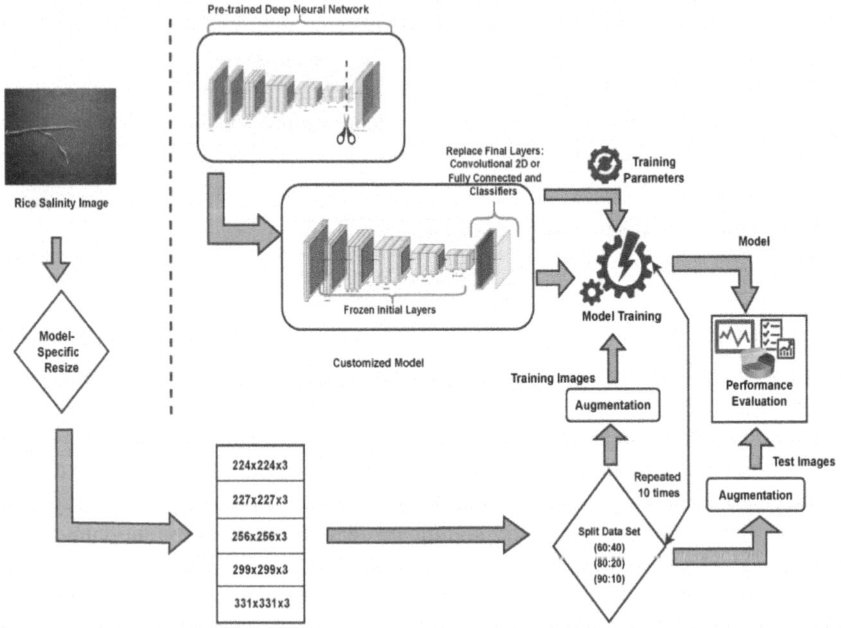

Fig. 2. Proposed Architecture

the process, the following subsections give details on each step. The suggested technique is, all things considered, a potent instrument for recognizing and categorizing pests and illnesses in rice crops, which may lead to an increase in both production and quality. A detailed flowchart of the suggested technique, which makes use of transfer learning on pre-trained CNNs, is depicted in Fig. 2.

3.2 Dataset Collection

Samples of experimental rice seedlings were gathered after consulting with the Central Coastal Agricultural Research Institute (ICAR; located in Old Goa, India). In September of 2022, we use a high-definition camera to take images straight from the field. The image dataset consists of the first, third, fifth, seventh, and ninth grades.

When images are scaled down to 224 × 224, an augmentation method is used to create the final product. There is a split between training and testing data of 70%, with each using 30%. First, we use integers to label the classes, and then, using one-hot encoding, Table 1 shown below.

Table 1. Split Train and Test Datas

Models	Count
Training image (70%)	630
Testing image (30%)	220
Total Images	850

Data Pre-processing: The dataset consists of images from various places. For optimal performance with the used deep learning architectures, all input images were converted to JPEG and scaled down to 224 × 224 pixels. Image enhancements were made by manipulating the luminance, rotation, breadth, height, vertical flip, zoom, and shear. Augmented Table 2 and Fig. 3 shown below.

Table 2. Augmented Data

Models	Count
Training image	630
Testing image	220
Augmented images	5950
Total images	6800

For magnification purposes, the image is shrunk by picking a value from [256, 480] for its shorter side. The per-pixel mean is subtracted from a randomly selected 224 ×

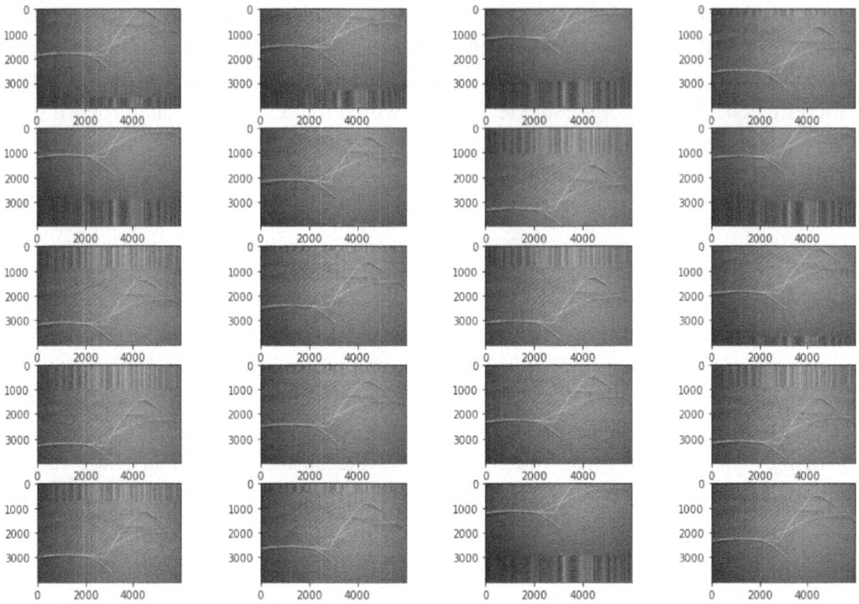

Fig. 3. Augmented Images

224 area of an image or its horizontal flip. Models may be trained for up to 60 × 10000 iterations, during which time the learning rate begins at 0.1 and is split by 10 when the error plateaus. They use a 0.9 momentum and 0.0001 weight decay.

4 Convolutional Neural Network

A ConvNet is a multi-layer network whose activation volumes are switched between at each layer by use of a differentiable function. Three main types of layers are overlaid to build a ConvNet architecture: the Convolutional Layer, the Pooling Layer, and the Fully-Connected Layer (Stanford, a). Initially, a convolution layer computes the output of neurons related to local areas in the input. Next, an activation function is used, like ReLU (Stanford, a), which employs a zero-threshold maximum (0, x) activation. Subsequently, a pooling layer performs a downsampling process in both the horizontal and vertical spatial directions. Class scores are determined by the Fully-Connected layer, which is the last layer in a typical neural network.

4.1 VGG16 and VGG19

The model's key improvement was the use of compact 3×3 convolutional filters. Max-pooling was used across a 2-by-2-pixel window with a 2-pixel strive to perform the pooling. There are 143 million parameters in the model, and it has a depth of 16. Model depth of 19 in VGG19 (Fig. 4).

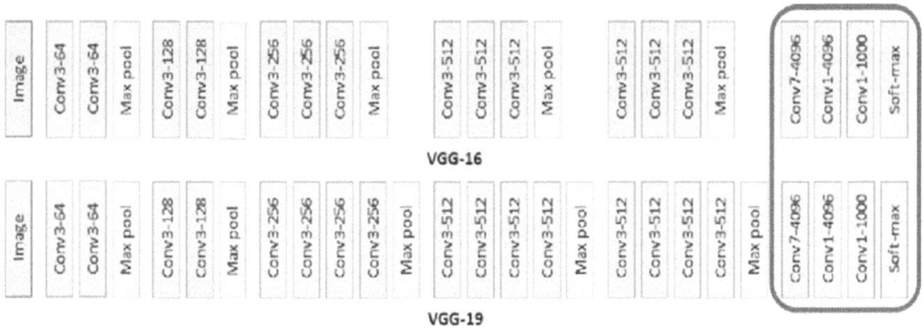

Fig. 4. VGG Architecture

4.2 Inception-V3

The "going deeper" contribution was made by Szegedy et al. in 2012 when they developed the inception model. The suggested model included 27 levels of complexity, not including conception. It consists of 159 layers, and its parameters total 23 million (Fig. 5).

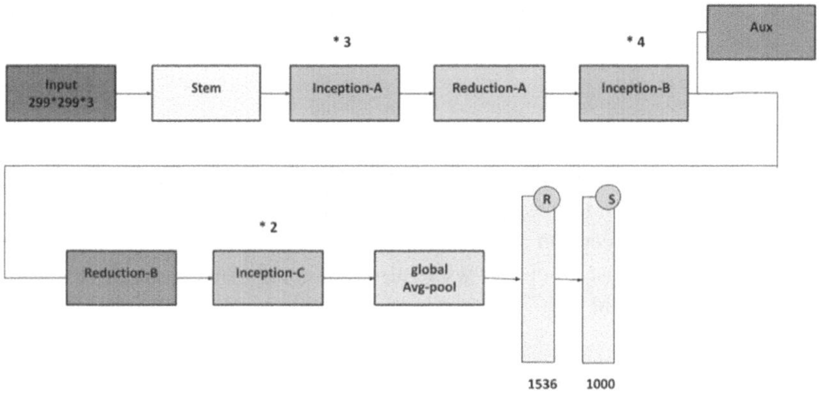

Fig. 5. Inception V3 Architecture

4.3 Xception

The Xception model. The term "extreme inception" describes a scenario in which the inception principle is taken to an extreme. Only depth-separable convolution layers are used in this neural network design. In Inception, the original input was compressed using

1×1 convolutions, and then various types of filters were applied to each of the depth spaces based on the compressed inputs. However, Xception flips that order around, first applying filters and then compressing (Fig. 6).

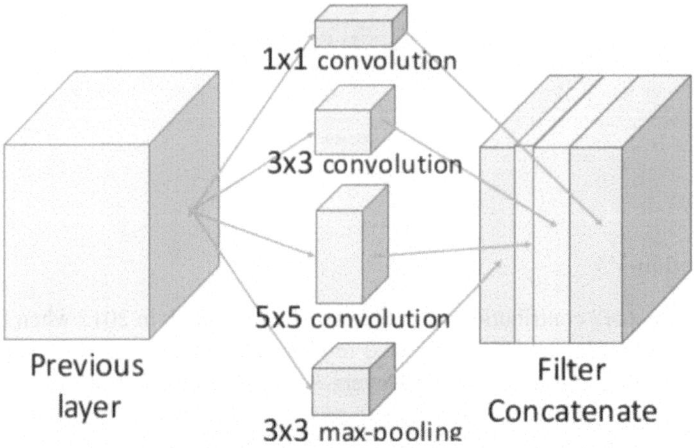

Fig. 6. Xecption Architecture

4.4 InceptionResNetv2

Building on the Inception family of architectures, InceptionResNetv2 was developed in 2016 by Szegedy et al. to integrate residual connections by replacing the filter concatenation step of the Inception architecture. Shortcuts in the model made possible by residual connections boost efficiency while also making the Inception blocks easier to understand and implement (Fig. 7).

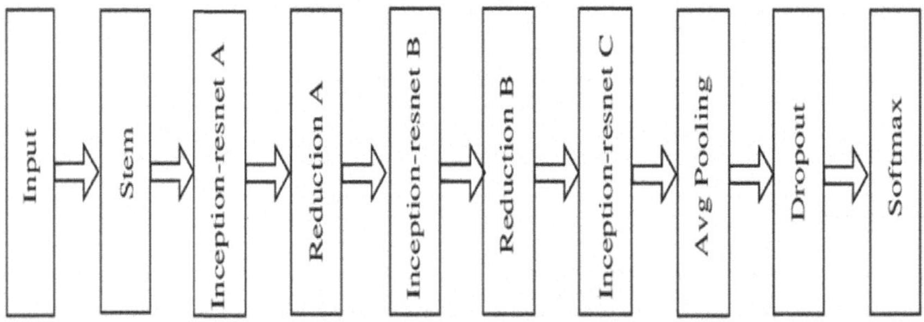

Fig. 7. InceptionResNetv2 Architecture

VGG16 has the shallowest architecture with just 23 layers, whereas the Xception model has the fewest parameters at only 2291480 (Table 3).

Table 3. Transfer Learning Modeling parameters

Model	Size	F1-Score	Accuracy	Parameters
Xception	87 MB	0.780	0.948	22,899,412
VGG19	536 MB	0.752	0.897	148,568,120
InceptionResnet-v2	200 MB	0.801	0.958	54,589,458
Inception-v3	90 MB	0.769	0.924	22,789,158
VGG16	520 MB	0.725	0.915	149,249,861
AgriNet	450 MB	0.713	0.902	48,267,02
ResNet 50 / 100 / 101V2	480 MB	0.710	0.908	11,028,03
MobileNetV2	450 MB	0.700	0.956	10,587,563
ImageNet121	487 MB	0.695	0.912	130,156,02

4.5 Resnet 50 and 50v2 and 100 V2

Resnet50 and 50V2 architecture borrow heavily from the preceding model but with one key modification. When it became clear that training the layers would take too long, the building block was reworked into bottleneck architecture. In this case, three layers were stacked instead of two. To remedy this, the bottleneck blocks in Resnet100 were all upgraded to three layers in Resnet50. The 100-layer ResNet model can't compare to the precision of this (Fig. 8).

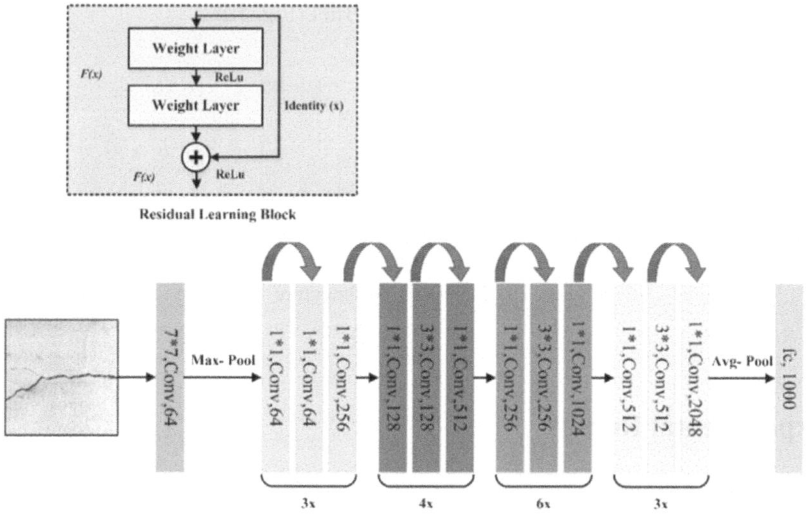

Fig. 8. ResNet Architecture

4.6 MobileNet

MobileNet is a lightweight, deep convolutional neural network architecture that combines depthwise separable convolutions to build an effective model for mobile and embedded vision applications (Fig. 9).

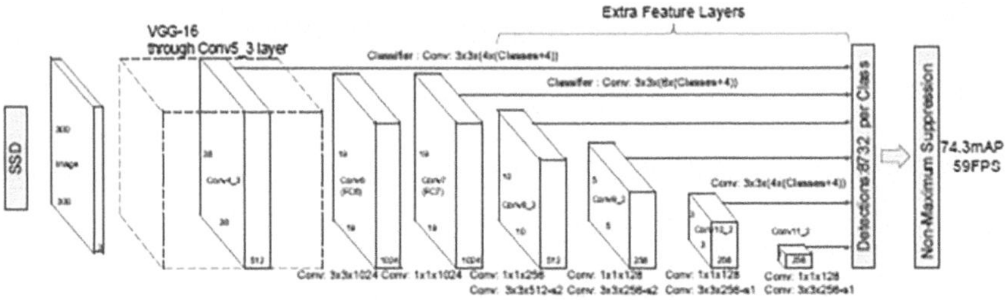

Fig. 9. MobileNet Architecture

4.7 DenseNet

With the novel connection method described by DenseNet, each network layer is linked to the layers below it, allowing the current layer to use the feature map outputs from all the layers below it as input features. The issue of gradient disappearance may be mitigated to some degree by using a connection of this kind. Because each layer is linked to its predecessors, the same features may be mined again and over to produce further feature maps with a smaller convolution kernel (Fig. 10).

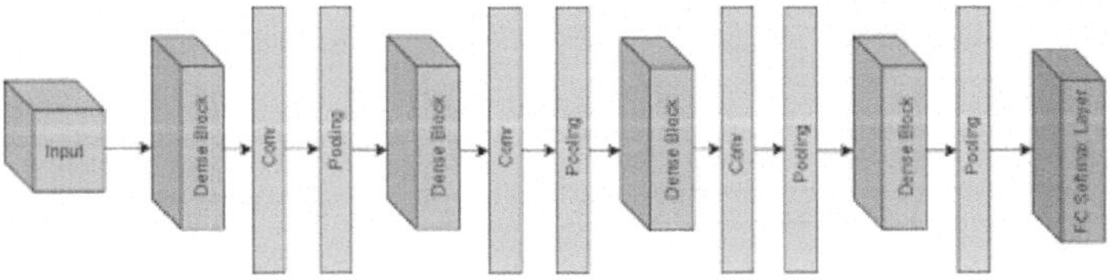

Fig. 10. DenseNet Architecture

5 Experimental Results

5.1 Transfer Learning

In Transfer Learning the parameters that determine which pre-trained model is the best for image classification include task specificity, computing limitations, and dataset size. We need to try out several models and tuning strategies to determine which works best for our application.

The purpose of this research is to examine the feasibility of using modified and retrained advanced CNN models for disease classification in rice leaf photos. Different from one another in the input size, construction, and computational effectiveness, the 9 pre-trained CNN models were compared in this research. This research set out to figure out how best to use such models to the problem of classifying diseases in rice salinity samples.

5.2 Pretraining Models

Since current ConvNets are labor-intensive, it is common practice to provide final ConvNet checkpoints to aid in fine-tuning activities. For instance, training a ConvNet on ImageNet using several GPUs may take up to two months. ImageNet was used as the generic dataset and AgriNet as the target dataset in the first round of transfer learning on pretrained models (Fig. 2). After AgriNet models were trained, their stored architectures and weights were presented as pretrained models for use in additional agricultural categorization tasks.

As can be seen in Table 4, all of the models used in this study were trained using the same set of hyperparameters. Before being fed into the various pre-trained CNNs, all images of rice salinity diseases were resized to meet the input size requirements listed in Table 5. All images were in full color (RGB, or red, green, and blue). Data augmentation [33] was used to enlarge the pool of available training data. Overfitting may be avoided and model generalization can be enhanced through data augmentation. It strengthens models by providing more data for training them. Parameters such as image rotation, reflection, and shear are included into this technique. Figure 11 displayed a few examples of ROI Extraction enhanced images. Table 6 displays the Augmenter's data values. To allow for horizontal reflection of images, true was included into RandXReflection. The images were translated across a narrow range (3 to 3) thanks to the settings of RandXTranslation (Range of horizontal translation) and RandXTranslation (the range of vertical translation). Shear in the imagegraphs was controlled by adjusting the RandXShear (Range of horizontal shear) and RandYShear (Range of vertical shear) from 30 to 30. This method may enrich the dataset used to train the model by adding new types of data.

Table 4. Parameters for Training

Function Name	Value
Optimizer	Adam
MiniBatchSize	32
MaxEpochs	30
ValidationFrequency	30

(continued)

Table 4. (*continued*)

Function Name	Value
VerboseFrequency	30
ExecutionEnvironment	gpu
Verbose	true
LearnRateDropFactor	0.1
LearnRateDropPeriod	8
LearnRateSchedule	none
Shuffle	Every-epoch

Table 5. Input Size for pre-trained models.

Pre-Trained CNN Model	Parameters (Millions)	Input Image Size
InceptionV3	25.6	224 × 224 × 3
Xception	44.6	224 × 224 × 3
VGG 16	7	224 × 224 × 3
VGG 19	138	224 × 224 × 3
Resnet 50	1.4	224 × 224 × 3
Resnet 100	5.3	224 × 224 × 3
Resnet 50V2	3.5	224 × 224 × 3
MobileNetV2	5.3	224 × 224 × 3
DenseNet121	20	224 × 224 × 3
ResNet101V2	1.24	227 × 227 × 3
AgriNet	61	227 × 227 × 3

Table 6. Data Agumentations

Properties	Values
RandXReflection	True
RandXTranslation	− 3 to 3
RandYTranslation	− 3 to 3
RandXShear	− 30 to 30
RandYShear	− 30 to 30

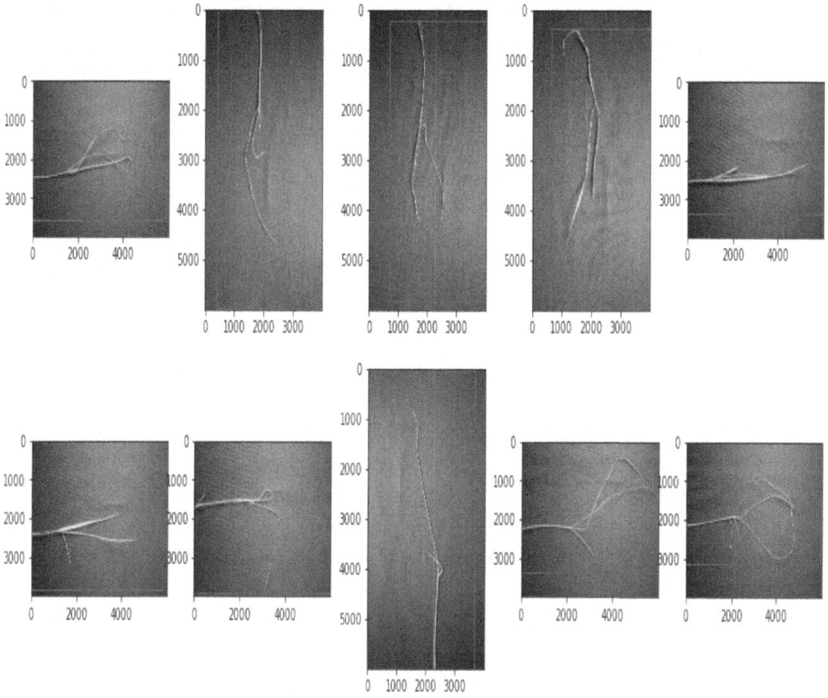

Fig. 11. ROI Extraction mages

6 Results and Discussion

6.1 Framework for Assessment

In our studies, we make use of pre-trained transfer learning models on the Imagenet dataset's one thousand classes. The initial and final layers of these models, including fully connected, convolutional, softmax, and classification output layers, are modified such that they are suitable for our intended use. Upper layers of pre-trained models often catch task-specific data whereas lower levels capture generic characteristics. The pre-trained model may be modified to new tasks by altering the higher layers, drawing on the information gained from the original work. Following this, once all models were trained with the identical set of hyperparameters. Training lasted 30 epochs, the number of which was established by comparing the models' training and validation results. Network training using the Adam optimizer and a learning rate of 0.0001. Several data partitioning strategies were used to assess the models' capacity to generalize to a larger testing set. The first strategy used a 60/40 split between the dataset's training and validation sets. In the second way, only 80% of the photos were utilized for training, but in the third method, 90% of the images were used [37].

The models were developed and tested on an DELL Laptop, which is equipped with 64 GB of RAM, a 4.00 GHz Intel(R) Xeon(R) W-2125 central processing unit (CPU) with 4 cores, a high-performance NVIDIA GeForce GTX 1080 Ti graphics card, and 8 logical processors. Throughout construction and testing, Jupyter Python was used.

6.2 Evaluation Metrices

In machine learning, a model's precision and efficiency are measured by its performance measures. Metrics may vary widely from issue to problem and algorithm to algorithm. We employed the accuracy, precision, recall, f1-score specificity, Matthews Correlation Coefficient (MCC), and false positive rate (FPR) classification criteria in this study. The term "accuracy" refers to the rate at which illness samples are accurately identified relative to the overall amount of test datas. Accuracy refers to how many "yes" forecasts turn out to be correct. True positive rate (also known as recall) is the fraction of cases when a result is indeed positive. The F1 Score is the mathematical mean of the recall and accuracy scores. A model's specificity is quantified by how well it can identify non-positive samples. Measuring consistency between observations and predictions, MCC provides a number between −1 and 1 that varies depending on the proportion of true and erroneous positives and negatives. FPR It's a metric for how frequently a test falsely concludes that a target event has happened when in fact it hasn't. The results Eqs. 1, 2, 3, 4, 5 and 6 listed below.

$$\text{Accuracy} = \frac{T_{+ve} + T_{-ve}}{T_{+ve} + T_{-ve} + F_{+ve} + F_{-ve}} \tag{1}$$

$$\text{Precision} = \frac{T_{+ve}}{T_{+ve} + F_{+ve}} \tag{2}$$

$$\text{Recall} = \frac{T_{+ve}}{T_{+ve} + F_{-ve}} \tag{3}$$

$$\text{F1 Score} = 2 \times \frac{(\text{Precision} \times \text{Recall})}{(\text{Precision} + \text{Recall})} \tag{4}$$

$$\text{Specificity} = \frac{T_{-ve}}{F_{+ve} + T_{-ve}} \tag{5}$$

$$MCC = \frac{(T_{+ve} \times T_{-ve}) - (F_{+ve} \times F_{-ve})}{\sqrt{(T_{+ve} + F_{+ve})(T_{+ve} + F_{-ve})(T_{-ve} + F_{+ve})(T_{-ve} + F_{-ve})}} \tag{6}$$

We performed studies to determine whether or not CNN models could accurately identify the illness kind in rice leaves. The training procedure was performed 10 times to account for variations in the random data partition of images into training and validation sets. The average was used to determine the final tally. We recorded the timespan of each CNN model's training and validation processes, as well as their respective performance assessment criteria.

There were three different approaches to data partitioning used (i.e., 60/40, 80/20, and 90/10), and these may reveal whether or not the models were underfitting or overfitting the data. Pretrained deep neural network models' mean overall accuracy, precision, recall, F1-score, specificity, MCC, and FPR for the initial data partition scheme (60/40) are listed below. Along with their superior performance across the board, all models displayed remarkable accuracy. With a mean F1 score of 98.21% and a specificity of 99.81%, InceptionResNetV2 outperformed all other models. AgriNet's mean F1 score of 86.03% and specificity value of 98.49% made it the worst performing model. Confusion matrices for the best and worst performing models are shown in Fig. 12 and Table 7, validating the reported performance levels.

Table 7. Using 60% of the training data, the mean general performance measure values

Model	Accuracy(%)	Precision(%)	Recall(%)	F1-Score(%)	Specificity(%)	MCC	FPR
InceptionV3	99.64	98.22	98.2	98.21	99.81	0.98	0.017
Xception	99.61	98.13	98.08	98.08	99.78	0.978	0.018
VGG 16	99.51	97.57	97.56	97.55	99.73	0.972	0.024
VGG 19	99.47	97.53	97.38	97.38	99.71	0.971	0.024
Resnet 50	99.47	97.39	97.36	97.35	99.7	0.97	0.026
Resnet 100	99.35	96.77	96.76	96.75	99.64	0.964	0.032
Resnet 50V2	99.27	96.4	96.34	96.35	99.59	0.959	0.035
MobileNetV2	99.19	96.06	95.94	95.94	99.55	0.955	0.039
DenseNet121	99.18	96.18	95.89	95.89	95.54	0.955	0.038
ResNet101V2	98.98	95.22	94.92	94.95	99.43	0.955	0.047

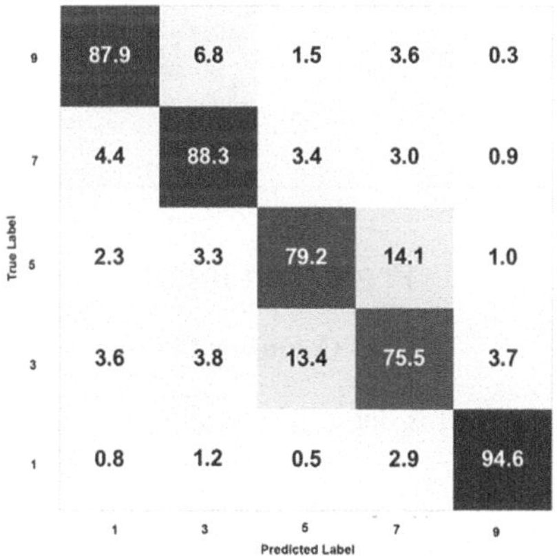

Fig. 12. Confusion Matrix

Performance metrics for 9 pre-trained deep learning models over 10 runs utilizing 80% of the data is shown in Table 8.The confusion matrices, shown in Fig. 13, also reveal the models' best and worst performances.

Table 8. Using 80% of the training data, the mean general performance measure values

Model	Accuracy (%)	Precision (%)	Recall (%)	F1-Score (%)	Specificity (%)	MCC	FPR
InceptionV3	99.69	98.48	98.45	98.46	99.82	0.982	0.015
Xception	99.59	97.99	97.96	97.94	99.77	0.977	0.020
VGG 16	99.54	97.82	97.71	97.72	99.74	0.974	0.021
VGG 19	99.53	97.71	97.66	97.65	99.74	0.974	0.022
Resnet 50	99.51	97.58	97.56	97.56	99.73	0.972	0.024
Resnet 100	99.49	97.57	97.46	97.46	99.71	0.972	0.024
Resnet 50V2	99.41	97.16	97.06	97.02	99.67	0.967	0.028
MobileNetV2	99.41	97.09	97.01	96.99	99.66	0.967	0.029
DenseNet121	99.32	96.93	96.61	96.64	99.62	0.963	0.030
ResNet101V2	99.14	95.93	95.72	95.69	99.52	0.953	0.040

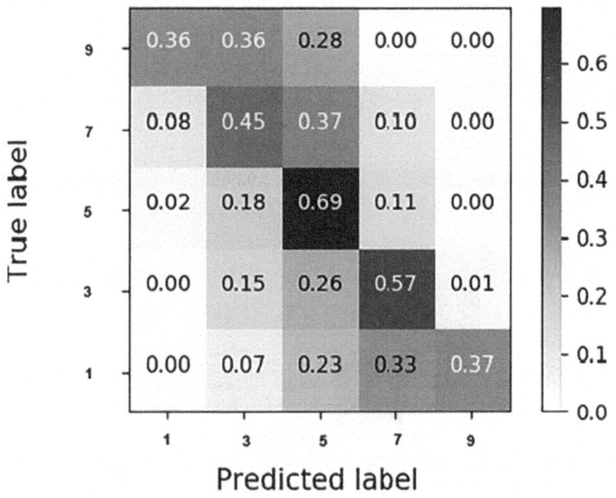

Fig. 13. Confusion Matrix

Some of the results observed is depicted in Table 9.

Table 9. Using 90% of the training data, the mean general performance measure values

Model	Accuracy (%)	Precision (%)	Recall (%)	F1-Score (%)	Specificity (%)	MCC	FPR
InceptionV3	99.74	98.72	98.7	98.7	99.85	0.985	0.012
Xception	99.68	98.42	98.4	98.4	99.82	0.982	0.015

(*continued*)

Table 9. (*continued*)

Model	Accuracy (%)	Precision (%)	Recall (%)	F1-Score (%)	Specificity (%)	MCC	FPR
VGG 16	99.64	98.23	98.2	98.21	99.8	0.98	0.017
VGG 19	99.64	98.21	98.2	98.2	99.8	0.98	0.017
Resnet 50	99.64	98.24	98.2	98.19	99.8	0.98	0.017
Resnet 100	99.52	97.67	97.6	97.61	99.73	0.973	0.023
Resnet 50V2	99.48	97.75	97.4	97.44	99.71	0.972	0.022
MobileNetV2	99.46	97.34	97.31	97.29	99.71	0.97	0.026
DenseNet121	99.22	96.34	96.1	96.09	99.56	0.957	0.036
ResNet101V2	99.14	95.96	95.7	95.73	99.52	0.953	0.04

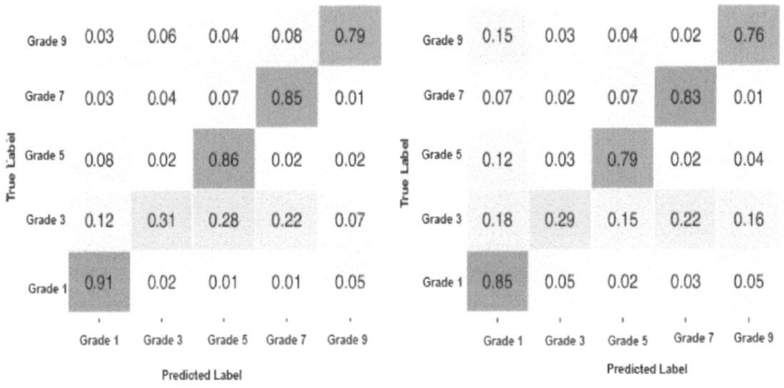

Fig. 14. Confusion Matrix

The results showed that the AUC for InceptionResNetV2 was 99.94%, for Xception it was 99.92%, for ResNet50 it was 99.91%, and for AgriNet it was 97.70%. The CNN's ability to accurately identify each class is shown by visual representations of confusion matrices, receiver operating characteristic (ROC) curves, area under the curve (AUC), and overall model accuracy (Fig. 14 and Table 10).

The scoring class is determined for each image. The algorithm's prediction for the Score 9 sample is represented in Fig. 16 along an X-axis of Class Names and a Y-axis of likelihood (Figs. 15 and 17).

7 Constraints on the Undertaking

There are gaps in the present literature on the use of transfer learning for the categorization of rice leaves. The difficulty in generalizing these models is a major obstacle to employing this method. Images of plants of differing kinds or captured in varied lighting situations might hinder the models' performance. This restriction reduces the model's credibility and may cause misdiagnosis of illnesses.

Table 10. Generalized Models of Transfer Learning Performances

Model	Accuracy (%)	Precision (%)	Recall (%)	F1-Score (%)	Specificity (%)	MCC	FPR
InceptionV3	99.64	98.23	98.21	98.20	99.80	0.98	0.02
Xception	99.61	98.12	98.06	98.07	99.78	0.98	0.02
VGG 16	99.56	97.93	97.80	97.82	99.75	0.98	0.02
VGG 19	99.55	97.77	97.74	97.73	99.75	0.97	0.02
Resnet 50	99.53	97.78	97.68	97.68	99.74	0.97	0.02
Resnet 100	99.48	97.43	97.40	97.40	99.71	0.97	0.03
Resnet 50V2	99.41	97.13	97.05	97.05	99.67	0.97	0.03
MobileNetV2	99.40	97.12	96.99	97.00	99.66	0.97	0.03
DenseNet121	99.35	96.89	96.75	96.73	98.31	0.96	0.03
ResNet101V2	99.09	95.70	95.45	95.46	99.49	0.95	0.04

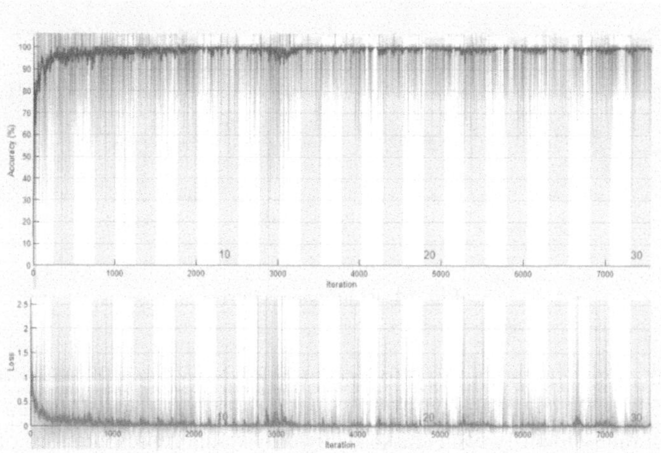

Fig. 15. Accuracy and loss

Furthermore, dietary inadequacies may be incorrectly diagnosed as illnesses by the models. Inadequate soil nutrient levels may lead to rice plant nutritional deficits. Nutritional deficits in rice plants include visual symptoms that resemble those of illnesses, but there are important distinctions. Nutritional deficits often manifest as yellowing or slowed development, whereas illnesses typically show dark rings or lesions on the leaves. Plant diseases, on the other hand, may quickly spread from one plant to another, whereas nutrient deficits are often the result of an overall deficiency in the soil. Since the therapy for rice illnesses and nutritional deficiencies is different, it is essential to distinguish

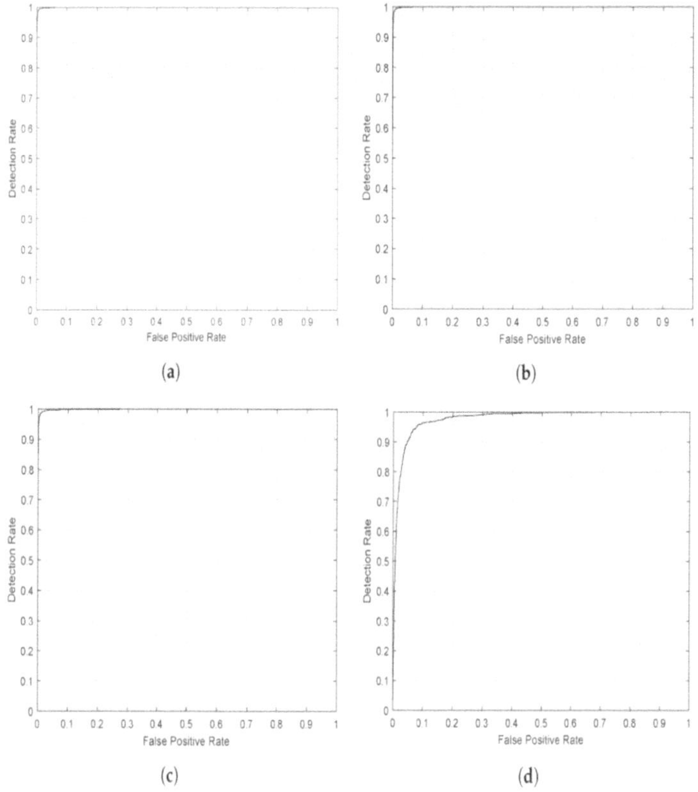

Fig. 16. ROC Curve

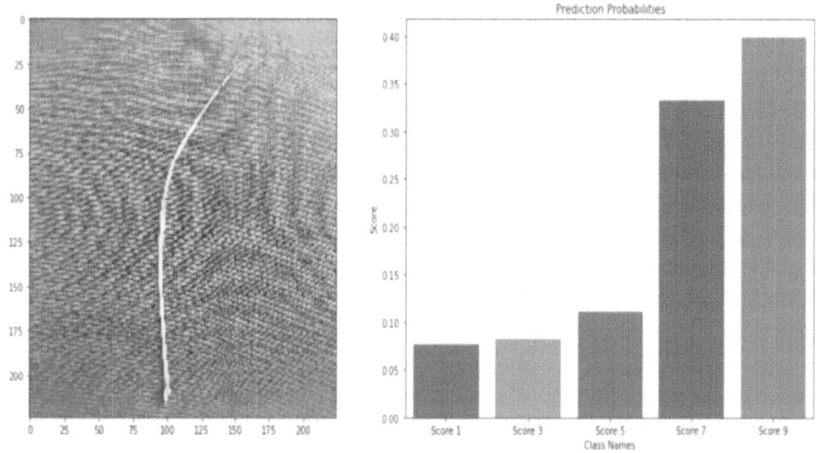

Fig. 17. Score of Grade 1, 3,5,7,9

between the two. Soil amendments may fix nutritional deficits, but only so much, and diseases need targeted treatments to stop their spread and heal the plants they infect.

While transfer learning offers a practical answer for identifying rice leaf diseases, its limitations must be taken into account. Improving a model's performance and making it more dependable for real applications may be achieved by investigating and testing its constraints.

8 Conclusion

More than half of the world's population relies on rice for sustenance, making this grain crucial to ensuring the safety of the world's food supply. However, farmers might lose a lot of money due to rice plant illnesses. To counteract this, scientists are working on innovative strategies to fight these illnesses. Machine learning and deep learning are two cutting-edge technologies that show promise for detecting rice salinity stress. The purpose of this research is to automate the process of detecting illnesses in rice leaves by applying transfer learning to a set of 9 previously-trained convolutional neural network (CNN) models. Compared to AlexNet, the findings demonstrate that the InceptionV3 model is the most successful one, with an average accuracy of 99.64 percent. Because genetic, morphological, and physiological differences across rice varieties might influence the expression of illness symptoms, it is not always easy to identify whether a given approach is appropriate for all rice types.

A plant pathologist should be consulted, along with additional symptoms including symptom location and intensity, as well as environmental variables like weather and soil conditions, to confirm a diagnosis in such circumstances. Microscopy, PCR, and culture of bacteria or fungi, as well as other laboratory procedures, may also be used to validate the identity of the pathogen.

Symptoms of illnesses affecting rice plants may appear anywhere on the plant. Water-soaked lesions on leaves are a telltale sign of bacterial leaf blight, whereas lesions on the leaf sheath and collar are indicative of sheath blight. Blast disease is another factor that might diminish the quantity and quality of the panicles and grains. Therefore, the diagnosis could be incomplete if just leaf symptoms are considered. And this may cause early signs of the illness to be missed. This may be remedied by taking into account the disease's manifestations in various plant organs throughout the diagnostic process.

Artificial intelligence and machine learning frameworks are useful in many areas of modern agriculture and food production. Since rice salt stress may result in a significant drop in rice yield, these frameworks can be utilized to automatically and more effectively identify and categorize the various rice crops involved in the score evaluation process. Agricultural scientists and researchers may benefit from the information gained from this work to evaluate and speed up breeding by selecting potential genotypes at early seedling stages required for salinity tolerance tests, even though the images in the proposed work are insufficient to assign each image a specific score. To make the automatic standard assessment system more effective, additional photos might be gathered as part of the study's future aims.

References

1. Gregorio, G.B., Senadhira, D., Mendoza, R.D.: Screeningrice for salinity tolerance, IRRI discussion paper series no. 22 (1997). http://www.knowledgebank.irri.org/ricebreedingcourse/documents/Screening_manual.pdf
2. Gregorio, G.B., Senadhira, D., Mendoza, R.D., Manigbas, N.L., Roxas, J.P., Guerta, C.Q.: Progress in breeding for salinity tolerance and associated abiotic stresses in rice. Field Crops Res. 76, 91–101 (2002). http://dx.doi.org/10.1016/S0378-4290(02)00031-X
3. Gregorio, G.B.: Tagging salinity tolerance genes in rice using amplified fragment length polymorphism. Ph.D. Dissertation, University of the Philippines Los Baños. Laguna, Philippines, p. 118 (1997)
4. International Rice Research Institute (IRRI): Annual report for 1967. IRRI, Los Banos, Philippines, p. 180 (1967)
5. IRRI: Rice breeding course, Module 4 (2006). http://www.knowledgebank.irri.org/ricebreedingcourse/Breeding_for_salt_tolerance.htm
6. IRRI: Annual Report for 1997. IRRI, Los Banos, Philippines, p. 308 (1997)
7. IRRI: Standardization evaluation system for rice. International Rice Research Institute, Philippines, p. 34 (2013)
8. Islam, M.M.: Mapping salinity tolerance genes in rice (Oryza sa-tive L.) at reproductive stage. Ph.D. Dissertation, University of the Philippines Los Banos, Laguna, Philippines, p. 150 (2004)
9. Barbedo, J.G.A.: Impact of dataset size and variety on the effectiveness of deep learning and transfer learning for plant disease classification. Comput. Electron. Agric. **153**, 46–53 (2018)
10. Shrivastava, V.K., Pradhan, M.K., Minz, S., Thakur, M.P.: Rice plant disease classification using transfer learning of deep convolution neural network. Int. Arch. Photogramm. Remote Sens. Spat. Inf. Sci. **3**, 631–635 (2019)
11. Chowdhury, M.E., et al.: Automatic and reliable leaf disease detection using deep learning techniques. AgriEngineering **3**, 294–312 (2021)
12. Li, L., Zhang, S., Wang, B.: Plant disease detection and classification by deep learning—a review. IEEE Access **9**, 56683–56698 (2021)
13. Malathi, V., Gopinath, M.: Classification of pest detection in paddy crop based on transfer learning approach. Acta Agric. Scand. Sect. B Soil Plant Sci. **71**, 552–559 (2021)
14. Sethy, P.K., Barpanda, N.K., Rath, A.K., Behera, S.K.: Deep feature based rice leaf disease identification using support vector machine. Comput. Electron. Agric. **175**, 105527 (2020)
15. Haridasan, A., Thomas, J., Raj, E.D.: Deep learning system for paddy plant disease detection and classification. Environ. Monit. Assess. **195**, 120 (2023)
16. Yang, L., et al.: Goog-LeNet based on residual network and attention mechanism identification of rice leaf diseases. Comput. Electron. Agric. **204**, 107543 (2023)
17. Latif, G., Abdelhamid, S.E., Mallouhy, R.E., Alghazo, J., Kazimi, Z.A.: Deep learning utilization in agriculture: detection of rice plant diseases using an improved CNN model. Plants **11**, 2230 (2022)
18. Daniya, T., Vigneshwari, S.: Exponential Rider-Henry gas solubility optimization-based deep learning for rice plant disease detection. Int. J. Inf. Technol. **14**, 3825–3835 (2022)
19. Bari, B.S., et al.: A real-time approach of diagnosing rice leaf disease using deep learning-based faster R-CNN framework. PeerJ Comput. Sci. **7**, e432 (2021)
20. Islam, M.A., Shuvo, M.N.R., Shamsojjaman, M., Hasan, S., Hossain, M.S., Khatun, T.: An automated convolutional neural network based approach for paddy leaf disease detection. Int. J. Adv. Comput. Sci. Appl. **12**, 280–288 (2021)
21. Wang, Y., Wang, H., Peng, Z.: Rice diseases detection and classification using attention based neural network and Bayesian optimization. Expert Syst. Appl. **178**, 114770 (2021)

22. Rahman, C.R., et al.: Identification and recognition of rice diseases and pests using convolutional neural networks. Biosyst. Eng. **194**, 112–120 (2020)
23. Chen, J., Zhang, D., Nanehkaran, Y.A., Li, D.: Detection of rice plant diseases based on deep transfer learning. J. Sci. Food Agric. **100**, 3246–3256 (2020)
24. Feng, L., Wu, B., He, Y., Zhang, C.: Hyperspectral imaging combined with deep transfer learning for rice disease detection. Front. Plant Sci. **12**, 693521 (2021)
25. Thenmozhi, K., Reddy, U.S.: Crop pest classification based on deep convolutional neural network and transfer learning. Comput. Electron. Agric. **164**, 104906 (2019)
26. Upadhyay, S.K., Kumar, A.: A novel approach for rice plant diseases classification with deep convolutional neural network. Int. J. Inf. Technol. **14**, 185–199 (2022)
27. Hossain, S.M.M., et al.: Rice leaf diseases recognition using convolutional neural networks. In: Yang, X., Wang, C.D., Islam, M.S., Zhang, Z. (eds.) ADMA 2020. LNCS, vol. 12447, pp. 299–314. Springer, Cham (2020). https://doi.org/10.1007/978-3-030-65390-3_23
28. Chen, S., et al.: An approach for rice bacterial leaf streak disease segmentation and disease severity estimation. Agriculture **11**, 420 (2021)
29. Stephen, A., Punitha, A., Chandrasekar, A.: Designing self attention-based ResNet architecture for rice leaf disease classification. Neural Comput. Appl. **35**, 6737–6751 (2022)
30. Li, D., et al.: A recognition method for rice plant diseases and pests video detection based on deep convolutional neural network. Sensors **20**, 578 (2020)
31. Rice Leaf Diseases Dataset. https://www.kaggle.com/datasets/vbookshelf/rice-leaf-diseases. Accessed 17 Jan 2023
32. Rice Leaf Disease Image Samples. https://data.mendeley.com/datasets/fwcj7stb8r/1. Accessed 17 Jan 2023
33. Nanni, L., Paci, M., Brahnam, S., Lumini, A.: Comparison of different image data augmentation approaches. J. Imaging **7**, 254 (2021)
34. Thangaraj, R., Anandamurugan, S., Kaliappan, V.K.: Automated tomato leaf disease classification using transfer learning-based deep convolution neural network. J. Plant Dis. Prot. **128**, 73–86 (2021)
35. Noor, A., Zhao, Y., Koubâa, A., Wu, L., Khan, R., Abdalla, F.Y.: Automated sheep facial expression classification using deep transfer learning. Comput. Electron. Agric. **175**, 105528 (2020)
36. Kaya, A., Keceli, A.S., Catal, C., Yalic, H.Y., Temucin, H., Tekinerdogan, B.: Analysis of transfer learning for deep neural network based plant classification models. Comput. Electron. Agric. **158**, 20–29 (2019)
37. Fraiwan, M., Faouri, E., Khasawneh, N.: Multiclass classification of grape diseases using deep artificial intelligence. Agriculture **12**, 1542 (2022)
38. Shiragudikar, S.K., Bharamagoudar, G., Manohara, K.K., Malati, S.Y.: Insight analysis of deep learning and a conventional standardized evaluation system for assessing rice crop's susceptibility to salt stress during the seedling stage. SN Comput. Sci. **4**(3), 262 (2023)
39. Das, B., Manohara, K.K., Mahajan, G.R., Sahoo, R.N.: Spectroscopy based novel spectral indices, PCA- and PLSR-coupled machine learning models for salinity stress phenotyping of rice. Spectrochimica Acta Part A: Mol. Biomol. Spectrosc. **229**, 117983 (2020). ISSN 386-1425. https://doi.org/10.1016/j.saa.2019.117983

Multi-camera HD Pedestrian Dataset for Person Detection and Re-identification

Prashant Narayankar(✉), Vishwanath P. Baligar, Prajwalkumar A. Gouli, Naresh Illur, Pooja Doddannavar, and Pooja M. Malagund

School of Computer Science and Engineering, KLE Technological University, Hubballi 580031, India
{Prashant_narayankar,Vpbaligar}@kletech.ac.in

Abstract. People's detection and tracking methods are susceptible to occlusions and overlapping problems in a multi-camera environment. Such issues are persistent when you mount the camera at a low height. Detection and tracking performance can be improved by combining visual inputs from synchronized camera setups.

In this paper, we provide a novel large-scale, high-resolution multi-camera dataset. The dataset has been captured with four stationary cameras with the common region of interest in a university campus where a swarm of spontaneous people are standing and walking. Including frames, we present extrinsic and intrinsic calibration parameters and four sequences of annotated frames for detection at two frames per second. As a result, almost 150 bounding boxes delimit each person present in the region of interest.This dataset can be used for multiple computer vision tasks like pedestrian detection, classification and person reidentification etc.

Keywords: Mult-camera · Multi-people · ReIdentification · Person Detection · Person Tracking

1 Introduction

Detection of pedestrians is challenging in computer vision and artificial intelligence. It has applications in numerous fields, such as automation, robotics, self-driving cars, security and surveillance. The differences in height, walking speed, the ray of light, and contour lines of pedestrians increase this problem's complexity. However, the availability of single-camera databases of labelled pedestrians on a colossal scale and advanced detection algorithms has considerably increased the accuracy of pedestrian detectors in recent years. Furthermore, contemporary deep learning-based detection methods allow us to develop portable discriminative features across datasets. Deep learning-based monocular sensors have recently achieved human-level performance on conventional benchmark datasets.

On the other hand, many cases of practical significance necessitate detection in heavily packed and cluttered surroundings. In these cases, severe occlusions render monocular pedestrian detection ineffective. Fortunately, image inputs from several cameras with

T. Shreekumar et al. (Eds.): ISCComm 2023, CCIS 2231, pp. 309–317, 2025.
https://doi.org/10.1007/978-3-031-75605-4_20

overlapping fields of interest are frequently available in real-world applications. The cameras are usually placed slightly higher than the typical human height. As a result, building pedestrian detectors that take advantage of numerous viewpoints and the scene's geometry will result in accurate detection estimates in congested environments.

No large-scale, high-quality public dataset replicates this setting by contemporary standards. PETS 2009, the most often used, is merely a few hundred frames long. The camera calibration is poor, making it challenging to exploit geometrical limitations simultaneously. Furthermore, it was captured in an "actor setup," which means that only a few different people appear throughout the sequence. The WILDTRACK dataset presents a high resolution and large-scale dataset captured with seven static cameras. The camera frames also provide extrinsic and intrinsic calibrations as an accurate joint. All other standard datasets are substantially smaller and sparsely populated, feature a minimal scenario, or Use cameras that aren't overlapping.

The unavailability of such a dataset severely restricts the advancement of multiple camera detection people algorithms. Recent community improvements need the creation of a more realistic and rigorous benchmark for comparing multi-camera detection techniques. Deep learning-based detectors that use multi-camera input improve classification accuracy and confidence. However, the lack of a large-scale dataset on which to train without over-fitting severely limits these techniques. As a result, most known joint systems integrate information gathered from pre-trained monocular detectors utilizing ad-hoc methods.

We propose a large-scale people dataset gathered employing four static cameras with overlapping regions of interest to aid research on techniques that use multi-camera setups. We capture an unscripted scene in which pedestrians frequently obstruct one other. Camera location in the 3D plane is represented as extrinsic calibration and focal length, and the optical centre of the camera plane is described as intrinsic calibration. Synchronize frames from four viewpoints and four series of labelled frames for detection and re-identification at two frames per second. This results in segmenting every person in the region of interest, resulting in over 150 people. Individual track annotations are provided as 3D ground plane positions and 2D segmentation projected in the four perspectives. We also offer annotations for evaluating camera calibration methods as a final point. Many person detection dataset are available but the disadvantage of those dataset is that they don't provide intrinsic and extrinsic parameters and many dataset available are not labelled dataset. Advantage of our dataset over such dataset is that we provide intrinsic and extrinsic parameters and also our dataset is prelabelled data with groundtruth information.

Dataset generation and gathering has involved many challenges, Those challenges are Occlusion, proper light, camera synchronization and peoples consent. We could able to manage all challenges by proper camera set up on day light and also a standard template for people consent was established.Synchronisation of all camera was taken care by GoPro remote.

The following is a breakdown of the paper's sections: Sect. 2 describes the study of several datasets which focus on multi-camera person detection and tracking.

Section 3 describes the new dataset's features, including our camera calibration procedure. Section 4 compares state-of-the-art multi-camera detection systems, and Sect. 5 discusses impending future research possibilities.

2 Related Work

In this section, we studied several datasets focusing on multi-camera datasets. N. Dalal et al. [1] provide a more complex dataset encompassing over 1800 labelled human photos with various stance variations and backdrops. The new approach gives near-perfect separation from the original MIT pedestrian database.

SALSA [2] is a dataset that facilitates multimodal and Synergetic sociAL Scene Analysis. This dataset automates social interaction analysis; research and makes significant advancements. SALSA records social interactions between 18 individuals for over 60 min in a natural, indoor setting. Low-resolution photos, illumination fluctuations, multiple occlusions, repercussions, and interfering sound sources are obstacles to this registration. The author enables multimodal analysis by recording the social interplay using four static surveillance cameras and sociometries to solve these challenges. Throughout the event, the paper includes annotations about individuals' personalities and their position, head, body orientation, and F-formation information.

The PETS 2009 [3] workshop included a crowd image analysis challenge described in the paper. The challenge's goal was to employ new or existing technologies in a real-world setting to i) count and estimate crowd density, ii) monitor individual's inside a crowd, and iii) recognize discrete flows and specific crowd activities. The situations in the dataset were shot using various cameras and featured many actors. All three datasets were registered at the workshop at the Whiteknights Campus, University of Reading, U.K. Multi-sensor sequences with crowd scenarios with increasing scene complexity make up the datasets. Dataset D1 specifies the person to count and density estimation. Dataset D2 solves issues relating to people tracking. Dataset D3 describes flow analysis and event recognition.

In T. Chavdarova et al., [4] a new large-scale, high-resolution dataset was presented. It was captured in a public open place with spontaneous dense groups of pedestrians standing and walking utilizing seven static cameras synchronously with ~ 50 ms. They provided an extrinsic and intrinsic calibration and seven series of 400 annotated frames for detection at a rate of 2 frames per second, along with the camera frames. Over 40,000 bounding boxes delimited every person present in the area of interest, totalling over 300 people.

In WILDTRACK [4], because of its broad use and support in several libraries, including OpenCV, they utilized the Pinhole camera model to calibrate the cameras. The authors first used camera calibration in terms of intrinsic and extrinsic parameters, and for the latter, they employed hand-measured distances between places on the ground. The calibration is modelled as a nonlinear least-squares problem with an error equal to the squared L2 norm of the difference between the observed feature location and the projected 3D point on the camera's image plane. Using the camera calibration provided, they then estimated the point's 3D location as an intersection of the rays defined by those projections. Finally, the 3D point is projected in the remaining viewpoints.

In WILDTRACT [4], the authors took advantage of precise camera calibration. An annotation task is defined as altering the position of a 3D cylinder on the ground so that its projections (rectangles) overlay the person the annotator is marking from all viewpoints.

3 Dataset

Hardware: Four HD cameras were placed in a static position to capture the data. The layout of all four cameras and the area of interest (AOI) is witnessed in Fig. 1. We used four GoPro cameras; all four sequences have a 1920 x 1080 pixels resolution and were captured at the rate of 60 fps. The synching between the four cameras was accomplished with a precision of 10 ms.

Camera Placement and Layout: The cameras are placed at the average height of humans, and the camera architecture, i.e., the region of interest, is extensively overlapping. Overlap between the camera's fields has been taken from a top-view perspective. In **Fig. 1**, A dark grey region states AOI; the deeper the colouring, the more cameras record that area. To make this diagram, we counted how many cameras have a given point in their field of vision by looking at points on the ground plane. The circles in the diagram show the placements of the cameras.

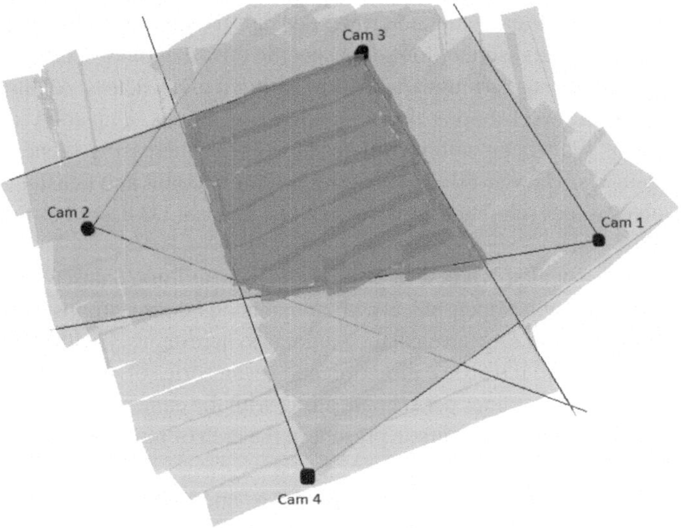

Fig. 1. The layout and area (AOI) captured by all four cameras

The mathematical model's objective is to describe a person's projection into the image mathematically. The model is generally represented by an equation that has coordinates (X, Y) of a pedestrian into image coordinates (x′,y′). Image coordinates are dependent on the camera calibrations. Figure 1 shows the layout of the cameras, namely C1, C2, C3 and C4. The distance and angle of the cameras are shown as follows.

Distance between all Cameras:

C1 to C2 is: 68ft (20726.4 mm), C1 to C3 is: 48ft (14630.4 mm), C1 to C4 is: 55ft (16764 mm), C2 to C3 is: 54ft (16459.2 mm), C2 to C4 is: 53ft (16154.4 mm) and C3 to C4 is: 75ft (22860 mm).

Angles between each camera:

Camera 1: C3C1C4 has 110 Ds, Camera 2: C3C2C4 has 110°, Camera 3: C2C3C1 has 140° and Camera 4: C2C4C1 has 120 ,

Considering the layout, we have two triangles, C1C3C2 and C1C4C2.

The area of both triangles is given by the

C1C3C2 = ½*D*H1

C1C4C2 = ½*D*H2

Where D is diagonal, H1 is the height of the C1C3C2 triangle, and H2 is the height of the C1C4C2 triangle.

Both triangles give the area of the quadrilateral [5] as

C1C2C3C4 = Sum of are of ΔC1C3C2 and ΔC1C4C2

Thus, the area of quadrilateral

C1C2C3C4 = 1/2*D*H1 + 1/2*D*H2.

C1C2C3C4 = ½*D*(H1 + H2).

Area of Interest (AOI), which is included in the quadrilateral C1C2C3C4's computed area, is once more determined using the overlapped regions of all cameras. AOI region is used for people detection and tracking.

Camera Specifications

- GoPro Hero 8 camera
- 4K videos + 12MP photos
- levels of stabilization: On, Hight, and Boost; for smooth footage
- Digital Lenses: Narrow, Linear, Wide, SuperView
- Resolution: 1080 X 1920
- Capture rate: 30 fps, 60 fps, 90 fps, 120 fps

Data Acquisition and Preparation On a warm sunny day, data acquisition has taken place ahead of the C-Lite building of the KLE Technological University, Hubballi in India. We captured unscripted students standing and walking on the way to their classes. Four sequences of videos were shot for 11 min and 47 s in synchronization. In the videos captured, we observed that until the first four minutes, the student's movement in the region of interest was minimal.

Frames Extraction We obtained 60 frames per second from the videos captured by all four cameras. However, for the dataset, only two frames per second are retained. The frames are retained manually such that there is considerable people count in the retained frame. Six hundred ten frames are extracted from each camera. Table 1 desribes camera specifications. Statistics of proposed data (Table 2).

Table 3 specifies the the number of people present in the resulting frame number 256 and also people reidentified details. Total frames considered 610 X 4 = 2,440 frames, we have discarded the duplicate frames and considered 610 distinct frame for our system (Figs. 2,3,4,5 and 7).

Table 1. Camera Specifications

Sl.No	Camera	Resolution	FPS(frames per second)	Angle	Total Frames (D)	Length of the videos
1	GoPro Hero 8-C1	1920x 1080	60 fps	110° Wide angle	610	6 min 11 s
2	GoPro Hero 8-C2	1920x 1080	60 fps	110° Wide angle	610	6 min 11 s
3	GoPro Hero 8-C3	1920x 1080	60 fps	140° Wide angle	610	6 min 11 s
4	GoPro Hero 8-C4	1920x 1080	60 fps	120° Wide angle	610	6 min 11 s

Table 2. Camera to camera Distance

Camera 1	Camera 2	Distance
C1	C2	68 ft -20726.4 mm
C2	C3	54ft-16459.2 mm
C3	C1	48ft-14630.4 mm
C4	C1	55ft-16764 mm
C2	C4	53ft-16154.4 mm
C3	C4	75ft-22860 mm

Table 3. Frame and people details.

Camera	Frame Number	No.of People present in the frame	Person Reidentified
GoPro Hero 8-C1	Resulting 256th frame	11	12
GoPro Hero 8-C2	Resulting 256th frame	8	8
GoPro Hero 8-C3	Resulting 256th frame	4	4
GoPro Hero 8-C4	Resulting 256th frame	10	11

4 Person Detection and Re-identification

Associating photos or videos of the same individual captured from different perspectives and cameras is known as re-identification (PID). In a wide range of applications, from crowd traffic management to personalized healthcare, person re-identification is a helpful tool for non-invasive biometric validation, monitoring, and human-robot interaction.

We employ the YOLOv3 algorithm for person detection and re-identification in this project. You Only Look Once, Version 3 (YOLOv3) [6, 7] is a real-time object recognition

Fig. 2. GoPro camera 1 captured the 105th frame of video.

Fig. 3. GoPro camera 2 captured the 105th frame of video.

Fig. 4. GoPro camera 3 captured the 105th frame of video.

Fig. 5. GoPro camera 4 captured the 105th frame of video.

system that detects specific objects in movies, live feeds, and pictures. YOLO uses features learned by a deep convolutional neural network to detect an item. The baseline architecture used is ResNet 18 as the backbone. Additionally, the system predicts a person that a bounding box may contain while using a multiple-label classification. In the end, the network will predict a Person's ID One of the problems regarding the older versions of the YOLO model was the detection of small objects. The YOLOv3 model improves this detection.

This model locates small objects with high accuracy due to the use of multi-scale predictions, which leads to a better and more efficient detection system which made us use the YOLO v3 for person Reid. The input for this model is the video that has been scaled down to 480 X 360.

5 Results

The results of the detection and re-identification are shown in the below figures. we have randomely choosen frame number 256 from all cameras. In Fig. 6, the person detected with the pink bounding box is assigned a personal identification(PID) number. In Fig. 8 and Fig. 9, the same person in a different frame with a pink bounding box is reidentified by assigning the exact identification again. The re-identification results are accurate on average for three different views at the same timestamp.

The Results give a holistic view of the total person in the video captured. In our context, there are altogether 17 persons identified, and they are assigned PIDs. The figure also shows the number of occurrences of each person in input frames. Assigning the IDs and counting each person's occurrence is our project's primary goal, achieved with some limitations.

Fig. 6. Results of person re-identification for view 1

Fig. 7. Results of person re-identification for view 2

Fig. 8. Results of person re-identification for view 3

Fig. 9. Results of person re-identification for view 4

6 Conclusion and Future Scope

We have prepared a multi-camera dataset for this project and performed person re-identification on the dataset. The proposed system involved two parts, dataset capturing, which further involves significant data pre-processing and the extractions of the frames from the captured multi-camera videos. At the same time, we evaluated our captured dataset by detection and re-identification. Our result says person detection and re-identification are valid for only three input frames. YOLOv3 is achieving these results for detection and re-identification. There is a limitation where a person's exact match is not seen across all frames. To improvise this, work in progress for person detection and re identification and looking after all four frames for the exact PID match of a each person.

This work can also be extended to explore more research possibilities and be deployed in real-time surveillance systems. The research challenges like person ID switching and occlusions can be addressed in such systems.

References

1. Dalal, N., Triggs, B.: Histograms of oriented gradients for human detection. In: 2005 IEEE Computer Society Conference on Computer Vision and Pattern Recognition (CVPR'05), 2005, vol. 1, pp. 886–893. https://doi.org/10.1109/CVPR.2005.177
2. Alameda-Pineda, X., et al.: SALSA: a novel dataset for multimodal group behavior analysis. In: IEEE Transactions on Pattern Analysis and Machine Intelligence, vol. 38, no. 8, pp. 1707–1720, 1 Aug. 2016. https://doi.org/10.1109/TPAMI.2015.2496269
3. Ferryman., Shahrokni, A.: PETS2009: dataset and challenge. In: Proceedings of the 12th IEEE International Workshop on Performance Evaluation of Tracking and Surveillance (Winter-PETS), pp. 1–6 (2009)
4. Chavdarova, T., et al.: WILDTRACK: a multi-camera HD dataset for dense unscripted pedestrian detection. IEEE/CVF Conf. Comput. Vis. Pattern Recogn. **2018**, 5030–5039 (2018). https://doi.org/10.1109/CVPR.2018.00528
5. https://www.cuemath.com/quadrilateral-area-formulas/
6. Patil, P., Meena, S.M.: Optimization in artificial intelligence-based devices and algorithms for efficient training: a survey. In: Kaiser, M.S., Xie, J., Rathore, V.S. (eds.) Information and Communication Technology for Competitive Strategies (ICTCS 2020). Lecture Notes in Networks and Systems, vol. 190. Springer, Singapore (2021). https://doi.org/10.1007/978-981-16-0882-7_79
7. Narayankar, P., Baligar, V.P.: A systematic review of video analytics using machine learning and deep learning—a survey. In: Information and Communication Technology for Competitive Strategies (ICTCS 2020) Intelligent Strategies for ICT, pp. 659–666 (2021). https://doi.org/10.1007/978-981-16-0882-7_58

Fire Detection System Using Deep CNN

Sireesha Vikkurty[1(✉)], P. Nagaratna Hegde[1], Vennela Preethi Chinthakrinda[1], G. P. Hegde[2], and Sudheer Shetty[3]

[1] Department of CSE, Vasavi College of Engineering, Hyderabad, Telangana, India
{v.sireesha,nagaratnaph}@staff.vce.ac.in

[2] SDM Institute of Technolog, Ujire, India

[3] Department of ISE, Alvas Institute of Engineering and Technology, Moodbidri, Affiliated to Visvesvaraya Technological University, Belagavi, Karnataka, India

Abstract. The occurrence of fire should be reported by a fire alarm at the earliest. With the advent of convolutional neural networks, this has become more reliable and advanced. The detection of fire conditions as early as possible is a prerequisite of a Fire Detection System, and to give sufficient time for effective counter actions by Automated Fire personnel. Fire detection and forecast by conventional fire detection systems was achieved by using the by-products of fire such as smoke, temperature and flame which take a considerable amount of time to produce a level of fire which will trigger the smoke sensors and heat sensors. In this project, the proposed system is building a deep learning solution using convolutional neural networks. The proposed system is fine-tuned to stabilize the efficiency and accuracy by keeping in mind the nature of the target problem and fire data. The effectiveness of the proposed system is revealed and validated by the experimental results on benchmark fire datasets.

Keywords: Fire · Alert System · Convolutional Neural Network · Deep Learning Solution · Heat Sensor · Smoke Sensor

1 Introduction

This fire detection system is used to detect and warn people through alarm systems when fire is present. Good solutions to this problem could prevent fire accidents in buildings particularly at risk such as universities, schools, airports, train stations, business and factory premises, hospitals and residential homes for the elderly.

Convolutional neural networks (CNNs) play a significant role in image classification and other computer vision tasks. The ecological and social effects, caused by fires, can be reduced by the application of CNN in fire detection systems. This will substantially improve fire detection accuracy and will in the end, diminish fire disasters. However, due to increased memory consumption and high computational requirements for inference, the biggest concern with CNN- based fire detection systems is their application in real-world surveillance networks. Traditional fire alarms and vision sensor-assisted fire detection are the two widespread categories of approach in fire detection.

T. Shreekumar et al. (Eds.): ISCComm 2023, CCIS 2231, pp. 318–325, 2025.
https://doi.org/10.1007/978-3-031-75605-4_21

Traditional fire alarm systems use sensors like infrared and optical, which demand proximity to get triggered. These sensors are not well equipped to handle life- threatening circumstances and require human interference to validate a fire in the case of an alarm and often involve a visit to the place of the fire. Moreover, such systems cannot normally give details about the size, location, and burning degree of the fire. To overcome these drawbacks, numerous vision sensor-based methods have been explored by researchers in this field. These have the benefits of less human interference, faster response, affordable cost, and larger surveillance coverage. Additionally, such systems can authenticate a fire without mandating a visit to the place of fire and provide detailed information about the fire including its size, location and degree.

So, in this paper, we propose an original, energy friendly, and computationally efficient CNN architecture for fire detection, localization, and semantic understanding of the scene of the fire. This paper is further organized as follows. Literature survey included in Sect. 2; the Proposed system in Sect. 3. Implementation details included in Sect. 4 and Results and Discussion in Sect. 5. Finally, the Conclusion and future scope included in Sect. 5.

2 Literature Survey

Multiple CNNs were investigated for improving the accuracy for flame detection and to decrease the rate of false warnings in [1]. Vision-based sensors are widely used, as a replacement for traditional sensors, which provide numerous advantages, like cheaper costs, faster response times, greater coverage of surveillance areas, and least to no human intervention, thus, avoiding the necessity of visiting the location, from where the fire alert was activated [2].

This method uses the premise that fire is the only thing that has an irregular shape. However, that is not always true as there are other moving objects which also modify their shape [3]. In [4], there is a proposal for a fire detection method which compares a regular image with its color information for tunnels. This model works only in the case of fire situations that are static in nature because it uses multiple ad- hoc parameters.

Considering the features that are heuristic in [5], the writers in [6] demonstrated a detection method for fire, based on rules and decisions, using fire and actively analyzing it along with RGB/HSI color space.

Later, another probable solution was proposed, which dictates that the shape of fire is changed constantly, which enabled in differentiating it from other rigid objects in motion. An example of this method is demonstrated in [7], where a vector for a feature is pulled out using the optical flow and the characteristics of fire that are physical in nature. This can in turn, separate the fire from rigid objects in motion. A similar model which is built on constantly changing textures and shapes is presented in [8].

3 Proposed System

Most of the studies from the last ten years were concentrated on traditional methods for features extraction for flame detection. The major issue with these methods is that they are extremely time-consuming in processing of features engineering and have low performance for flame detection. A high number of false alarms are also generated by such methods, especially in surveillance with shadows, fire-colored objects, and varying lightings.

To cope with the above problems, we studied and analyzed various architectures for deep learning to detect a flame sooner. Furthermore, we explored multiple convolutional neural networks to enhance the accuracy for flame detection and reduce the rate of false positives generated. An outline sketch for the detection of fire/ flame in surveillance cameras is shown in the below figure.

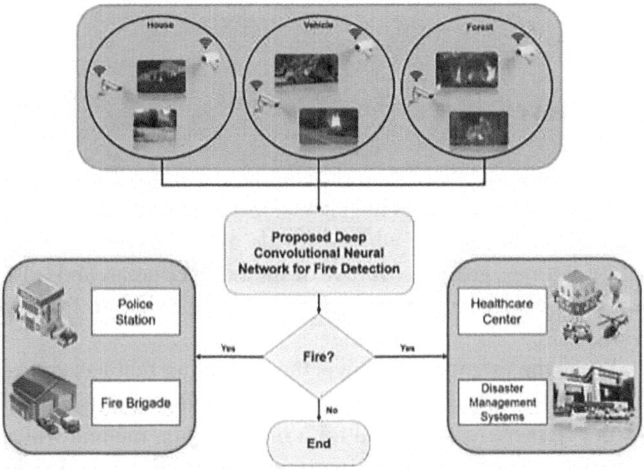

Fig. 1. Architecture

3.1 Algorithm Description

The proposed algorithm consists of the following steps:

1. Dividing video into frames using OpenCV.
2. Extracting the label from each frame name and resizing each frame to a fixed size of 128 x 128.
3. Perform one – hot encoding on the labels so that they get converted into binary labels
4. Partitioning the data into training (75%) and testing (25%).
5. Performing Image Augmentation.
6. Building layers like average pooling, dense, flatten and dropout layers to form a network or model.
7. Loading the model for testing the video.

8. Detection of fire from input videos.
9. Adding label to the frames/video.
10. Message is sent out to the fire station after detecting fire.

3.2 Flow Chart

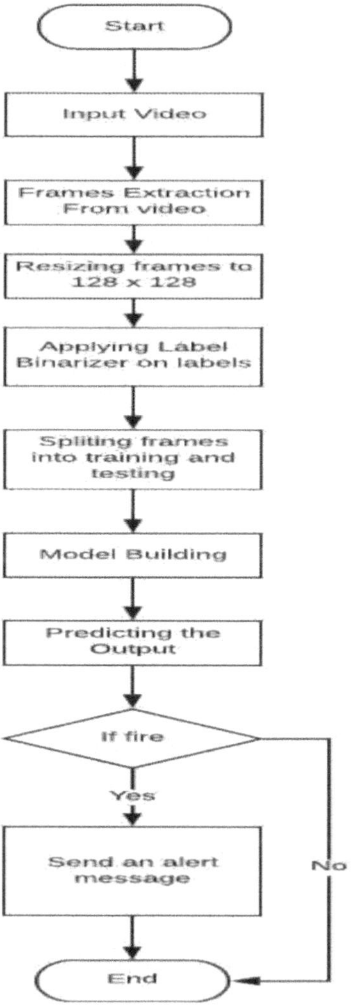

Fig. 2. Flow chart of the proposed system

4 Implementation

The implementation of this project consists of the following modules:

1. Frames Extraction
2. Model building
3. Video testing

4.1 Frames Extraction

Frames Extraction is used to convert videos into frames i.e., images. The path where the video is present sent as an argument to the function. Using OpenCV, the video is captured from the provided path. A variable for the path is considered which is used for storing the frames extracted from the video. Then the video is read, and frames are extracted using imwrite inbuilt function from cv2 library and are stored in the path given. Frames are saved by the name of the label i.e., fire or non-fire.

```
import cv2
def FrameCapture(path, i):
    pathout = "C:/Users/TOSHIBA/Desktop/major_project/sports-Type-classifier/data/no_fire"
    vidObj = cv2.VideoCapture(path)
    count = 0
    success = 1
    while success:
        success, image = vidObj.read()
        cv2.imwrite(pathout + str(i)+"frame%d.jpg" % count, image)
        count += 1
if __name__ == '__main__':
    for i in range(1,2):
        FrameCapture("C:/Users/TOSHIBA/Desktop/major_project/example_clips/test"+str(i+2)+".mp4", i)
```

4.2 Model Building

For implementing the model, path to input dataset, number of epochs to train our model, path to output serialized model, path to output label binarizer and path to output loss/accuracy plot are given as command line arguments.

```
ap = argparse.ArgumentParser()
ap.add_argument("-d", "--dataset", required=True,
    help="path to input dataset")
ap.add_argument("-e", "--epochs", type=int, default=25,
    help="# of epochs to train our network for")
ap.add_argument("-m", "--model", required=True,
    help="path to output serialized model")
ap.add_argument("-l", "--label-bin", required=True,
    help="path to output label binarizer")
ap.add_argument("-p", "--plot", type=str, default="plot.png",
    help="path to output loss/accuracy plot")
args = vars(ap.parse_args())
```

4.3 Video Testing

After Building the model, Video Testing module is used for detecting whether fire is present in video or not. If it detects fire, then an alert will be sent to the fire station. To execute this, paths to the trained serialized model, label binarizer, input video and the output video and the size of queue for averaging are send out as command line arguments.

```
ap = argparse.ArgumentParser()
ap.add_argument("-m", "--model", required=True,
    help="path to trained serialized model")
ap.add_argument("-l", "--label-bin", required=True,
    help="path to  label binarizer")
ap.add_argument("-i", "--input", required=True,
    help="path to our input video")
ap.add_argument("-o", "--output", required=True,
    help="path to our output video")
ap.add_argument("-s", "--size", type=int, default=128,
    help="size of queue for averaging")
args = vars(ap.parse_args())
```

5 Results

When a video is passed after building the model to detect fire, label gets added to the frames which is extracted from the test video. If fire is present, then it adds a label "fire" and if there is no fire then it adds a label "neutral" to each frame and the output video is saved with these labels (Figs. 1, 2, 3, 4, 5 and 6).

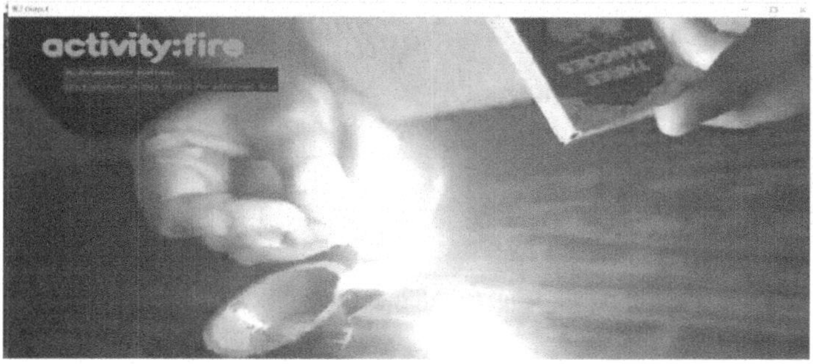

Fig. 3. Detection of Video Frame

If the model detects the fire in the video, then it will immediately send an alert message to the respective person or nearby fire station.

Fig. 4. Detection of Fire in video frame

Fig. 5. Non-fire frame

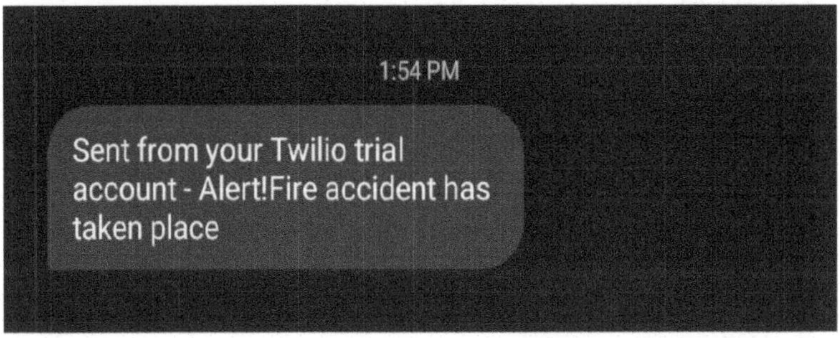

Fig. 6. Alert Message

6 Conclusion and Future Scope

Smart devices are showing promising results due to the recent improved processing capabilities, for detection of various abnormal events in surveillance systems such as fire, accidents, and other emergencies. Extreme catastrophes can be experienced if fire is not put out in time. This urges us to come up with fire detection systems that trigger an alarm while the fire is still in its early stages. Therefore, in this project we used deep learning approach.

The designed model is a relatively cheaper fire detection CNN architecture for cctv cameras. VGG16 architecture inspired this model. This prototype is fine-tuned with extra focus on computational complexity and detection accuracy. It has achieved an accuracy of around 89% with all the networks used which has better accuracy than the other existing features-based hand-crafted fire detection methods as well as the AlexNet, ResNet models' architecture fire detection methods which has an accuracy of 76%. In the future, by using GoogleNet or FireNet, the fire detection system can achieve higher accuracy.

References

1. Muhammad, K., Ahmad, J., Baik, S.W.: Early fire detection using convolutional neural networks during surveillance for effective disaster management. Neurocomputing **288**, 30–42 (2018). Available:https://www.sciencedirect.com/science/article/pii/S0925231217319203
2. Muhammad, K., Hamza, R., Ahmad, J., Lloret, J., Ge Wang, H.H., Baik, S.W.: Secure surveillance framework for IoT systems using probabilistic image encryption. In: IEEE *Transactions on Industrial Informatics*, to be published (2018). https://doi.org/10.1109/TII.2018.2791944
3. Phillips Iii, W., Shah, M., da Vitoria Lobo, N.: Flame recognition in video. Pattern Recogn. Lett. **23** 319–32 (2018)
4. Mueller, M., Karasev, P., Kolesov, I., Tannenbaum, A.: Optical flow estimation for flame detection in videos. IEEE Trans. Image Process. **22**, 2786–2797 (2013)
5. Liu, C.-B., Ahuja, N.: Vision based fire detection. In: Proceedings of the 17th International Conference on Pattern Recognition, ICPR 2004, pp. 134–137 (2004)
6. Guha-Sapir, D., Vos, F., Below, R., Penserre, S.: Annual disaster statistical review 2015: the numbers and trends, 2015. http://www.cred.be/sites/default/files/ADSR _ 2015.pdf
7. Toulouse, T., Rossi, L., Akhloufi, M., Celik, T., Maldague, X.: Benchmarking of wildland fire colour segmentation algorithms. IET Image Process. **9**, 1064–1072 (2015)
8. Foggia, P., Saggese, A., Vento, M.: Real-time fire detection for video-surveillance applications using a combination of experts based on color, shape, and motion. IEEE Trans. Circuits Syst. Video Technol. **25**, 1545–1556 (2015)
9. Sathishkumar, V.E., Cho, J., Subramanian, M., et al.: Forest fire and smoke detection using deep learning-based learning without forgetting. Fire Ecol. **19**, 9 (2023). https://doi.org/10.1186/s42408-022-00165-0

Deep Learning Classification Model for Analysing and Predicting Road Accidents

Prakash Kumar Udupi[1](✉) and Manju Jose[2]

[1] Middle East College, Muscat, Oman
prakash@mec.edu.com
[2] Majan University College, Muscat, Oman

Abstract. The Road traffic accidents are one of the major concerns in developing and developed countries, primarily due to fatalities. Rapid urbanisation, improved economic conditions, better roads, large number of people owing the vehicles resulted into more number of accidents in the recent past. Despite the implementation of various safety measures, whether in automobiles, while driving or in the surrounding traffic environment, accidents are unavoidable due to human errors, adverse weather conditions or influence of other factors such as alcohol consumption, rash driving or wrong judgements. Capturing of various accident parameters in terms of data, identifying and evaluating the patterns or pattern recognition are carried out, which are hidden in these data can help to understand various influencing factors for causing accidents. Machine learning is a technique in computer science, that facilitates the machine to learn from the input data using algorithms and predict the results. Deep learning is a subset of machine learning and artificial intelligence, used to automate the prediction of results using network models, that imitates the functioning of human brain. This research evaluates accident dataset and explores deep learning classification algorithms to predict the type of accidents.

Keywords: Machine Learning · Artificial Intelligence · Deep Learning · Classification Algorithm · Pattern Recognition

1 Introduction

Road accidents are primarily due to human errors and causes economic losses to individuals, families as well as to the nations. According to WHO, each year approximately 1.3 million people die in road accidents [1]. Approximately 93% of road accident death occurs among low income countries and middle income countries. As per the world health rankings and according to WHO data for the year 2020, approximately 3.55% of the total deaths in Oman are due to road accidents. Road accidents causes negative impact on public heath due to disabilities, injuries and deaths [2]. Rapid growth in the economy enabled growth of industrialization, especially in the automobile sectors. In addition, rapid urbanisation, improved infrastructures and roads along with increased buying power of urban populations enabled possession of large number of vehicles by

T. Shreekumar et al. (Eds.): ISCComm 2023, CCIS 2231, pp. 326–333, 2025.
https://doi.org/10.1007/978-3-031-75605-4_22

the peoples, who are living in urban areas. Increase in the possession of vehicle leads to increase in the usage of vehicles and leads to increase in accidents. Many reasons are observed as causes of road accidents. Human errors, driving under the influence of alcohol, speed driving, rash driving, unsafe road infrastructure, negligence, unsafe vehicles, distracted driving, enforcement of poor traffic laws are few reasons for the road accidents. Capturing of various road accident information in terms of data, analyzing these accident data, identifying the patterns in the available accident data can help to identify the reason for accidents, which can help to draw strategies to prevent accidents. Road accident data comprises of diversified information and consists of complex data structures.

2 Study of Existing Machine Learning and Deep Learning Framework

Machine learning is a subset of artificial intelligence that facilitates the machines to learn from the available data using various algorithm and predict the outcome as results. Machine learning enables the machine to learn from data automatically without exclusively being programmed [3]. Supervised learning, unsupervised learning, reinforcement learning are some of the categories, which are commonly used as a part of machine learning techniques for predicting the outcomes [4]. Unsupervised learning facilitates the machine to discover the patterns available in unlabeled data [5]. Clustering techniques are most commonly used unsupervised machine learning methods used in data science. Similarly, Supervised learning are task driven algorithms, uses labelled data. The machine learns from these labeled data and predicts the outcomes [6]. Classification and regression techniques are the most common supervised learning techniques used in machine learning [7].

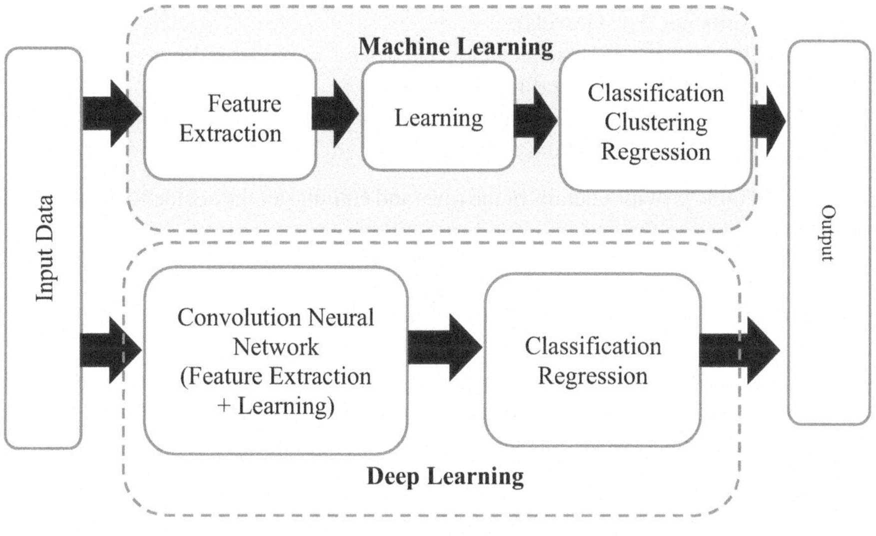

Fig. 1. Machine learning and deep learning framework.

Deep learning is a subset of machine learning and artificial intelligence, that facilitates learning through data, primarily derived from artificial neural network techniques[8]. Deep neural network using supervised learning techniques and unsupervised learning techniques are used for classification and regression applications (Fig. 1).

Accident data can be applied with machine learning techniques or deep learning techniques and various different outcomes such as type of accidents, accident categories, accident outcomes or any other accident information can be predicted based on the availability of different attributes in the accident dataset. Hence, patterns hidden in accident data under difference circumstances can be detected, identified, evaluated and different predictions of accidents can be obtained using machine learning or deep learning techniques [9].

3 Designing and Description of Dataset

In order to evaluate the deep learning framework, random data are assumed for simulations by following relevant deep learning framework requirements. Deep learning and data structure rules are followed during the construction of dataset and deep learning algorithm requirements are studied during construction of dataset. Accident prediction dataset consists of 8 column variables comprises of total 150 row random dataset. Following variables are used for analyzing classification of accidents.

- Speed: Continuous data describes speed of the vehicle
- Brake Power: Discrete data represents the brake power applied at the time of accident.
- Driver's Age: Discrete data represents age of the vehicle driver
- Vehicle Life: Discrete data represents life of the vehicle
- Road Type: Discrete data represent type of road. (1 = Highway, 2- Urban Road, 3 –Cross Roads)
- Road Surface: Discrete data represents road surfaces (1 = Pavements, 2 = Thin Asphalt Pavements, 3 = Gravels)
- Weather: Discrete data represents weather types (1 = Dry, 2 = Sunny, 3 = Cloudy, 4 = Wet, 5 = Rainy, 6 = Drizzling, 7 = Windy)
- Accident Types: Nominal data represents type of accident. (Head-on Collision, Interstate accidents, Intersection accidents)

Following table provides details of the rows and columns of the accident dataset along with data displayed in the first and last 3 rows (Table 1).

4 Evaluation of Deep Learning Classification Algorithm

Deep learning facilitates solutions through classification and regression methods. Both classification and regressions are supervised machine learning techniques and uses labelled data for evaluation.

Table 1. Description of accident dataset

Speed	Brake Power	Driver's Age	Vehicle Life	Road Type	Road Surface	Weather	Accident Types
124.2	56	22	10	1	3	4	Head-On Collision
112.7	58	29	5	3	2	3	Head-On Collision
118.7	60	28	10	2	1	7	Head-On Collision
-	-	-	-	-	-	-	-
-	-	-	-	-	-	-	-
-	-	-	-	-	-	-	-
49.8	42	58	14	3	1	3	Intersection accidents
35.1	49	61	11	1	2	2	Intersection accidents
78.6	46	64	14	2	2	4	Intersection accidents

4.1 Neural Network Classification Algorithm

Neural network classification algorithm can be used for featuring the categorization of output results. Artificial neural network comprises of inter connected networks along with one input layer, one output layer and one or more intermediate layers [10]. Hence, at the end of classification process, neural network provides categorization of data along with identifiable groups.

4.2 Classification Metrics

The process of neural network classification algorithm provides one hidden layer with 8 node network for the given input accident dataset. Artificial neural network algorithm is applied on the accident dataset and following results obtained after the deep learning process (Table 2).

Table 2. Neural Network Classification Result Table

Hidden Layers	Nodes	N (Train)	N (Validation)	N (Test)	Validation Accuracy	Test Accuracy
1	8	96	24	30	0.958	1.000

The classification result table demonstrates the deep learning process splits the data into training, testing and validation dataset. The machine learns the patterns in the training dataset [11–14], validates the learning and predict the outcomes for the 30 test dataset with the validation accuracy of 0.958 and test accuracy of 100 percent. Hence, for the given accident dataset, deep learning neural network classification algorithm provides the best possible predictions in terms of classification. The process uses artificial neural network algorithm for leaning the patterns in the dataset and subsequently predicting the results.

4.3 Confusion Matrix

Confusion matrix measures the performances of deep learning classification techniques [11]. Confusion matrix is visualized in the form of a table and predicted versus observed data are displayed in the table format (Table 3).

Table 3. Confusion Matrix Table

		Predicted		
		Head-On Collision	Intersection accidents	Interstate accident
Observed	Head-On Collision	14	0	0
	Intersection accidents	0	8	0
	Interstate accident	0	0	8

The confusion matrix table demonstrates three observed classification versus three predicted outcomes as shown in the table. From the table, we observe that, observed and predicted data are same and provides 100 percent accuracy.

4.4 Evaluation Metrics

Evaluation metrics of deep learning classification outcome is in the form of a table, comprises of various evaluation results such as accuracy, precision, recall, correlation coefficient etc. (Table 4).

The important information such as accuracy demonstrate the efficiency of the predictions resulted by applying deep learning algorithm on the accident dataset. From the table, we observe that, all the thee classification of accidents predicted perfectly. Hence, 100 percent accuracy achieved.

Table 4. Evaluation Metrics Table

Evaluation Metrics	Head-On Collision	Intersection accidents	Interstate accident	Average / Total
Support_ad	14	8	8	1.000
Accuracy_ad	1.000	1.000	1.000	1.000
Precision_ad (Positive_Predictive_Value)	1.000	1.000	1.000	0.000
Recall_ad (True_Positive_Rate)	1.000	1.000	1.000	0.000
False_Positive_Rate_ad	0.000	0.000	0.000	1.000
False_Discovery_Rate_ad	0.000	0.000	0.000	1.000
F1_Score_ad	1.000	1.000	1.000	0.813
Matthews_Correlation_Coefficient_ad	1.000	1.000	1.000	1.000
Area_Under_Curve_ad_(AUC)	0.500	1.000	0.938	1.000
Negative_Predictive_Value_ad	1.000	1.000	1.000	0.000
True_Negative_Rate_ad	1.000	1.000	1.000	0.000
False_Negative_Rate_ad	0.000	0.000	0.000	∞
False_Omission_Rate_ad	0.000	0.000	0.000	1.000
Threat_Score_ad	∞	∞	∞	1.000
Statistical_Parity_ad	0.467	0.267	0.267	1.000

4.5 Neural Network Structure Plot

Neural networks provides multi layer network formation of neurons, which are used for classifying and predicting the results (Fig. 2).

During the learning process, for the seven different input variables of accident dataset, one hidden layer is created. Neural network technique interconnects various nodes. By observing the lines connected between input layer, hidden layer and output layer, we can observe how data travels through these connections.

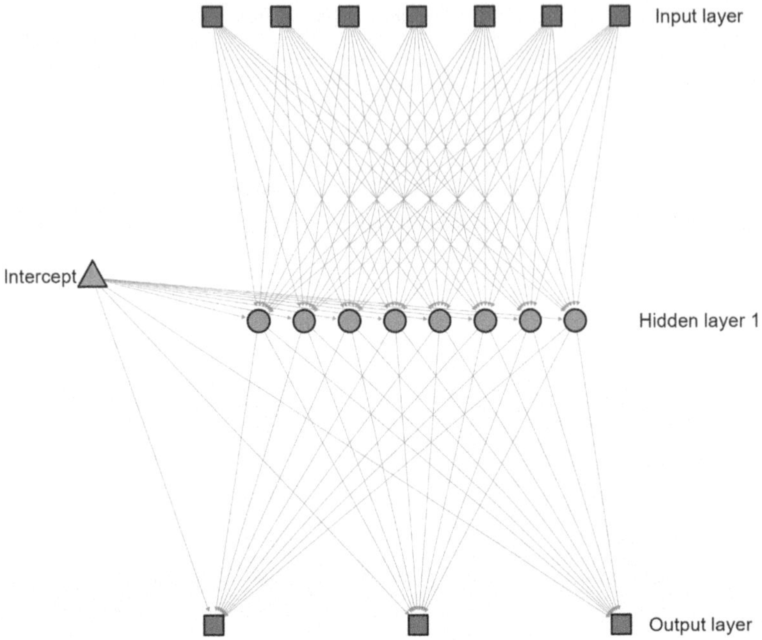

Fig. 2. Neural Network Model for Accident Dataset

5 Conclusion

Deep learning facilitate neural network approaches to classify and predict the outcomes. Hence, deep learning enables robust learning process for the input dataset. Further, deep learning process outcomes are more accurate and can handle large dataset. Accident dataset consists of diversified and complex data structures. For the sample dataset, we observed that the deep learning approach provides best results for predicting and classifying different types of accidents. The same classification process can be further extended to predict various other attributes such as weather, type of roads, type of road surfaces or any other relevant variables. Similarly, deep learning regression techniques can also be used for predicting vehicle speed, driver's age, brake power, vehicle life or any other relevant parameters. Hence, by predicting such information related to road accidents will help to make proper decision to monitor, control and reduction of road accidents.

References

1. Peden, M., et al.: World Report on Road Traffic Injury Prevention. WHO, Geneva (2004)
2. Government of India: Road accidents in India 2020. Ministry of Road, Transport and Highways, Transport Research Wing (2020)
3. Sarker, I.H.: Machine learning- algorithms, real-world applications and research directions. SN Comput. Sci. **2**, 160 (2021). https://doi.org/10.1007/s42979-021-00592-x
4. Said, A.S., Muhammed, T.A., Ashraf, M.E., Khaled, S.: Machine learning and deep learning techniques for cybersecurity-a review. In: Proceedings of the International Conference on Artificial Intelligence and Computer Vision (AICV2020), pp 50–57 (2020)

5. El Bouchefry, K., de Souza, R.S: Learning in big data: Introduction to machine learning. Knowledge Discovery in Big Data from Astronomy and Earth Observation. Elsevier Science, United States, pp 225–249 (2020)
6. Singh, A., Thakur, N., Sharma, A.: A review of supervised machine learning algorithms. In: 2016 3rd International Conference on Computing for Sustainable Global Development (INDIACom), pp. 1310–1315 (2016)
7. Li, J.: Regression and classification in supervised learning. In: Proceedings of the 2nd International Conference on Computing and Big Data (ICCBD 2019), Association for Computing Machinery, New York, NY, USA, 99–104 (2019) https://doi.org/10.1145/3366650.3366675
8. Sarker, I.H.: Deep learning: a comprehensive overview on techniques, taxonomy, applications and research directions. SN Comput. Sci. **2**(6), 420 (2021) https://doi.org/10.1007/s42979-021-00815-1. Epub 2021 Aug 18. PMID: 34426802; PMCID: PMC8372231
9. Devi, S.S., Parthasarathy, G., Asif, L.M., Dylan, P.N.C., Parasuraman, B., Vinoth, R.: Prediction of accident prone-zone and alert messages to the hospitals using machine learning and internet of things. Eng. Technol. Res. **3**(2), 025–031 (2020)
10. Bala, R., Kumar, D.: Classification using ANN: a review. IJCIR **13**(7), 1811–1820 (2017)
11. Udupi, P.K.: Practical Machine Learning: No-coding, Easy and Fast Track using JASP – A Free and Open Source Software (2023). amazon.com
12. Pandey, K., Udupi, P.K., Vishal, D.: Artificial Intelligence in Supply Chain and Logistics (2023). amazon.com
13. Sharma, N., Udupi, P.K., Vishal, D.: Artificial Intelligence in Human Resource Management (2023). amazon.com
14. Udupi, P.K., Artificial Intelligence in Fashion Industry (2023). amazon.com

Web Scraping System for Recommendation of Mobiles from Multiple E-commerce Websites

Shivananda V. Seeri(✉), K. P. Sonia Peal, Sandeep Seetaram Naik, and A. Aneesh Kumar

Mangalore Institute of Technology and Engineering, Mangaluru, Karnataka, India
seeri123@gmail.com

Abstract. Recent advancements in Internet technologies have led to the proliferation of online shopping portals, offering customers the convenience of purchasing a diverse array of products from the comfort of their homes. However, navigating through these online portals can be time-consuming, as they often feature numerous similar products with varying price tags and brands. Analyzing customer comments manually to pinpoint the top products can consumes significant amount of time. Automated recommendation systems have emerged as a valuable solution to help customers make informed online purchases that align with their preferences and needs. These systems leverage technology to assist users in finding the best-branded products in an efficient manner. The continuous innovation and development in Information Technology have led to the proliferation of numerous E-commerce websites, motivating customers to explore and purchase products online without the need to visit physical stores. However, in E-commerce portals, customers are often required to manually sift through the vast product offerings to identify the best-branded products. The paper proposed a novel automated recommender system that leverages web crawling and scraping methods to find the best deals on E-commerce websites. The scraping dynamically scan HTML elements/tags and process queries without storing information in a local database, thereby improving storage efficiency and processing power.

Keywords: Web-scrap · Recommendation · Polarity · E-commerce · Requests Beautiful-Soup

1 Introduction

The web page encompasses a diverse range of textual and non-textual data. The wealth of data present on web pages can encompass both advantages and disadvantages aspects. The sheer magnitude of data generated on web portals inspires the development of new techniques and technologies for swift and convenient data access. The efficiency of information retrieval relies on both user queries and the characteristics of the consumer-data stored in the database. E-commerce holds a pivotal role in contemporary business by eliminating the need for scheduled shopping and overcoming geographical constraints. Within E-commerce portals, users can effortlessly access product details and brand

T. Shreekumar et al. (Eds.): ISCComm 2023, CCIS 2231, pp. 334–341, 2025.
https://doi.org/10.1007/978-3-031-75605-4_23

information at their fingertips. This convenience allows customers to make purchases without the necessity of visiting physical markets or shops, ultimately saving time and energy. As more businesses transition online, the a surplus of shopping options are available for buyers to quickly find items, even though it may require additional time to locate their preferred products.

Web-scraping is used for extracting large amounts of unstructured-data (HTML) from web-portals. This method transforms unstructured website data into structured data tailored to user interests. Web scraping emerges as a powerful method for data mining, extracting valuable insights from extensive web data.

This paper introduces a novel recommender system based on web-scraping, designed to identify the best offers on mobile products from E-commerce websites. Therefore, the proposed system assists customers in purchasing smart phones based on their preferences, from either Amazon or Flipkart portals.

2 Literature Survey

Sanjay Kumar and Rizvi [1] have investigated web mining based on web-usage mining, web-scrapping, and semantic annotation. The experimental results demonstrated the success of the proposed method.

Osmar Castrillo-Fernández [2] has presented different techniques used for web scraping and web crawling. The paper demonstrates the need for web-scraping for websites, PDF documents, and web-scraping tools.

Glez-Pena et al. [3] have discussed the importance of methods to scrape data from websites in the current context and proposed a novel way of building a web data scraper. The paper implemented various web-scraping techniques using web-scraping APIs.

Giulio Barcaroli et al. [4] conducted a survey paper on the text-mining and web-scraping processes, focusing on contributions by various authors in the field of ICT across multiple enterprises.

Umit Turkut and colleagues [5] proposed an internet-based recommendation system utilizing deep learning specifically designed for textile products. They employed a convolutional-neural- network (CNN) to classify diverse patterns and preferences of both users and designers. The deep learning model suggests various strategies for recommending color-compatible textile products.

Mana and Sasipraba [6] have implemented AI models for product and service recommendation systems. The authors have developed two prototype models: (1) for recommending banking services and (2) for recommending movies. Machine learning techniques and various types of scraped data filtering are adopted in the proposed-novel recommendation system.

Kiran et al. [7] developed product recommendation system based on sentiment analysis of product reviews and ratings. Each product has a rating system, and the ratings or reviews given for the product are considered for the sentiment analysis. Each comment has been tagged with a score based on its polarity value: positive, neutral, or negative. The total product rating is computed by combining the individual feature scores. This approach is advantageous for customers seeking particular features in a product.. The system recommends a product that has mostly positive reviews.

Nan Hu et al. [8] have developed a novel technique for the decision support system to recognize the interdependence among sales, review observations, and product ratings. It is found that the most favorable and recent consumer reviews impact product sales. The paper also presented validations of the opinion or polarity ratings, which were calculated using human opinion on a subset of the feedback.

Venkata Rajeev and Smrithi Rekha [9] have proposed a product recommendation system based on opinion mining of product reviews and features given in online portals. The helpfulness score, Customer-Age, star rating, reviews, and review polarity are used for recommending products to customers. This model graphically depicts which of the two products is superior based on the aforementioned criteria. Natural-language-processing was adopted for reviews to ascertain their sentiment. The subjective and objective parameters are combined to calculate the product score for the Flipkart website's mobile phone reviews.

Aciar Silvana et al. [10] have suggested a recommendation system grounded in a prioritizing mechanism for consumer product reviews on digital cameras. The consumer reviews are provided in a blended of structured and unstructured data. Ontology techniques are used to convert reviews into a structured format, and the product is graded based on user comments. Consumer remarks on each product's features are used to rank the product's quality. The digital camera is recommended based on the consumer's feedback and the product ranking."

Parneet Kaur [11] introduces an innovative method for conducting sentiment analysis on live news data by employing web scraping techniques and leveraging supervised-machine-learning algorithms such as Naïve Bayes, and Logistic Regression. This approach involves scrutinizing sentiments within diverse news headlines and statements, aiming to assess the impact of positivity, negativity, or neutrality in the content.

Saurabh Sahu et al. [12], Proposed is a novel approach employing long short-term memory (LSTM) and a combination of convolutional neural network with LSTM (CNN-LSTM) models for sentiment analysis. The data undergoes preprocessing techniques, including stop word removal, punctuation removal, and tokenization, resulting in a high accuracy rate of 97.8%.

3 Methodology

The proposed methodology for web scraping details of mobile phones from E-commerce websites consists of the following stages:

1. Selecting E-commerce websites (Flipkart and Amazon)
2. Gather the mobile features: product name, price, ratings, battery power, memory, etc.
3. Perform Sentiment analysis on the consumer reviews
4. Compare the product features available on different E-commerce websites and recommend the mobile product.

The proposed web scraping system is implemented using Python, with experimentation conducted on two E-commerce websites: Flipkart and Amazon. Web scraping is performed using the Beautiful Soup tool, while sentiment analysis of consumer ratings (reviews) was conducted employing natural-language-processing (NLP) tools. The system searches for different brands of mobile products on both websites and displays their details of comparison for potential buyers. The stages of the proposed system are described below.

A. Working procedure

The proposed system is implemented using the following steps:

1. Import the Python libraries.
2. Fetch the URL using Python Requests module
3. Parse the HTML and convert it into Beautiful Soup format.
4. Scrap product label, price, specification, and image.
5. Analyze the product's details.
6. View scraped data according to details and then compare

B. Web Scraping

The proposed system is implemented using Python Requests and Beautiful Soup libraries. Python Requests is utilized for making HTTP requests to specific URLs, ensuring reliable communication with web services and facilitating connection pooling. This library automatically handles parameter encoding and interprets responses in Unicode. The Python Requests module supports various built-in methods for making HTTP requests, including POST, PUT, GET, PATCH, and HEAD requests.

Beautiful Soup (Beautifulsoup3) is employed to extract data from HTML and XML files. It creates a parse tree from the source code of a web page, allowing data extraction in a hierarchical and more readable manner. This library provides a parse tree and functions to navigate, search, or modify this parse tree. It generates a parse tree based on the Document Object Model (DOM). Filter functions are used to locate specific tags, strings, attributes, or combinations thereof.

To begin parsing, the document is passed as an argument to the Beautiful Soup function, which then converts the document to Unicode and its elements to Unicode characters. By default, Beautiful Soup uses the HTML parser for text files formatted in HTML, but other parsers can also be employed with Beautiful Soup.

C. Proposed Methodology

The block diagram of the proposed system is presented in Fig. 1. The system comprises the following stages: selecting a website, obtaining the HTML content, extracting mobile features and comment data, storing and analyzing data, and displaying comparison data. The proposed system scrapes mobile details from both Flipkart and Amazon websites and displays them. Furthermore, consumer reviews (ratings/comments) that are extracted undergo sentiment analysis. The block diagram depicting sentiment analysis is displayed in Fig. 2.

D. Sentimental Analysis

Sentiment analysis, a sub-field of Natural Language Processing (NLP), measures people's opinions as Positive, Negative, or Neutral by analyzing comments or feedback provided on social media or online shopping portals. This type of data is typically unstructured.

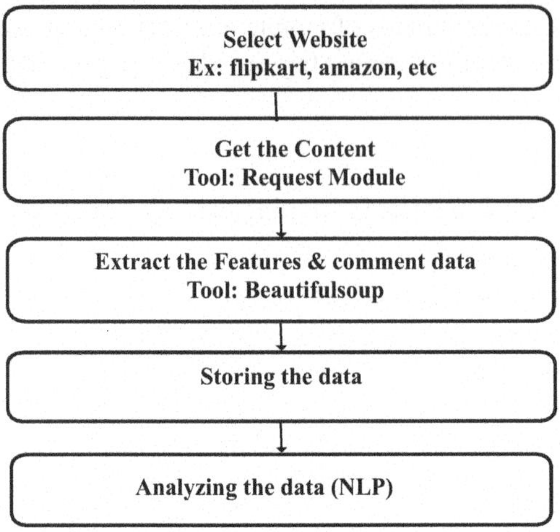

Fig. 1. Block diagram of proposed system

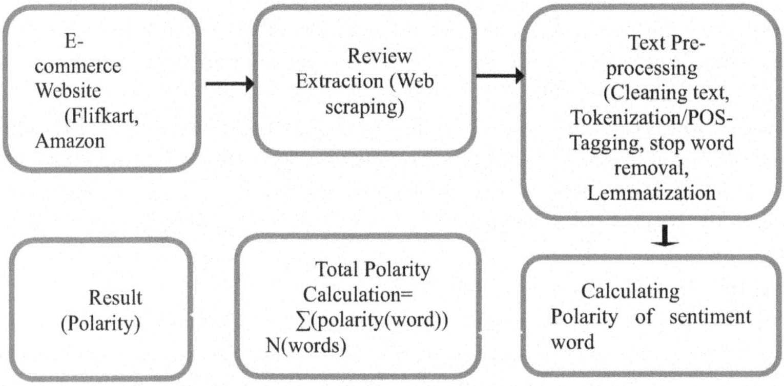

Fig. 2. Block diagram of sentimental analysis

Sentiment analysis can be implemented using two main methods: Rule-based and Machine Learning-based. In the proposed system, Rule-based sentiment analysis is employed. The rule-based approach is a practical method that uses a predefined set of rules to label text as positive, negative, or neutral.

Data preprocessing steps in NLP:

1. Removing punctuations like., ! $() * % @
2. Tokenization
3. POS tagging
4. Removing Stop words
5. Obtaining the stem words

In the cleaning step, special characters and numbers are removed from the text using regular expression operations. In the tokenization step, the text is decomposed into small parts known as tokens. In the POS tagging step, each token is converted into tuples, with each tuple comprising (word, tag). POS tagging is also applied to maintain the context of the word for lemmatization.

In the fourth step, stop words are eliminated. Stop words in English are frequently used words, such as 'a,' 'the,' 'is,' and 'are,' which convey little meaningful information for extracting the essence of the text. Removing stop words enables us to concentrate more on crucial information.

In the last step, words/tokens are reduced to their root forms by eliminating unnecessary characters, aiding in the preprocessing of text, words, and documents for text normalization. Finally, sentiment analysis is applied to calculate the polarity. Based on the polarity value, comments are tagged as positive, negative, or neutral.

4 Results

The Python programming language is used to implement the suggested model and tested on both Flipkart and Amazon web portals for various brands of mobiles, each with multiple features and comments. Table 1 represents the details fetched from the Flipkart website, while Table 2 represents the details fetched from the Amazon website. Basic features such as product name, RAM, ROM, price, and ratings of each mobile are considered for extraction.

The proposed system extracts mobile features for the same brand from both Flipkart and Amazon web portals, enabling users to effortlessly compare products prior to finalizing their purchase choice. Finally, comments for each mobile are fetched from the websites, and sentiment analysis is applied to compute a polarity factor based on the reviews or comments. If the mobile has received positive reviews, the polarity factor will be greater than zero, and the system recommends buying the product. If the polarity is zero or less than zero, the system displays a neutral or negative status, respectively.

Finally, the inference module compares the product details in Tables 1 and 2 and makes inferences using the following criteria: If the polarity is positive, and the rating is greater than 3, the corresponding E-commerce site (Amazon/Flipkart) is recommended for mobile purchases based on the lower price of the product. Table 3 displays the

Table 1. Flipkart details of Samsung Galaxy mobile

SAMSUNG Galaxy							
Model	Colour	RAM	ROM	Price	Ratings	Polarity Factor	Polarity
M42	Prism Dot Gray	8 GB	128 GB	20948	4.3	0.405731	Positive
A52s 5G	Awesome Black	8 GB	128 GB	32499	4.2	0.353662	Positive
A22 5G	Gray	8 GB	128 GB	21999	4.1	0.423458	Positive
A22 5G	Mint	8 GB	128 GB	19949	4.1	0.402819	Positive
A22 5G	Violet	8 GB	128 GB	19999	4.1	0.411323	Positive
S20 FE	Cloud Navy	8 GB	128 GB	49999	4	0.417166	Positive
S20 FE	CLoud Mint	8 GB	128 GB	38990	4	0.419961	Positive

recommendations of the system for different brands of mobile products based on ratings and the polarity factor.

Table 2. Amazon details of Samsung Galaxy mobile

SAMSUNG Galaxy							
Model	Color	RAM	ROM	Price	Ratings	Polarity Factor	Polarity
M42	Prism Dot Gray	8 GB	128 GB	24999	4.1	0.202920	Positive
A52s 5G	Awesome Black	8 GB	128 GB	32499	4.2	0.191371	Positive
A22 5G	Gray	8 GB	128 GB	20999	4.0	0.171051	Positive
A22 5G	Mint	8 GB	128 GB	20899	3.6	–0.006786	Negative
A22 5G	Violet	8 GB	128 GB	17849	3.6	–0.006796	Negative
S20 FE	Cloud Navy	8 GB	128 GB	39999	4.4	0.206847	Positive
S20 FE	Cloud Mint	8 GB	128 GB	42920	3.4	0.419961	Positive

Table 3. Recommendation based on the ratings and polarity factor

SAMSUNG Galaxy				
Model	Color	Price (In Rs.)	Ratings	Recommended Ecommerce website
M42	Prism Dot Gray	20948	4.1	Flipkart
A52s 5G	Awesome Black	32499	4.2	Flipkart/Amazon
A22 5G	Gray	20999	4.0	Amazon
S20 FE	Cloud Navy	39999	4.4	Amazon

5 Conclusion and Future Scope

These days, an increasing number of individuals favor online shopping to traditional in-person shopping. With the increasing number of online E-commerce websites, competition between them is on the rise. This often leaves customers in a state of confusion when choosing a website for shopping, one that offers maximum benefits and quality products. Web scraping techniques can significantly reduce the effort customers need to put into online searches, helping them easily find and purchase products from various websites offering high-quality items with attractive features. In this paper, we introduce a novel web scraping system designed for searching mobile phones on both Flipkart and

Amazon websites. The goal is to identify widely accepted branded mobiles with excellent features. Additionally, sentiment analysis is applied to comments provided about mobiles on both Flipkart and Amazon websites to recommend suitable mobiles for purchase by customers. The proposed system can be easily scaled up to include additional websites and can be generalized for different types of products.

References

1. Malik, S.K., Rizvi, S.A.M.: Information extraction using web usage mining, web scrapping and semantic annotation. In: 2011 International Conference on Computational Intelligence and Communication Networks, pp. 465–469. IEEE (2011)
2. Castrillo-Fernandez, O.: Web scraping: applications and tools. European Public Sector Information Platform Topic Report 2015 (2015)
3. Glez-Pena, D., Lourenc¸o, A., Lo´pez-Ferna´ndez, H., Reboiro-Jato, M., Fdez-Riverola, F.: Web scraping technologies in an API world. Briefings Bioinform. **15**(5), 788–797 (2014). https://doi.org/10.1093/bib/bbt026
4. Barcaroli, G., Nurra, A., Scarno‘, M., Summa, D.: Use of web scraping and text mining techniques in the istat survey on information and communication technology in enterprises. In: Proceedings of Quality Conference, pp. 33–38 (2014)
5. Turkut, U., Tuncer, A., Savran, H., Yılmaz, S.: An online recommendation system using deep learning for textile products. In: 2020 International Congress on Human-Computer Interaction, Optimization and Robotic Applications (HORA), pp. 1–4. IEEE (2020)
6. Mana, S.C., Sasipraba, T.: A machine learning based implementation of product and service recommendation models. In: 2021 7th International Conference on Electrical Energy Systems (ICEES), pp. 543–547. IEEE (2021)
7. Kiran, M.V.K., Vinodhini, R.E., Archanaa, R., Vimalkumar, K.: User specific product recommendation and rating system by performing sentiment analysis on product reviews. In: 2017 4th International Conference on Advanced Computing and Communication Systems (ICACCS), pp. 1–5. IEEE (2017)
8. Hu, N., Koh, N.S., Reddy, S.K.: Ratings lead you to the product, reviews help you clinch it: the dynamics and impact of online review sentiments on products sales (2013). Decision Support Systems
9. Rajeev, P.V., Rekha, V.S.: Recommending products to customers using opinion mining of online product reviews and features. In: 2015 International Conference on Circuits, Power and Computing Technologies [ICCPCT-2015], pp. 1–5. IEEE (2015)
10. Aciar, S., Zhang, D., Simoff, S., Debenham, J.: Recommender system based on consumer product reviews. In: 2006 IEEE/WIC/ACM International Conference on Web Intelligence (WI 2006 Main Conference Proceedings)(WI'06), pp. 719–723. IEEE (2006)
11. Kaur, P.: Sentiment analysis using web scraping for live news data with machine learning algorithms. Mater. Today Proc. Science Direct, **65**, Part 8, 3333–3341 (2022)
12. Sahu, S., Divya, K., Rastogi, N., Yadav, P.K., Perwej, Y.: Sentimental analysis on web scraping using machine learning method. J. Inform. Comput. Sci. **12**(8), 24–29 (2022). ISSN: 1548–7741.

A Machine Learning Approaches with Special Emphasis for Identification of Plant Leaf Diseases in Crop Yield Prediction

Ch Ravindra Babu[1(✉)], E. V. N. Jyothi[1], K. Naga Prassana[2], T. Siva Jayanth[2], Sk. Sandani[2], S. Veera Venkata Gopi Chand[2], and R. J. Shwetha[3]

[1] Pace Institute of Technology and Sciences, Ongole, AP, India
ravindra.chimmiri@gmail.com

[2] IV B. Tech CSIT, Pace Institute of Technology and Sciences, Ongole, AP, India

[3] Global Payex Technologies Pvt. Ltd., Mumbai, India

Abstract. Productivity in farming is very essential to the Indian economic system. When a plant gets any disease it can significantly reduce production, cost money and reduce the quality and quantity of farm produce. To prevent agriculture from losing quantity and quality, it is essential to recognise diseases of plants. Plant disease identification is receiving more and more attention these days since large crops area is being monitored. Monitoring plant health and identifying diseases is critical for sustainable agriculture. This requires a lot of work, knowledge of plant diseases, and a lot of processing time. Utilizing Machine Learning algorithms, To predict increased agricultural yields, crop yield projections are made. One of the difficult problems in the agricultural industry is this. Given the growing significance of agricultural yield prediction, with a concentrate on agricultural yield prediction, it is important to apply machine learning techniques to crop yield predictions. Techniques like decision tree, support vector machine, Bayesian Network, and random forest, among others, are used in Machine Learning support in the automatic identification of plant disease based on the plant's visual symptoms. This research presents an analysis of several machine learning approaches for plant disease prediction. The outcomes demonstrate that this technique conducted illness diagnosis with success and enhanced diffusion pattern analysis and disease progression level detection accuracy. Thus machine learning approaches with special emphasis for identification of plant leaf diseases in crop yield prediction results show that plant diseases can be accurately classified.

Keywords: crop yield prediction · Random Forest (RF) · Support Vector Machine (SVM) · Decision Tree (DT) · Bayesian Network · Machine Learning (ML)

1 Introduction

One of the most popular occupations in India is farming. All countries economies depend heavily on agriculture. Due to food shortages or scarcity caused by an increasing global population, the agricultural industry has to be upgraded in order to sustain the current

T. Shreekumar et al. (Eds.): ISCComm 2023, CCIS 2231, pp. 342–353, 2025.
https://doi.org/10.1007/978-3-031-75605-4_24

scenario [1]. In order to guarantee agricultural development and production in a timely and dependable way, there are interconnected effects from an increasing population, soil loss, natural climatic changes, and demand practises that change with the climate. Additionally, it ought to increase the viability of agricultural food production [2]. These needs show that agricultural yield estimation, crop protection, and land value are becoming more crucial to the world's food production. As a result, accurate crop output estimation will be dependent on the country's policymakers obtaining desirable export and import valuations in order to improve national food security.

Both bacterial and fungal infections can harm crops. The impact on farmers output is significant. Crops need to be healthy in order to yield their maximum. Diagnosing diseases with our eyes always remains a complicated process. The farm must be continuously watched in order to achieve success. When the farm is big, it is also very expensive. Agricultural experts are also struggling to diagnose diseases and find solutions due to this challenge. Farmers would benefit greatly from a machine that could recognize plant diseases. This method can inform farmers in a timely manner so they may take the appropriate procedures.

Therefore, it is challenging to forecast agricultural productivity due to a number of complicated aspects. Crop output is affected by landscapes, soil quality, insect infestations, genotypes, water quality and availability, meteorological conditions, harvest scheduling, and other variables. Time-specific and essentially nonlinear processes and approaches are used to predict crop yields [16]. These approaches are also complex due to the inclusion of a wide range of associated parts that non-arbitration and outside sources can specify and impact.

Plant diseases are caused by many organisms such as infections, germs, and fungus. The automated disease detection procedure improves in the early detection of plant pathology. Early disease diagnosis has a positive impact on plant health. Most disease symptoms show on the leaves, branches, and fruits. Plant leaf indicators are used to diagnose illness faster, more accurately, and at a cheaper cost. In general, farmers use visual inspection to identify plant diseases, which enables disease detection and diagnosis. When working with large farms, this requires a significant number of specialists and continuous plant monitoring, both of which have costs. Farmers in certain countries, however, without the necessary resources or knowledge on how to approach experts. Professional consultation is therefore both expensive and time-consuming.

It seems like the best way to deal with this problem is through Machine Learning. There are several ML algorithms that can classify and diagnose plant diseases automatically. The leaves, stem, seeds, and fruits are only a few of the elements of a plant that are affected by diseases. The diseases have varying effects on different areas of the plant's body [18]. The process of photosynthesis can only be accomplished with the help of leaves, despite the fact that the central component of a plant is commonly referred to as its leaves.

A plant's life cycle is immediately affected when it suffers from a foliar disease. To fight these diseases effectively, a method for automatically detecting and classifying diseases must be established. ML is a potential solution to this problem. Several ML methods have recently been proposed for the identification and classification of plant diseases.

2 Literature Survey

R. Reda, B. Ilham, T. Saffaj, O. Saidi, L. Brahim, K. Issam, and E. M. El Hadrami et al. [3] Soil samples were taken from four agricultural fields in Morocco, and the total nitrogen (TN) and soil organic carbon (SOC) were estimated using machine learning (ML) techniques. In contrast to conventional chemical procedures, near-infrared spectroscopy data are used since they require less processing time and use less resources. In contrast to back-propagation neural networks (BPNN) and alternative regression models, the ensemble learning modelling approach performs the best.

Y. Cai, D. Lobell, T. Xu, K. Guan, J. Peng, S. Wang, B. Potgieter, S. Asseng, A. Y. Zhang, L. You, and B. Peng et al. [4] Over fourteen years, ML algorithms were used to estimate wheat yield in Australia utilizing two data sources, meteorological data and satellite data (NN, SVM, and RF). According to the simulation's outcomes, when compared to satellite data for yield prediction, meteorological data provide different information. Crops planted on time are more productive and suffer less financial loss, which is exactly timing is so crucial.

Mohsen Azadbakht, Soheil Radiom, Hossein Aghighi, Davoud Ashourloo, and Abbas Alimohammadi et al. [5] Performance was examined for to identify wheat leaf rust, use the high, medium, and low leaf area index (LAI) levels. Four methods namely random forest regression (RFR), gaussian process regression (GPR), v-support vector regression (v-SVR), and boosted regression tree (BRT) were used to describe the performance of rust detection. To evaluate performance, an experiment was carried out on 7000 hectares of wheat farming in North West Iran. For the same purpose, hyperspectral reflectance data collected under various environmental circumstances is used. The results of a comparison between ML and the Spectral Vegetation Index (SVI) demonstrate that ML exceeds SVIs.

Kerim Karada˘g, Mehmet Emin Tenekeci, Ramazan Ta¸saltın, and Ay¸sin Bilgili et al. [6] A technique has been proposed to aid in the detection of fusarium wilt in pepper. The spectroradiometer can learn more about the plant by observing how it reflects light. Between 350 and 2500 nm ought to be the wavelength. There are two layers of processing. 1. A feature vector; 2. A categorization of feature vectors. Three methods are used for classification purpose namely KNN, Artificial Neural networks (ANN), and Naïve Bayes (NB). As shown in the table, while ANN and Naïve Bayes have success rates of 97.5 and 90 percent, respectively, KNN has a success rate of 100 percent. Performance in classification is hampered by large data sizes. Large data dimensions decrease the performance of categorization. Reduced dimension is achieved through the use of Wavelet Transformation (WT). For this, Sym5, db2, and Haar are utilized. In ANN, the backpropagation algorithm is trained for classification. Using cross-validation on 10 fields, its performance is evaluated. For the purpose of implementing the classification method, Mat lab R2015b is utilized. Sym5, Haar, and db2 have success rates of 100%, 100%, and 98% for KNN when $K = 2$. KNN offers excellent results when compared to other algorithms. Out of 2150 characteristics, wavelet decomposition decreases 75.

Cramer S., Freitas A. A., Kampouridis M.and A. K. Alexandridis et al. [7] shown that seven ML systems were efficient in predicting rainfall. According to statistical outcomes, radial-basis function neural network (RBFNN) performs better than other advanced algorithms.

Vijai, S., A.K.Misra et al., [8] utilized genetic algorithm for plant disease detection. The only disease detection technologies now in use include plant disease specialists inspecting leaves in the field. This is a very costly operation to carry out for larger farms since it requires a significant number of specialists and frequent plant evaluation. In some countries proper agricultural facilities are not available. Hence, it is imperative for them to hire professionals. However, this results in time-consuming and costly activities. In such cases, this advice strategy is useful in preventing crop overproduction. Plant illness can be automatically detected by examining the state of the leaves, which is less expensive and more efficient. This machine vision provides robot guidance, inspection and automatic process control based on images. It takes a lot of effort and work to visually identify leaf diseases. However, this is not particularly reliable and can only be done in limited locations. However, it is quicker, faster reliable, and takes less time to detect plant diseases automatically rather than manually.

K. L. Chong, K. D. Kanniah, C Pohl. and K. P.Tan et al. [9] Comprehensive examination of the many applications of remote sensing for palm oil farms is included, such as the counting of trees, estimating age, detecting changes, identifying pests and diseases, estimating above-ground biomass (AGB) and carbon production, and computing yield. They offered proposed fixes and pointed out potential research holes. The emphasis on palm oil yield by utilizing machine learning algorithms is not, highlighted in this assessment.

A. Morellos, G. Tziotzios, X. T. Alexandridis, E. Pantazi, R. Bill, D. Moshou, R. Whetton, J. Wiebensohn, and A. M. Mouazen et al. [10] In the Premslin, Germany field, TN, SOC, and Moisture Content (MC) were also determined using visible and infrared spectroscopy. A predictive Machine Learning (ML) model is constructed using the spectroscopy dataset to predict the three soil properties. When comparing multivariate principal component regression with partial least square regression techniques, residual prediction deviation (RPD) is better for cubist and least square support vector machine (LS-SVM) machine learning techniques With RMSEs of 0.062 and 0.457 and RPDs of 2.24 and 2.20, SOC and MC are predicted using LS-SVM.

S. K. Singh, A. Porwal, R. Saxena, and S. S. Ray et al. [11] Hail storms in February and March 2015 alone resulted in an 8.4% drop in the country's wheat production, according to an evaluation of the impact of hail rains on India's wheat harvest. Correct weather forecasting can be a lifesaver for farmers, especially those who are economically disadvantaged and live in countries like India where it is difficult to store harvested products for long periods of time. When frequently used on a system, ML models serve as feedback controls. We are able to predict the variables that impact crop production with the use of precise ML models. Therefore, preventative measures can be taken before a crop production imbalance ever happens.

Arivazhagan, S., Ananthi, S., Newlin S., Vishnu V et al. [12] developed a technique that involves colour transforming an input colour image. Using a threshold value, the image's green pixels are hidden and removed. The segmentation process is then carried out. All reliable segments are used to calculate the texture statistics. Disease is categorized using the next classifier.

Revathi, P., Hemalatha, M et al. [13] created an innovative method for identifying plant visual disorders. The plant is captured on camera and prepared for digital imaging.

Techniques for extraction are used, including edge detection, colour space, and textural components. The classifiers get the extracted features. The main goal of this investigation is to identify the contaminated portion of the cotton leaf using image processing.

Camargo, A., Smith et al. [14] showed a colour image analysis technique that reproduces the visual signs of plant illnesses. Starting with a grayscale image of the diseased plant, the constructed algorithm transforms it into colour transformations. Using the histogram's intensity distribution, the transformed image is first segmented. This method works especially only when concentrating on a single image with a large range of brightness. Following the segmentation process, the acquired region is adjusted to eliminate pixel sections outside the target area. The outcomes were accurate in locating unhealthy areas in the plant and matched all of the specifications. Additionally, they put out an algorithm that would use image processing to identify unhealthy spots in plants. The colour adjustment is made after the image has been taken. The colour change is done after the initial picture capture. The Gaussian filter is used to improve photographs that have been transformed. The region of interest is then obtained through segmentation. The location of the optimum threshold separates segments. Then, using the SVM classifier, the segmented regions are classified and labeled as diseased or healthy.

R. Kaundal, A. S. Kapoor, and G. P. S. Raghava et al. [15] When support vector machines (SVM) are utilized to develop plant disease prediction models based on weather, both the amount of training time and the incapacity to predict the values of unknown data points are reduced. It is based on comparisons with neural networks, multiple regression, and other modelling techniques like ANN architecture and Generalized Regression Neural Network (GRNN) technology. When it comes to general-purpose supervised estimation, SVM is a very effective approach. This research was done to find out the way SVM models could forecast rice blast severity dependent on weather. The absence of local minima, sparsity of the solution, use of kernel-induced feature spaces, and simplicity of fitting non-linear relationships and complicated variable interactions are further benefits of utilising SVM.

3 A Machine Learning Approaches with Special Emphasis for Identification of Plant Leaf Diseases in Crop Yield Prediction

The block diagram of identification of plant diseases in crop yield by utilizing Machine Leaning approaches with special emphasis is represented in Fig. 1.

Gather pictures of healthy and diseased plants. The training time was calculated using the Open Source Computer Vision Library (OpenCV) framework using a Python script that automatically cropped the photos. In order to identify the diseases, 20,639 photos of healthy and diseased plant leaves were categorized into 15 classifications from plant village dataset. C-1 Pepper, Bacterial Spot, Diseased, and 997 images are among the 15 categories. 1478 shots of pepper; 1000 photos of potato; 152 photos of potato; 1000 photos of potato; 1000 photos of potato; 1000 photos of early blight; Target Location, C-6, Tomato, Illness, 373 images, 1401 images Mosaic Virus, C-7, Infected Tomato virus causing yellow leaf curl, 3209 images of a C-8 infected tomato and 2127 images of a C-9 tomato with a bacterial spot 1000 images, C-10, Tomato, Infected, Blight Early Images represent a tomato from 1591 that was in good health, a tomato from 1909 that was disease, and The tomato illnesses C-13, C-14, and C-15 (where C stands for classification) include of two spotted spider mite (1676 photographs), leaf mould (952) and Septoria leaf spot (1771 images).

A sample collected in the initial step and then sample is pre-processed and labelled. The labelled part is transferred to the input image. Then the segmentation is done with the feature extraction is done. Augmentation process is done after the feature extraction. The augmentation process is divided into regression and classified. Then finally get the result Whether leaf is having disease or not.

A Python script that makes use of the OpenCV framework included photos that were automatically scaled in the dataset to reduce training time. Scaling the data points from [0, 255] (the image's minimum and maximum Red Green Blue (RGB) values) to [0, 1] pre-processes the input data. The plant's leaf's picture is photographed. The picture is in RGB format.

A training part and a testing segment have been created from the dataset. The dataset has 20% for testing and 80% for training. 16,511 photos make up the training dataset, whereas 4,128 images make up the testing dataset. The test data set is hidden from view to test the model's accuracy while the model is trained using the training data set. The division of a picture into pixel-like characteristics is known as image segmentation. It mostly supports the visual interpretation process. It moves an image from a lower level to a higher level. Its capacity to successfully analyse an image is largely dependent on the precise the segmentation process. Contextual and non-contextual factors both have a role in segmentation. The segmentation procedure makes use of a number of algorithms.

A method to expand the quantity of photographs in a database is called data augmentation. In order to diversify our collection, several procedures such as moving, rotating, stretching and rotating are applied to the image datasets. Overfitting can be minimized during the training phase by expanding the dataset and introducing distortion to the images. In-place or real-time data augmentation are supported by Keras ImageDataGenerator class. They can make sure that our network learns about new deviations by providing this type of data. This allows us to come up with higher results using a smaller dataset.

Numerous prediction algorithms based on regression and classification have been utilised for predicting crop yield. The following methods are used to predict agricultural yield: k-nearest neighbours (K-NN), linear regression (LR), support vector machine (SVM), decision tree (DT), support vector regression (SVR), Extremely Randomised Trees (extra tree) (ERT), and random forest (RF).

The interaction between an independent variable and one or more dependent variables is represented by the LR model. Regression algorithms are used in a machine learning framework to minimise loss or error (RMSE or MSE), which is the way learning is achieved. Multiple Multiple Linear Regression (MLR) implementations, analysis have shown that multiple independent variables are more effective and dependable than a single individual element.

Regression estimation in MLR models is commonly performed using the least-squares method (LSM). By combining MLR techniques with the widely used Generalised Cross-Validation (GCV) index, which takes nonlinearity and variable correlations into consideration, the Multivariate Adaptive Regression Splines (MARS) model may

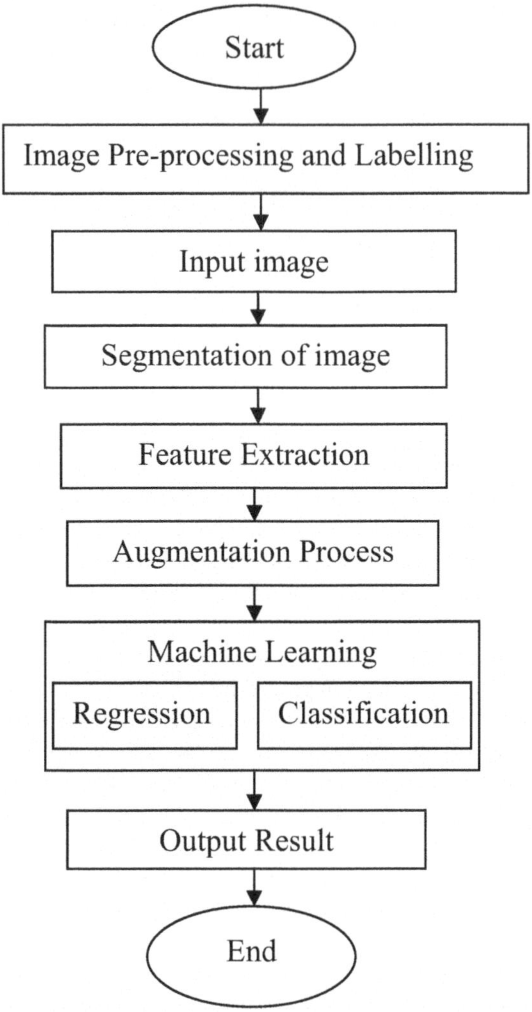

Fig. 1. Block Diagram of Identification of Plant Diseases in Crop Yield By Using Machine Leaning Approaches With Special Emphasis

be assessed. Adaptive splines with different slopes are separated out of the learning data. K-NN is used in classification and regression, which gives closer neighbours more weight in the prediction because they are more correlated with the mean than further neighbours. There are several distance calculations that may be used to determine the separation between two neighbours, including Euclidean, Minkowski, and Manhattan. A linear separating hyperplane is produced by the binary classifier SVM for the purpose of classifying data instances. SVM differs slightly from SVR in that the SVR approach can be utilized to solve regression issues.

Both classification and regression may be performed using the Supervised Machine Learning model DT. Two leaf nodes and a root node without an input edge make up these three nodes, and a decision node with input and output edges (no incoming edge). Each output node in a decision tree splits each attribute into two or more branches in accordance with the split criterion. To distribute the qualities, Classification and Regression Trees (CART) utilizes information gain, gain factor and the Gini Diversity Index (GDI). The use of RF in agricultural research is an effective tool for evaluating production.

When calculating regression, it produces an extensive range of regression trees derived from a huge number of decision trees. Because of its higher performance and ability to correct for bias owing to randomness, the RF decision tree outperforms all other decision trees.

The ensemble model known as extremely randomised trees (extra tree) uses unpruned decision trees which is similar to RF. The entire learning sample is included, and the nodes are divided along randomly chosen cut-points to create the trees

(without bootstrapping any samples). The nodes are divided into groups using the same number of trees and variables as in RF. By simulating a sophisticated nonlinear input–output relationship, Artificial Neural Networks (ANN) are a widely utilized Machine Learning technique for predicting agricultural yields. This system's three levels are the concealed layer, input layer, and output layer. An ANN's learning rate, training tolerance, and number of nodes in the hidden layer all impact its performance. In order for the predicted value to fall within a range that is suitable for the observed value, the weights need to change by some amount over several iterations, which is determined by the learning rate. These methods produce extremely accurate results.

4 Result Analysis

The main topic of this section is the performance of plant foliar disease detection in agricultural yield prediction by ML methods. The performance is evaluated based on the classified instances namely True Negative (TN), False Negative (FN), True Positive (TP), and False Positive (FP) which are defined as follows:

True Positive: If a sample is predicted correctly as positive and actually it is positive.

True Negative: When a sample's result is projected to be negative but turns out to be positive.

False Positive: If a testing is assumed to be negative but is really positive.

False Negative: If a testing is estimated to be positive but is actually positive.

For the purpose of evaluating the performance of the system being presented, based on these numbers, performance indicators such as F1 score, precision, and accuracy are computed.

Precision: All positive classes that the model correctly predicted are described by precision; how many of those actually have a positive outcome. The accuracy is calculated by dividing the total number of correctly classified instances by the total number of predicted positive examples. The formula for the equation is:

$$\text{Precision} = \frac{TP}{TP + FP} x100 \tag{1}$$

Accuracy: It is a performance parameter that expresses the system's capacity to produce right predictions.

$$\text{Accuracy} = \frac{TP + TN}{TP + FP + TN + FN} x100 \tag{2}$$

F1 Score: The F1-score provides an overall estimate of a test subject's precision and recall. It is the harmonic mean of a test value's precision and recall.

$$\text{F1 Score} = \frac{2xprecision\, x\, recall}{\boldsymbol{precision} + recall} \tag{3}$$

Table 1. Performance Comparison Table

Classifiers	Precision	Accuracy	F1 Score
SVM	97	99	97
RF	98	98.6	96
CNN	70	75	78

When compared to SVM, RF, and CNN, the given ML classifiers outperform them in related to precision, accuracy, and F1 Score.

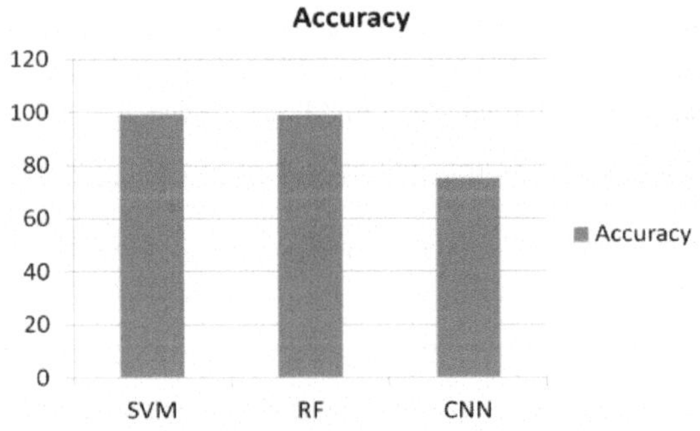

Fig. 2. Accuracy Comparison Graph.

A comparison graph of accuracy between SVM, RF, and CNN may be found in Fig. 2.

A comparison graph of precision between SVM, RF, and CNN may be identified in Fig. 3 (Fig. 4).

A comparison graph of F1 scores between SVM, RF, and CNN is shown in Fig. 2.

The agricultural yield forecast of plant leaf disease identification using SVM, RF, and CNN is time-comparison in Fig. 3. In comparison to CNN, the SVM and RF-based

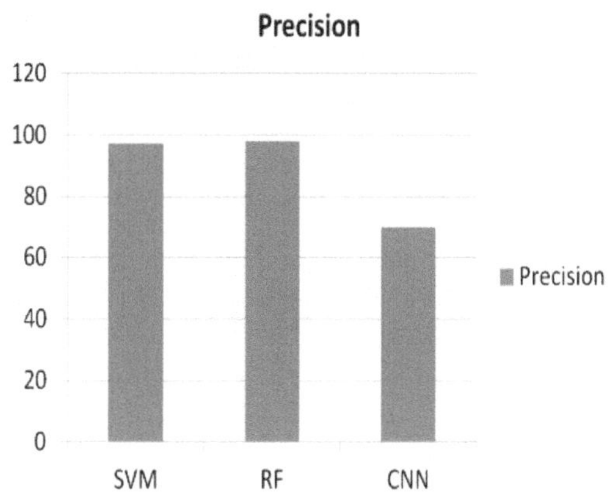

Fig. 3. Precision Comparison Graph

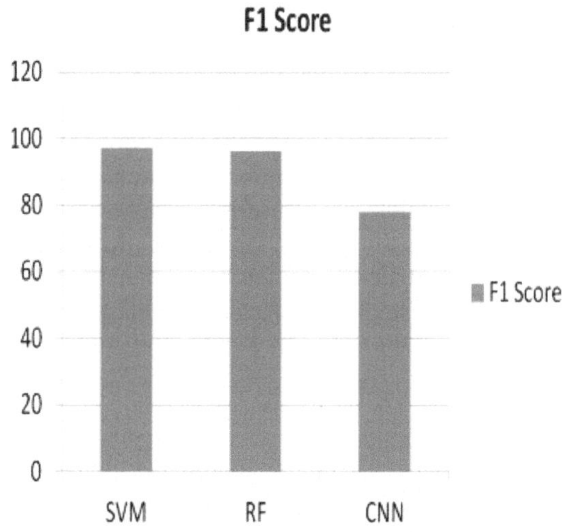

Fig. 4. F1-Score Comparison Graph

method predicts plant leaf disease in less time. The Y-axis represents the time in ms (milliseconds) and X-axis represents the classifiers (Fig. 5).

Therefore, when compared to other classifiers, ML techniques for identifying plant leaf disease perform better in related to precision, accuracy, and F1 Score.

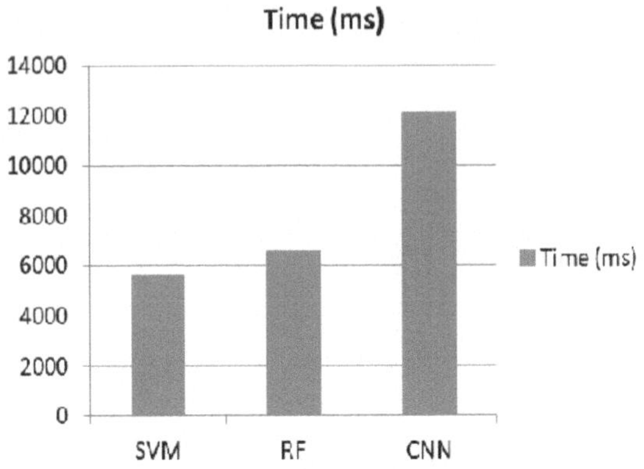

Fig. 5. Prediction Time Comparison Graph

5 Conclusion

In this analysis, identification of plant leaf disease in crop yield prediction by using ML approaches is presented. Here ML classifier of SVM and RF is used for the predication of plant leaf disease. Although identifying the most effective methods is challenging in light of current research, several ML algorithms are being used and showing a large amount of potential. For sustainable agriculture, it is essential to keep a check on the health of the plants and look for illnesses. This involves a lot of work, understanding of plant diseases, and reduced processing times. Finally the plant leaf disease is predicted as positive if a leaf has disease otherwise the leaf is predicted as normal. The outcomes demonstrate that the presented system outperforms other classifiers in terms of accuracy and prediction time, providing accurate and dependable results.

References

1. Elavarasan, D., Vincent, P.M.D.: Crop yield prediction using deep reinforcement learning model for sustainable agrarian applications. IEEE Access **8**, 86886–86901 (2020)
2. Huang, J., et al.: Assimilation of remote sensing into crop growth models: Current status and perspectives. Agricult. Forest Meteorol., vols. 276–277, Art. no. 107609, October 2019
3. Reda, R., et al.: A comparative study between a new method and other machine learning algorithms for soil organic carbon and total nitrogen prediction using near infrared spectroscopy. Chemometric Intell. Lab. Syst. **195**, Art. no. 103873, December 2019
4. Cai, Y., et al.: Integrating satellite and climate data to predict wheat yield in Australia using machine learning approaches. Agricult. Forest Meteorol. **274**, 144–159 (2019)
5. Azadbakht, M., Ashourloo, D., Aghighi, H., Radiom, S., Alimohammadi, A.: Wheat leaf rust detection at canopy scale under different lai levels using machine learning techniques. Comput. Electron. Agric. **156**, 119–128 (2019)
6. Karada˘g, K., Tenekeci, M.E., Ta¸saltın, R., Bilgili, A.: Detection of pepper fusarium disease using machine learning algorithms based on spectral reflectance. Sustainable Computing: Informatics and Systems (2019)

7. Cramer, S., Kampouridis, M., Freitas, A.A., Alexandridis, A.K.: An extensive evaluation of seven machine learning methods for rainfall prediction in weather derivatives. Expert Syst. Appl. **85**, 169–181 (2017)
8. Vijai, S., Misra, A.K.: Detection of plant leaf diseases using image segmentation and soft computing techniques. Inf. Process. Agric. **4**, 41–49 (2017)
9. Chong, K.L., Kanniah, K.D., Pohl, C., Tan, K.P.: A review of remote sensing applications for oil palm studies. Geo-Spatial Inf. Sci. **20**(2), 184–200 (2017)
10. Morellos, A., et al.: Machine learning based prediction of soil total nitrogen, organic carbon and moisture content by using VIS-NIR spectroscopy. Biosyst. Eng. **152**, 104–116 (2016)
11. Singh, S.K., Saxena, R., Porwal, A., Ray, S.S.: Assessment of hailstorm damage in wheat crop using remote sensing. Current Sci. **112**(10), 2095–2099 (2015)
12. Arivazhagan, S., Newlin, S., Ananthi, S., Vishnu, V.: Detection of unhealthy region of plant leaves and classification of plant leaf diseases using texture features. Agric. Eng. Int. CIGR **15**, 211–217 (2013)
13. Revathi, P., Hemalatha, M.: Classification of cotton leaf spot diseases using image processing edge detection techniques, 169–173 (2012). 10.1109/ INCOSET.2012.6513900J
14. Camargo, A., Smith: Image pattern classification for the identification of disease causing agents in plants. Comput. Electron. Agr., 121–125 (2009)
15. Kaundal, R., Kapoor, A.S., Raghava, G.P.S.: Machine learning techniques in disease forecasting: a case study on rice blast prediction. BMC Bioinform. **2006**(7), 1–16 (2006)
16. Wahal, A., Choudhary, S.S.S., Singh, N.: Leaf disease detection and crop prediction using machine learning. In: 2021 3rd International Conference on Advances in Computing, Communication Control and Networking (ICAC3N), Greater Noida, India, pp. 438–442 (2021). https://doi.org/10.1109/ICAC3N53548.2021.9725643
17. Kilaru, R., Raju, K.M.: Prediction of maize leaf disease detection to improve crop yield using machine learning based models. In: 2021 4th International Conference on Recent Trends in Computer Science and Technology (ICRTCST), Jamshedpur, India, pp. 212–217 (2022). https://doi.org/10.1109/ICRTCST54752.2022.9782023
18. Ramachandra, A.C., Rajesh, N., Aishwarya, K., Madugonda, N., Akash, R.A., Namratha, G.: Classification of Crop diseases through Remote-sensed data using Multi-class SVM. In: 2022 IEEE 2nd Mysore Sub Section International Conference (MysuruCon), Mysuru, India, pp. 1–5 (2022). https://doi.org/10.1109/MysuruCon55714.2022.9972622

State of the Art Recurrent Neural Network with Bidirectional Long Short-Term Memory for Cursive Handwriting Recognition

Manju Jose[1(✉)] and Prakash Kumar Udupi[2]

[1] Majan University College, Muscat, Oman
manju.jose@majancollege.edu.om
[2] Middle East College, Muscat, Oman

Abstract. Handwriting is a natural and one of the popular way of communication and it has always been a challenge to recognize it as it can be written in different ways by the same writer. With the advancement in artificial intelligence, convolutional neural networks can be deployed for handwriting recognition. The technical merits and advancement of artificial intelligence have made it widely accepted for pattern recognition and image processing. Wide range of research studies are evident for handwriting recognition in various languages using Convolutional Neural Network (CNN). But not much research is evident for English language cursive handwriting recognition in offline using IAM dataset. This research work aims to explore various convolutional neural networks which can be deployed for recognition of English language cursive handwritten characters. The methodology consists of collection and preprocessing of data, feature extraction, classification, evaluation, and enhancement of the model. Two models were deployed using GPU for the IAM open dataset. The first model using CNN with varying parameters yielded an accuracy of 92.6% and the second hybrid model which was deployed using CNN and Recurrent Neural Network (RNN), yielded an accuracy of 95.4%. The future direction is to enhance the accuracy of the model using various optimization techniques and increase the number of convolutional layers.

Keywords: Convolutional Neural Network · CNN · Recurrent Neural Network · RNN · LSTM · Long Short-Term Memory · Cursive handwriting recognition

1 Introduction

The handwriting recognition is the process by which handwritten characters are transformed into electronic form. English cursive handwriting is connected, slanted, and overlapped. This makes the handwriting character recognition challenging. The character recognition could be performed in two different modes such as online or offline. If the characters are recognized while it is being created, it is known as online handwriting recognition. The recognition of the scanned documents which are generally stored as images is known as offline handwriting recognition. In the case of online character

T. Shreekumar et al. (Eds.): ISCComm 2023, CCIS 2231, pp. 354–363, 2025.
https://doi.org/10.1007/978-3-031-75605-4_25

recognition, the time series coordinates which represent the movements of the pencil or pen-tip are captured. On the other hand, for offline recognition the features are extracted and identified. For online recognition the features are extracted while they are created. Therefore, enhanced results can be obtained for online recognition than the offline handwriting recognition.

Reading the doctors' prescriptions, postal applications, medical notes and extracting the data from different kinds of forms etc. are the most common applications of handwriting recognition. Highly proficient and accurate systems are compulsory for reading the bank check as the amount in digits must be read correctly along with the words. OCR (Optical Character Recognition) systems are used in digital libraries for converting textual and image details into digital form. A revolutionary performance is witnessed in handwriting recognition research using the deep learning algorithms. Extremely fast increase in the quantity of handwritten records and the convenience of massive processing potential demands tremendously efficient handwriting character recognition systems. English handwriting can be running cursive or mixed scripts. Mixed scripts are more common, and it is harder to recognize. Some research work is evident to recognize cursive handwriting. Currently, mixed, and cursive scripts are commonly used together. This leads to the requirement for advanced study in cursive handwriting character recognition.

As per the data of National Academies of Science Institute of Medicine, 7000 deaths are recorded in a year due to the sloppy handwriting of the doctors Education sector has incorporated technology to transform the algebraic equations which are sloppy in nature, class notes etc. into digital format. Computerization of many jobs such as medical prescriptions, tax returns etc. can be done using handwriting character recognition. The methods which were used previously were unproductive handcrafted feature approaches. Much research is available for online English cursive handwriting recognition using CNN. Unfortunately, not much widespread research is seen in literature for cursive offline handwriting recognition using CNN for English language. The novelty of this work is to study the deep convolutional networks to acquire best possible accuracy for cursive character recognition utilizing IAM dataset compared to the existing/peer works. This research work aims to study various feature extraction techniques on the classification precision of English offline character recognition and develop an enhanced method for character recognition using deep convolutional networks with improved accuracy.

2 Related Work

A writer independent handwriting recognition system [1] was developed using Hidden Markov Model (HMM). This system could segment the words into sub strokes. Later, the feature extraction is performed using discrete curve evolution. This HMM model was implemented using Baum Welch parameter algorithm which is used for spherical and linear correlated data. This system yielded high accuracy on Unipen-ICROW-2003 database. In addition, it performed well utilizing recurrent neural network and conditional random field on text which was writer independent. According to [2], a graph-based method using string representation and the relationship was calculated using Levenshtein distance which enhanced the accuracy by getting rid of the normalization procedures. This system was able to recognize the uppercase letters and images from the ETL-1

AIST database with an accuracy of 67.01% for alphabets and 84.69% for digits. [3] suggested a unique architecture using ensemble classifiers. The neural network, Support Vector Machine (SVM) and K Nearest Neighbor (KNN) algorithms were used for the recognition of cursive characters. Various architectures were used based on various ensemble classifier conformations. This study proved that for the C Cube data set the support vector machine-based classifier gave the maximum accuracy 92.03%. [4] used segmentation algorithm which calculated the ideal image distance and the intersection cutting point of images. Later the width and height of the images was used to derive the segmentation point. During the preprocessing step the slope correction was accomplished. This experiment yielded an accuracy of 97% for IAM dataset using support vector machine. The research work [5] used directional features of stroke orientation with Hough transform and Arnold transform, yielding an accuracy of 87.19 percentage using the support vector machine. [6] deployed hidden markov model and the k-nearest neighbor algorithms with signal to noise ratio intended for the scanned cursive image identification. This algorithm using a grayscale histogram and thresholding scale resulted in 90% accuracy.

The study by [7] deployed binarization technique for character recognition. The preprocessing consisted of noise removal, thinning, cropping and normalization. This study used back propagation with multilayered feed forward artificial neural network following the gradient descent principle achieved the accuracy 85.62%. This would have been further improved using improved classifier characteristics. [8] suggested deep convolutional RNN for mobile application with batch normalization. This system had an accuracy of 76% for training data and 72% validation accuracy. [9] focused on handwritten digits and characters of bank deposit slips. The proposed system using convolutional neural network with the three layers produced an accuracy of 82% for the recognition of digits and 80% for the characters. This study [10] focused on Arabic character recognition deploying regularization techniques along with deep neural network. Batch normalization was implemented to reduce overfitting. This model produced an accuracy of 94.8% and 97.6% for two different datasets. [11] implemented a fusion model using RNN and feed forward neural network for an extremelylarge corpus. This system was developed for Chinese handwriting recognition, gave an accuracy of 95%. [12] proposed a ConvNet architecture which is unique in nature with 7 layers followed by several pooling layers. This study focused on Malayalam handwriting recognition and yielded an accuracy of 96.46%. The research work [13] accomplished a high validation accuracy of 99.44% employing semi supervised generative adversarial network. This study was implemented for Bangla character recognition. The research work [14] accomplished an accuracy of 88% for Arabic characters written by children of age 7 to 12 using the convolutional neural networks. All these studies achieved good accuracy using convolutional neural networks. Unfortunately, all these studies focused on other languages than English.

3 Research Design

Figure 1 displays the research methodology used for this study. IAM, which is a popular sentence directory data set was used for the study [15]. It consisted of scanned text of 1539 pages. This had sentences from more than six hundred writers. Exploratory data

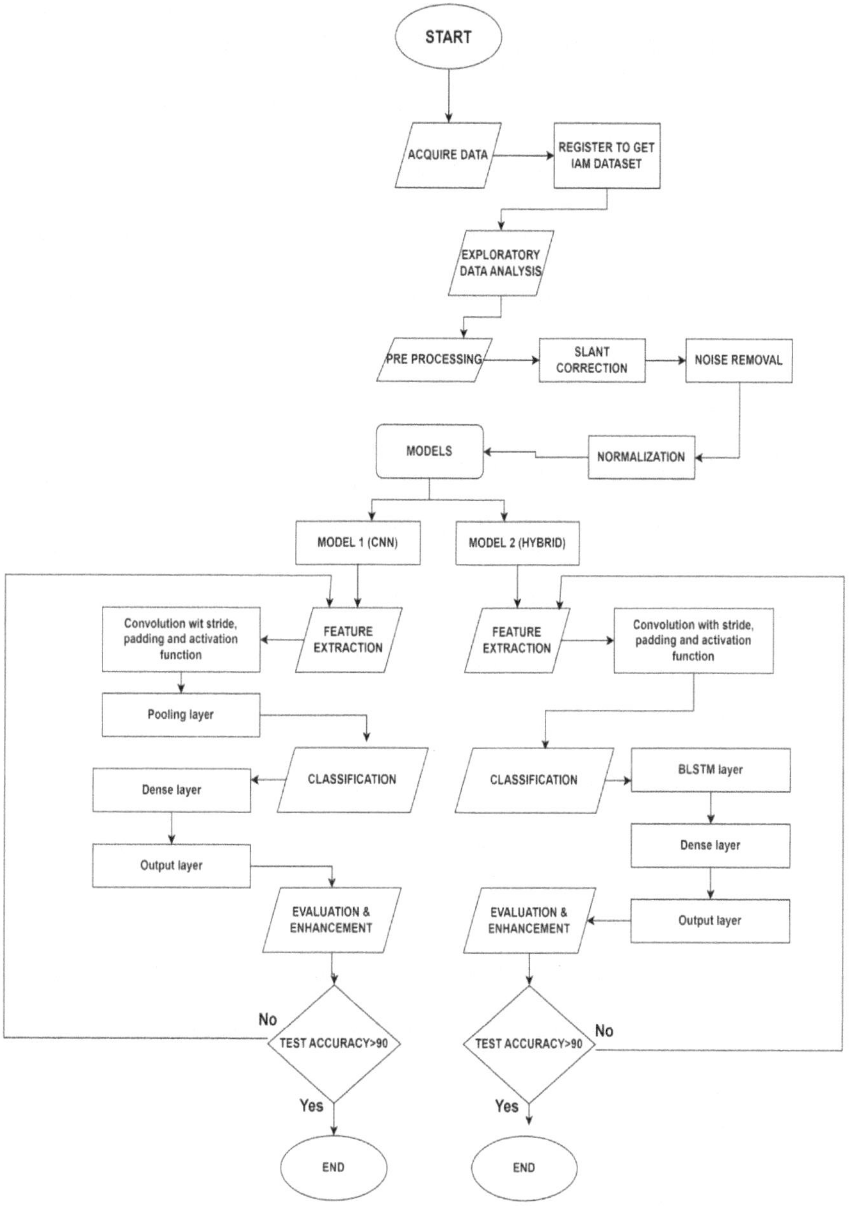

Fig. 1. Research Design

analysis was performed to study the data set. Preprocessing involved slant correction, size normalization, noise removal etc. to make the data ready to be fed into the model. The feature extraction was done to extract the vertical, horizontal, loops and curved lines. Two models were deployed during this research process. Model 1 using convolutional neural network was first deployed and tested for accuracy. Later it was improved with various techniques to improve accuracy. This model was implemented with suitable padding and stride. Appropriate activation functions were deployed with 3 pooling layers for extracting the features. This model had 2 fully connected dense layers. SoftMax activation was deployed for the final layer which is the output layer. Model 2 was a hybrid model which was implemented using RNN and BLSTM (Bidirectional Long Short-Term Memory) layers to achieve enhanced accuracy. This hybrid model with CNN and RNN architecture accomplished the feature extraction using the CNN layers. The dense layers along with RNN & BLSTM layer operated as the classifier.

4 Implementation and Evaluation

The curated IAM data set had the data arranged according to the writers. SoftMax function was used for reducing the dimensions of images for training. As per [16], the sentences were made into smaller size of 113x133. Model 1 used a generator function to scan through the pictures. The images were first converted into arrays (NumPy). The curvature, spacing etc. were obtained during the process of extraction of features. Various pooling and convolutions were deployed to accomplish this. The model was built using Tensor flow and Keras library with Python programming language. 8 epochs were deployed during the training of this model. First the weights were randomly assigned to have the output possibilities to be random during the first training. The total error was considered. Later back propagation technique was used to calculate the gradient of errors. Further to this, the parameters were optimized for appropriate classification of images. The Adam optimizer [16], was deployed to use the adaptive learning rate method for the optimization of the resultant model. The model accuracy was 84 percent with six epochs and two conversational layers. The accuracy was enhanced by using various permutations and combinations of fully connected layers, convolutional layers and epochs to achieve the desired accuracy.

The curated IAM data set consisted of data arranged according to the writers. SoftMax function was used for reducing the dimensions of images for training. As per [17], the sentences were made into smaller size of 113x133. Model 1 used a generator function to scan through the pictures. The images were converted into arrays. The feature extraction was implemented to obtain the curvature, spacing etc. Various pooling and convolutions were deployed to accomplish this. The model was built using Tensor flow and Keras library with Python programming language. 8 epochs were deployed for training. First the weights were randomly assigned to have the output possibilities to be random during the first training. The total error was considered. Later back propagation technique was used to calculate the gradient of errors. Further to this, the parameters were optimized for appropriate classification of images. The Adam optimizer [18], was deployed to use the adaptive learning rate method for the optimization of the resultant model. The model accuracy was 84 percent with six epochs and two conversational layers. The accuracy was enhanced by using various permutations and combinations of, convolutions, fully connected layers and different epochs in order to improve the accuracy. Figure 2 shows the performance given by model1. This model had 84.15% training accuracy for the initial epoch. Afterwards this enhanced to 99% for the rest of the epochs. The maximum accuracy for validation was for epoch 4 which was 96.48%. Later the accuracy decreased for the remaining epochs. Epoch 2 had the least validation accuracy of 95.44%. The model accomplished 92.6% test accuracy.

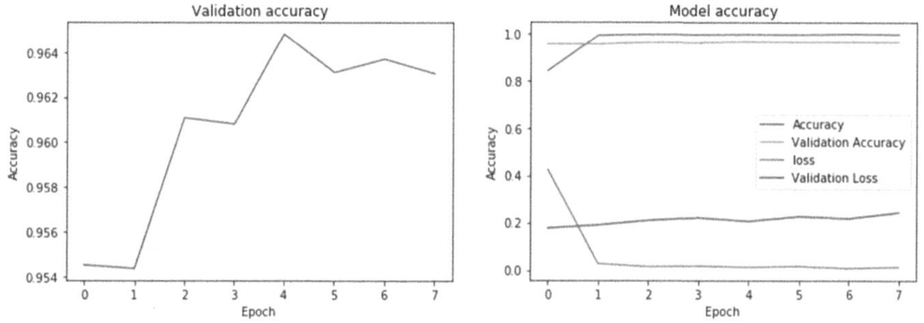

Fig. 2. Performance of model 1

The second model consists of CNN plus RNN architecture [19] used LSTM. The CNN sub model which functioned as the feature extractor had seven convolutions with padding and stride. Appropriate activation functions were also used. The RNN sub model which functioned as the classifier had 2 BLSTM layers along with a dense layer. For the noise removal the pictures were first transformed to a grayscale. House lines method [20], based slant correction algorithm was used. The extraction of the lines was done from every sentence by the sequential black coloured pixels. Further, contour tracking method was used to separate the words. The extraction of features was then done applying several convolutions of the Max pool layers to get the curvature of letters spacing etc. The output from the CNN sub model was fed into the RNN sub model using LSTM for feature prediction. The details of the model are given in Fig. 3. With the 5 convolutional layers the model had an accuracy of 88% with 100 epochs. So various combinations of RNN and convolutional layers were experimented with to achieve the required accuracy. The performance details of this hybrid model are shown in Fig. 4. An accuracy of 98% was achieved most of the time during the validation. The loss and accuracy of validation continued to be constant after the initial epochs. 95.4% was the test accuracy and 98.6% was the final accuracy of validation.

Layer (type)	Output Shape	Param #
input_1 (InputLayer)	(None, 32, 128, 1)	0
conv2d_1 (Conv2D)	(None, 32, 128, 64)	640
max_pooling2d_1 (MaxPooling2	(None, 16, 64, 64)	0
conv2d_2 (Conv2D)	(None, 16, 64, 128)	73856
max_pooling2d_2 (MaxPooling2	(None, 8, 32, 128)	0
conv2d_3 (Conv2D)	(None, 8, 32, 256)	295168
conv2d_4 (Conv2D)	(None, 8, 32, 256)	590080
max_pooling2d_3 (MaxPooling2	(None, 4, 32, 256)	0
conv2d_5 (Conv2D)	(None, 4, 32, 512)	1180160
batch_normalization_1 (Batch	(None, 4, 32, 512)	2048

Fig. 3. Details of hybrid model

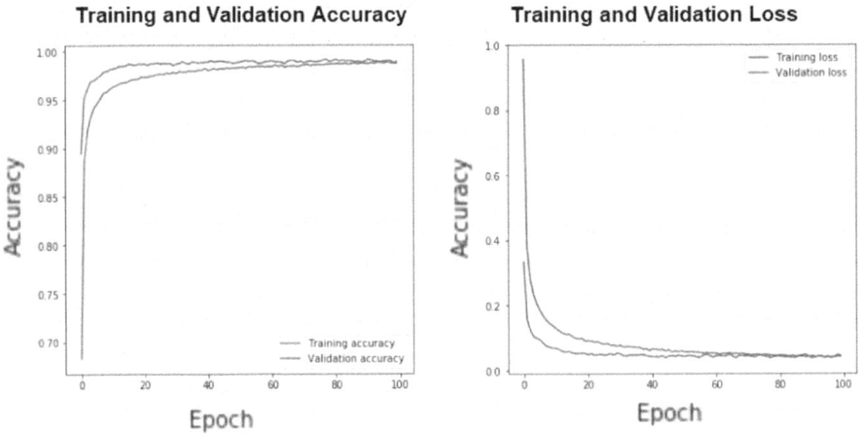

Fig. 4. Performance of model 2

5 Result Analysis

The model 2 which was the hybrid model performed better than model 1. The training was not overfitting for both model 1 and model 2. Table 1 shows the comparative analysis of the performance of the two models. The validation and training accuracies were almost close for model 1. The training and validation accuracies were almost similar in most of the epochs for model 2. Even though model 1 could also produce good accuracy in recognizing the cursive handwriting, the hybrid model outperformed. Model 2 could outperform because it had the sequence labeling using BLSTM [19]. The BLSTM was able to forget the previously deposited memory. Further it supplemented a portion of the fresh knowledge.

Table 1. Comparative performance analysis of the models

Models	Pre-processing Method	Major Characteristics	Accuracy of Training	Accuracy of Validation	Test Accuracy
CNN	Normalization, Slant correction	Adam optimizer + 3 convolutions	97%	95.5%	92.6%
Hybrid (CNN + RNN)	Line and word segmentation, Slant correction, Normalization	2 bidirectional LSTM + 7 convolutional layers	98%	98.6%	95.4%

There has been only one study [21], using the IAM database. This research focused on cursive handwriting character recognition of English language in offline. Unfortunately, this work used only 999 images which included the images from a local dataset and IAM database. The research work did not specify the number of images that were considered from the local data set and how many were taken from IAM data set. Also, this study had only the handwriting of 25 writers from both the data sets. The actual IAM data set had handwritings of 657 different writers. The recognition complexity surges with the increase in the number of writers because different writers have different styles of writing. So, this research [21], cannot be counted as an exclusive study of the IAM data set.

6 Conclusion and Recommendations

This research work provided an inspiring solution to recognize the handwritten text from bank slips, bank cheque, doctors' prescriptions etc. in order to improve the quality of service. Further this can be utilized for transforming historical documents into digital form providing a positive impact on society. This project does not have any ethical or legal consequences as an open source data set was deployed for the study. The future direction is to enhance the accuracy of the model using various optimization techniques and increase the number of convolutional layers. In addition, study the impact of substituting BLSTM with 2D-LSTM and extend this investigation on other datasets.

References

1. Samanta, O., Roy, A., Parui, S.K., Bhattacharya, U.: An HMM framework based on spherical-linear features for online cursive handwriting recognition. Inf. Sci. **441**, 133–151 (2018)
2. Putra, M.E.W., Supriana, I.: Structural offline handwriting character recognition using levenshtein distance. In: 2015 International Conference on Electrical Engineering and Informatics (ICEEI), Denpasar, Indonesia, pp. 31–36 (2015)
3. Rahman, A., Verma, B.: Effect of ensemble classifier composition on offline cursive character recognition. Inf. Process. Manage. **49**(4), 852–864 (2013). ISSN 0306-4573
4. Fitrianingsih, F., Madenda, S., Ernastuti, E., Widodo, S., Rodiah, R.: Cursive handwriting segmentation using ideal distance approach. Int. J. Electr. Comput. Eng. (IJECE) **75**, 2863 (2017)
5. Dasgupta, J., Bhattacharya, K., Chanda, B.: A holistic approach for Off-line handwritten cursive word recognition using directional feature based on Arnold transform. Pattern Recogn. Lett. **79**, 73–79 (2016). ISSN 0167-8655
6. Hepzi, J.L., Muthumani, I., Selvabharathi, S.: English Cursive Hand written Character Recognition. **117**, 57–63 (2017)
7. Choudhary, A., Rishi, R., Ahlawat, S.: Off-line handwritten character recognition using features extracted from binarization technique. Aasri Procedia **4**, 306–312 (2013)
8. Fajardo, L.J., et al.: Doctor's Cursive Handwriting Recognition System Using Deep Learning, pp. 6–11 (2019)
9. Gayathri, S., Mohana, R.S.: Optical character recognition in banking sectors using convolutional neural network. In: 2019 Third International conference on I-SMAC (IoT in Social, Mobile, Analytics and Cloud) (I-SMAC), Palladam, India, pp. 753–756 (2019). https://doi.org/10.1109/I-SMAC47947.2019.9032514
10. Etworks, N.: Arabic handwritten character recognition based on deep convolutional neural network. **33**, 25–27 (2017)
11. Wu, Y.-C., Yin, F., Liu, C.-L.: Improving handwritten Chinese text recognition using neural network language models and convolutional neural network shape models. Pattern Recogn. **65**, 251–264 (2017), ISSN 0031-3203
12. Sudarsan, D., Joseph, S.: A novel approach for handwriting recognition in Malayalam manuscripts using contour detection and convolutional neural nets. In: 2018 International Conference on Advances in Computing, Communications and Informatics (ICACCI), Bangalore, India, pp. 1818–1824 (2018). https://doi.org/10.1109/ICACCI.2018.8554592
13. Alom, M.Z., Sidike, P., Hasan, M., Taha, T.M., Asari, V.K.: Handwritten bangla character recognition using the state-of-the-art deep convolutional neural networks. Comput. Intell. Neurosci. **27**(2018), 6747098 (2018). https://doi.org/10.1155/2018/6747098.PMID: 30224913;PMCID:PMC6129853
14. Altwaijry, N., Al-Turaiki, I.: Arabic handwriting recognition system using convolutional neural network. Neural Comput. Appl. **33**, 2249–2261 (2021). https://doi.org/10.1007/s00521-020-05070-8
15. Fki.inf.unibe.ch.: IAM Handwriting Database — Computer Vision and Artificial Intelligence (2020). http://www.fki.inf.unibe.ch/databases/iam-handwriting-database
16. Tejas Reddy, G.H.: Handwriting Recognition using CNN. [online] IIT Joghpur. https://tejasreddy9.github.io/handwriting_cnn/
17. Bushaev, V.: Adam — latest trends in deep learning optimization. [online] Towards Data Science (2018). https://towardsdatascience.com/adam-latest-trends-in-deep-learning-optimization-6be9a291375c. Accessed 29 Sep 2020
18. Zhang, X., Chen, F., Huang, R.: A Combination of RNN and CNN for attention-based relation classification. Procedia Comput. Sci. **131**, 911–917 (2018). ISSN 1877-0509

19. Vinciarelli, A., Luettin, J.: A new normalisation technique for cursive handwritten words. Pattern Recogn. Lett. **22**, 1043–1050 (2001)
20. Greff, K., Srivastava, R.K., Koutník, J., Steunebrink, B.R., Schmidhuber, J.: LSTM: a search space odyssey. IEEE Trans. Neural Networks Learn. Syst. **28**(10), 2222–2232 (2017). https://doi.org/10.1109/TNNLS.2016.2582924
21. Fitrianingsih & Madenda, Sarifuddin & Ernastuti, Ernastuti & Widodo, Rini & Rodiah, Rodiah: Cursive Handwriting Segmentation using Ideal Distance Approach. Int. J. Electr. Comput. Eng. (IJECE) **7**, 2863 (2017). https://doi.org/10.11591/ijece.v7i5.pp2863-2872

BERT-Based Question-Answering for Narrative Text on Document Corpus

Ramesh Wadawadagi[1(✉)], Shrikant Tiwari[2], and Sanjay Hanji[3]

[1] Sai Vidya Institute of Technology, Bangalore 560064, Karnataka, India
rswlib@yahoo.co.in
[2] Galgotias University, Greater Noida 203201, India
[3] School of Business, MIT Vishwaprayag University, Solapur 413255, India
sanjay.hanji@mitvpu.ac.in

Abstract. Question-answering (QA) systems have gained immense popularity in the recent past due to their extensive usage in designing Chat- Bots and Kiosks. Developing interactive QA systems has always been a complex task in NLP. Time to look at QA systems now! These are available in different standards. However, this work focuses on designing an extractive QA system, which finds an answer-span from a con- text paragraph. Researchers have proposed many domain-specific QA systems to meet their needs. Yet high-quality QA systems are scarce for answering narrative text content from a collection of the document corpus. On the contrary, the BERT model has proved to be a powerful pre-trained language model and can be ne-tuned for developing efficient domain QA systems. In this paper, we develop an open-domain QA sys- tem based on BERT, termed Narrative QA-BERT (NQA-BERT). Particularly, the proposed NQA-BERT is pre-trained on the NarrativeQA dataset and ne-tuned on document corpus demonstrating the Indian epic stories such as Ramayana and Mahabharata. The effectiveness of the intended NQA-BERT model is measured with different performance metrics. The experimental results demonstrate that the ne-tuned BERT achieves high accuracy in extracting the answer-spans.

Keyword: Question-answering systems Transformer models Pre- trained BERT Narrative text content NarrativeQA dataset

1 Introduction

Question-answering (QA) has been a critical problem in NLP and continues to remain a premier problem in artificial intelligence. QA system aims to answer user's questions by identifying short text segments from the document corpus [1–3]. These QA systems can be broadly classified as either extractive or generative depending on the input question-type and the output results generated. The extractive models select the answer from a context provided as a paragraph [4]. On the contrary, generative models produce free text directly based on the con- text [5]. However, this work places emphasis on extractive QA task that accepts a question instance and relevant documents as input. The document

T. Shreekumar et al. (Eds.): ISCComm 2023, CCIS 2231, pp. 364–374, 2025.
https://doi.org/10.1007/978-3-031-75605-4_26

containing a text segment can be served as an answer to the question. This kind of machine reading comprehension is quite useful in building a wide range of applications, such as designing ChatBots, personal digital assistants (PDA) and Kiosks.

The literature study reveals that the end-to-end Neural Network architectures have achieved notable performance on extractive QA tasks. However, the most popular technique used in Neural Networks for finding the answer segments is to locate the 'begin' and 'end' points of the answers in the given documents. More particularly, pre-trained Bidirectional Encoder Representations from the Transformers (BERT) model trained with the SquAD dataset have proved to be sufficient for achieving high accuracy [6]. Despite incredible recent progress, high-quality QA systems are scarce for narrative content [7]. Most of the existing machine comprehension models are unsuccessful on narrative content, even though they employ a commonsense knowledge base [8]. However, few models are successful in extracting the right answers but in a context not relevant to the given input questions [9]. Moreover, this inconsistency is more critical when the number of occurrences of the answer segments increases within a document. To address this issue and return the most relevant answers, we propose a Narrative QA-BERT (NQA-BERT) that works based on the following idea. The proposed NQA-BERT model is a two-stage process. In the first stage, for each document relevant to the query identified by the Retriever engine, we compute the probability of each lexical term belonging to that document being considered in the context-span. Further, in the second stage, we perform the answer prediction task to figure out each word of the context-span to be included in the answer-span.

The significant contributions of this research work are as follows: i) Context- span prediction task that computes the probability of each term belongs to the context-span. Ii) Answer-span prediction task that computes the probability of each term in the context-span belongs to the answer-span. The remaining part of this article is structured as follows. Section 2 gives the literature review of relevant articles on narrative QA systems and pre-trained BERT language models. Meanwhile, the background knowledge and proposed methodology of NQA-BERT is discussed in Sect. 3. In Sect. 4, the experimental procedure is discussed, which includes the dataset preparation and selection of hyper-parameters. However, the analysis of experimental results is given in Sect. 5. Section 6 concludes the paper and identifies the future directions.

2 Literature Review

It is identified from the previous research works that narrative QA systems generally adopted a mixed-domain pre-training approach. However, general-domain pre-trained models are frequently employed domain-specific training on narrative texts. Some of the notable examples are included in the following paragraphs.

In [10], Xiong et al. Illustrated a large-scale QA dataset called TWEETQA dataset obtained from the social-media text. They applied deep language comprehension on QA pairs to differentiate them from standard QA datasets obtained from other text corpora. Further, they highlighted critical problems of QA systems for social-media data through performance deviation observed be- tween neural language models and human readers. Frermann [11] addressed the problem of QA on narrative text collected from book and movie script summaries. He designed predictive models for passage retrieval and answer-span prediction using heuristic datasets. Pre-trained BERT embeddings were used to introduce past knowledge into the system. This model proved to be efficient on both summary-level and narrative-level QA using the METEOR metric.

Boratko et al. [12] developed a new QA dataset to train and test the common sense reasoning abilities of AI models. A set of questions gathered from the FAMILY-F EUD reality game show served as the basis for creating the training set. More than 100 crowd workers are employed to answer each question collected, which serves as an evaluation set. A generative model has been pro- posed that throws a ranked list of prototypical answers for a given question. Their model proved to be surpassing many other competitive baseline models. An automated QA pair generation system based on the storybooks of eighth- grade students is proposed [13]. The performance of the proposed architecture is tested on an expert-annotated FairytaleQA dataset. The results show that their model outperforms other baseline systems.

Tian et al. [14] designs a safety hazard knowledge QA system for safety hazard management. The model uses a BERT combined with Bi-GRU and Self- attention mechanism for extracting the semantic features. Further, a Siamese Neural Network (SNN) is employed as an answer prediction method, which per- forms deep matching over the answer segments. The results are tested on hydraulic engineering construction and perform better than current QA models. Similarly, Mohamed Hassan et al. [15] proposed a model built using BERT to help identify the risky activities and potential hazards in construction tasks. They formulated a classification method for injury narratives based on related attributes like injury type and severity. The effectiveness of designed model is evaluated on the OSHA dataset.

However, in [16] Kierszbaum et al. employ a fine-tuned BERT model for QA tasks on the Aviation Safety Reporting System dataset consisting of text reports describing incident occurrences in an International aviation safety management. The model outperforms the contemporary DL baseline techniques. Interestingly, [17] and [18] addressed the task of predicting the correct answer-spans among multiple occurrences of the answers in a corpus, without annotating the correct answer-spans. Many researchers worked on domain-specific tasks including biomedical QA systems. For instance, [19, 20] proposed BioBERT trained on PubMed abstracts and full-length text. Further, they proved that domain-specific pre-trained models such as PubMed-BERT substantially outperform models with mixed-domain pre-training.

3 Proposed Methodology

The generic architectural diagram of the proposed NQA-BERT model is depicted in Fig. 1. The model comprises two major components, the retriever engine and the BERT reader. The task of the retriever engine is to identify relevant documents from text corpora consisting of the answer-spans. The candidate documents are then transferred to the BERT reader that performs two subtasks in a sequence, i.e., context prediction and answer prediction tasks. In the below paragraphs, we elucidate each of these subtasks in detail.

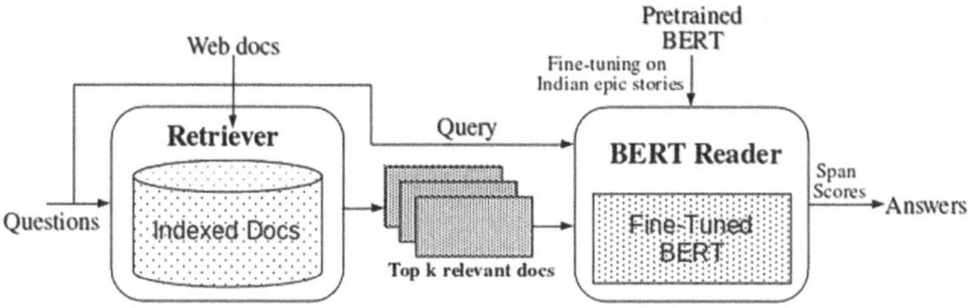

Fig. 1. BERT-based narrative Question-answering (QA) system.

The IR task that we take into account is known to be ad hoc retrieval, in which a user asks a retrieval system a question and the retrieval engine responds by returning an ordered set of relevant documents from the text corpus. To make search efficient, we use inverted indexing that helps to store valuable information such as document frequency and term count in each document. However, Chen et al. [21] claimed that indexing based on bigrams yields better results than unigrams. Additionally, we utilized an efficient hashing algorithm in the retriever engine, which improves document searching. Hence, the problem of document searching in IR is to find all documents di ∈ D that contain the terms qj ∈ Q. Further, to compute the similarity between the documents and the query, we employ cosine similarity that returns a score between 0 and 1. Hence, if any document di contains none of the query terms qj, it will receive a score of 0 and can be ignored.

However, this work mainly focuses on the second task, reading comprehension with BERT. Designing readers for narrative question-answering is much more difficult compared to conventional QA readers employing factoid answers. For example, given a question like "Who is Rama?" The passage containing the answers has different contexts, and a reader may return many possible answers such as Rama is a king of Ayodya, Rama is an incarnation of lord Vishnu, Rama married Sita, Rama is a brother of Lakshmana etc. This clearly indicates narrative questions can return multiple answers depending on the context, which is opposed to extracting factoid answers. Hence, the proposed model is designed to return top 'k' answers relevant to the query rather than returning a single answer. To alleviate this problem, we de ne context prediction as an auxiliary task, which explicitly drives the QA model to predict possible contexts. Then, from the contexts, our model can extract the answer-spans.

It is generally followed in span-based QA systems, presenting question as a first sequence and passage as the second sequence. Symbolically, we can denote this as H = NQA-BERT (Question, Passage). Figure 2 demonstrates the design of a BERT reader that performs both context prediction and answer-span prediction tasks. Each question token T_{qi} and passage token T_{pi} are input to BERT-Encoder as strings separated by a [SEP] token. This results in an encoding token embedding for both question and passage. Additionally, we have a linear layer trained in the ne-tuning phase for predicting the 'begin' and 'end' position of the answer-span. Two vectors S and E are included to hold start and end spans respectively, after learning in ne-tuning phase. However, the encoder that generates embedding is compatible with the attention model used in the reader. The following subsections illustrate the two tasks of the BERT reader, context prediction and answer prediction.

3.1 Context Prediction

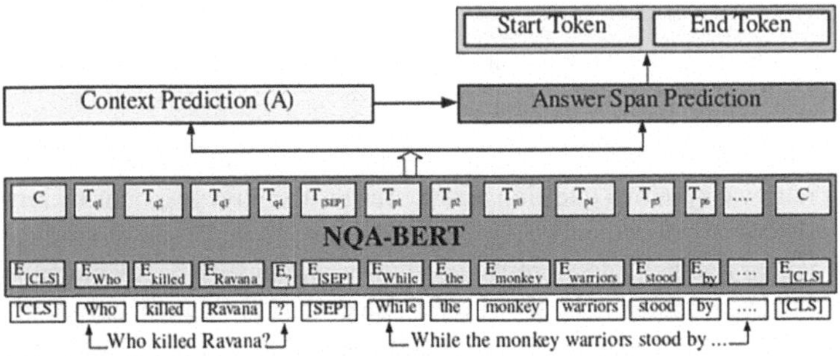

Fig. 2. Design of BERT reader for context and answer prediction.

We utilize the structure of answer-span prediction layer discussed in [23] for our context prediction task. The problem of context prediction is to estimate the begin and end pointers of context-span denoted as $p(l = B_c)$ and $p(u = E_c)$ and compute data locality of context-span $p(w_i \in C)$ through the cumulative distributions of $p(l = B_c)$ and $p(u = E_c)$. Further, $p(w_i \in C)$ is embedded in the attention layer for predicting the begin and end indices of context-span. The following formulae (Eq. 1 and Eq. 2) can be utilized for computing the begin and end indices.

$$p(l = B_c) = \frac{\exp\left(W_c^b.H_i\right)}{\sum_j \exp\left(W_c^b.H_j\right)} \tag{1}$$

$$p(u = E_c) = \frac{exp\left(W_c^e.H_i\right)}{\sum_j exp\left(W_c^e.H_j\right)} \tag{2}$$

In above formulae, H_i is an output transformer encoder vector, and H_j is an output vector of w_j, where W_b^c and W_e^c refers to span-begin and span-end embebeddings for context-span prediction layer respectively. Finally, $p(w_i \in C)$ is estimated as the product of the cumulative probabilities of words w_i appearing after B_c and before E_c as given in Eq. 3.

$$p(w_i \in C) = \sum\nolimits_{j \geq l} p(j = B_c) \times \sum\nolimits_{j \leq u} p(j = E_c) \tag{3}$$

3.2 Answer Prediction

For answer-span prediction, the BERT layer is similar to the layer used in the context-span prediction layer. We utilize context probability $p(w_i \in C)$ computed using Eq. 3 and multiplied to encoder output H for obtaining the begin and end indices of answer-spans with reference to context C. Mathematically, the begin and end pointers of answer-span can be computed using Eq. 4 and Eq. 5.

$$p(l = B_a) = \frac{exp\left(X_i W_a^b . H_i\right)}{\sum_j exp\left(X_i W_a^b . H_j\right)} \tag{4}$$

$$p(l = E_a) = \frac{exp\left(X_i W_a^e . H_i\right)}{\sum_j exp\left(X_i W_a^e . H_j\right)} \tag{5}$$

In Eq. 4 and Eq. 5, W_a^b and W_a^e refers to span-begin and span-end embeddings for answer-span prediction layer, and $X_i = p(w_i \in C)$. The loss function employed for computing the error of answer-span prediction task is given in Eq. 6.

$$L = -logP_{begin(i)} - logP_{end(i)} \tag{6}$$

4 Experimental Setup

In this section, we explain the experimental procedure and setup required for fine-tuning the NQA-BERT on Indian epic stories dataset, which is utilized for both context prediction and answer predicting tasks.

4.1 Data Preparation

The experiments are being carried out with four variations of the BERT model, namely, BERT basic, BERT Alberta, BERT Roberta, and SpanBERT, which are pre-trained on the NarrativeQA dataset [23]. The Narrative QA dataset has a total collection of 783 storybooks and 789 movie scripts, an average of 62K words per storybook or movie script. Furthermore, each document comprises of 30 pairs of question-answers created by human annotators with the help of natural language text. This dataset intuitively offers two modes of set- tings, which are defined as full-story settings and summary settings. The summary setting mode needs answering questions from Wikipedia book summaries, and the full-story setting mode needs answering questions from the story books. For narrative reading comprehension, the mode of a dataset is fixed to the full-story setting. However, the proposed NQA-BERT model is intended to carry out fine-tuning on document corpus demonstrating the Indian epic stories such as Ramayana and Mahabharata. Hence, the dataset containing 1200 documents depicting Ramayana and Mahabharata stories collected from various web resources is employed. Furthermore, every document has five to eight human-annotated question-answers.

4.2 Fine-Tuning the NQA-BERT Model

The important hyper-parameters such as batch size and learning rate are defined similar to those defined for pre-training BERT using the NarrativeQA dataset. We employed a single Nvidia GeForce GTX 1630 4 GB GPU to ne-tune the proposed NQA-BERT model. It is obvious that the ne-tuning process requires less computational time than pre-training BERT on the Narrative dataset. A batch size of 16, the learning rate of $3e^{-5}$, the number of training epochs set to 3, and the length specified for transformer encoder is 384. The model is trained using different optimizers with different λ values. But AdamW (PyTorch) optimizer with fixed weight decay proved to be efficient for fine-tuning BERT models. However, the recently developed model named Adafactor (PyTorch) proved to be better than other optimizers for training our dataset. Hence, we employed Adafactor (PyTorch) to fine-tune the proposed NQA-BERT model. Since our training data is less compared to the training data used in [22], the model took less time for training.

5 Results and Discussions

The performance evaluation of NQA-BERT is measured with the F1 Score based on ROUGH-L, which is defined by the number of overlapping tokens between predicted and ground truth answers [24]. The results are compared with other baseline language models based on BERT, including BERT ALBERT [25] BERT RoBERTa [26] and Span-BERT [27]. To overcome the computational limitations, we restrict the number of layers for BERT to only 12 layers. The results of the experiment are shown in Table 1. It is clearly implied that NQA-BERT outperforms other baseline BERT models for answer-span prediction.

Table 1. Results of answer-span prediction using NQA-BERT and baseline BERTs

Model	F1-Score
BERT Basic	65.5
BERT Albert	68.0
BERT Roberta	69.6
SpanBERT	70.3
NQA-BERT	71.2

5.1 Prediction Samples from NQA-BERT on Ramayana Text Scripts

The effectiveness of the proposed NQA-BERT for answer-span prediction task is illustrated with different test cases. Figure 3 and Fig. 4. Highlight the contexts and their corresponding answers extracted in response to the question being asked. The answers have been ranked according to their relevance scores considering 'k' value is equal to 5. i.e., the top five ranked answers have been returned by the model.

Test case-1: The input document for Test Case-1 is selected from the chapter presenting the context of war between Rama and Ravana. The question being asked is "Who killed Ravana?" The results of the query along with its context- spans is depicted in Fig. 3. It can be observed from the results, the proposed NQA-BERT can efficiently extract the answer-spans from the narrative content.

Test case-2: The input document for Test Case-2 is selected from the chapter presenting the context of who Sita is related to. However, this question seems to be a difficult one to answer. The actual question being asked is "Who is Sita?" The results of the query along with its context-spans are depicted in Fig. 4. It is noticed that the answers extracted are somehow related to the question being asked but not truly relevant. This is due to the fuzziness involved in the given context.

Query: Who killed Ravana?

Answers:

```
[ { 'answer': 'Rama',
    'context': 'caught by the death of Dushtabuddhi Rakasha, Surpnakha '
            'realized that only Rama could kill Ravana. So, when Laxman '
            'chopped off her nose, she provoked Ra'},
  { 'answer': 'Rama',
    'context': 'he she-demon died, than the gods in heaven rained lotus '
            'blossoms down on Rama blessing him. The three continued on '
            'their journey deeper into the fores'}
  { 'answer': 'humans',
    'context': '"Oh, Ravana. Khar and Dushan have been killed by humans. '
            'All their warriors are dead, too," Shurpanakha cried.'},
  { 'answer': 'unseen attackers',
    'context': "key army. Ravana's forces were merciless. Thousands were "
            'killed by unseen attackers. Seeing his forces in retreat, '
            'Hanuman charged on with a great cry'},
  { 'answer': 'Dushtabuddhi Rakasha',
    'context': 'a to avenge her husband's death. Surpanakha was first '
            'married to Dushtabuddhi Rakasha, who enjoyed great favors '
            'from Ravana, But his greed for more en'},
]
```

Fig. 3. Answer-span prediction for Test Case1.

Query: Who is Sita?

Answers:
```
[ { 'answer': 'my daughter, O Rama',
    'context': 'He placed her hand in his and said to Rama, "This'
           'is Sita, my daughter, O Rama, who is from today your'
           'partner in life. Accept her. Hold her ha'},
  { 'answer': 'beautiful wife',
    'context': 'His beautiful wife, Sita, and loyal brother, Lakshamana,'
           'voluntarily join him. The royal band journeys through the '
           'forest when their plans is drastica'},
  { 'answer': 'the she-demon',
    'context': 'like you live in this forest? Who are you?"Rama told her'
           'his story. Upon seeing Sita, the she-demon said, "That '
           'woman is not good enough for you."},
  { 'answer': 'one who is no more',
    'context': ' a mighty fallen warrior on the battlefield. Victory is'
           'the monopoly of none. Weep not for one who is no more.'
           'Rise, for we still have work to do."},
  { 'answer': 'Rama',
    'context': 'Rama responded, "And who, might I ask, is?"}
]
```

Fig. 4. Answer-span prediction for Test Case2.

6 Conclusion

In this paper, a domain-specific QA system based on BERT model named NQA- BERT is proposed for extracting answer-spans from narrative text content. The proposed work is carried out in two stages: (i) Identification of context-spans from the given document corpus relevant to the query being asked, and (ii) Development of an attention mechanism for extracting answer-spans from the context-spans. To achieve the above functionalities, NQA-BERT is designed, which can predicts both the context and answer segments from the input documents. The context-span prediction is achieved through determining the 'begin' and 'end' indices of contexts for input documents, whereas answer prediction is achieved through determining the begin and end indices of answer-span form given contexts. The NQA-BERT model is pre-trained on NarrativeQA dataset and ne-tuned on document corpus demonstrating the Indian epic stories such as Ramayana and Mahabharata. The experimental results shows that the NQA- BERT has an improved reading comprehension capability, and has proved to increase the performance gain for narrative question types.

References

1. Mishra, A., Jain, S.K.: A survey on question answering systems with classification. J. King Saud Univ. Comput. Inf. Sci. **28**(3), 345–361 (2016)
2. Wadawadagi, R., Pagi, V.: Sentiment analysis with deep neural networks: Comparative study and performance assessment. Artif. Intell. Rev. **53**, 6155–6195 (2020)
3. Wadawadagi, R., Pagi, V.: Polarity enriched attention network for aspect-based sentiment analysis. Int. j. inf. Tecnol. **14**, 2767–2778 (2022)

4. Wang, L., Zheng, K., Qian, L. Li, S.: A survey of extractive question answering. In: International Conference on High Performance Big Data and Intelligent Systems (HDIS), Tianjin, China, pp. 147–153 (2022). https://doi.org/10.1109/HDIS56859.2022.9991478
5. Huang, Y., Zhong, T.: Multitask learning for neural generative question answering. Mach. Vis. Appl. **29**, 1009–1017 (2018)
6. Pearce, K., Zhan, T., Komanduri, A., Zhan, J.: A comparative study of transformer- based language models on extractive question answering. CoRR arXiv:2110.03142 (2021)
7. Mou, X., et al.: Narrative question answering with cutting-edge open-domain QA techniques: a comprehensive study. Trans. Assoc. Comput. Linguist. **9**, 1032–1046 (2021)
8. Tikhonov, A., Samenko, I., Yamshchikov, I.P.: StoryDB: broad multi-language narrative dataset. In: Proceedings of the 2nd Workshop on Evaluation and Comparison of NLP Systems, pp. 32–39 (2021)
9. Seonwoo, Y., Kim, J., Ha, J., Oh, A.: Context-aware answer extraction in question answering. In: Proceedings of the 2020 Conference on Empirical Methods in Natural Language Processing, pp. 2418–2428 (2020)
10. Xiong, W., et al.: TWEETQA: a social media focused question answering dataset. In: Proceedings of the 57th Annual Meeting of the Association for Computational Linguistics, pp. 5020–5031 (2020)
11. Fremann, L.: Extractive NarrativeQA with heuristic pre-training. In: Proceedings of the Second Workshop on Machine Reading for Question Answering, Hong Kong, China, November 4, pp. 172–182 (2019)
12. Boratko, M., Li, X., Gorman, T., Das, R., Le, D., McCallum, A.: ProtoQA: a question answering dataset for prototypical common-sense reasoning. In: Proceedings of the 2020 Conference on Empirical Methods in Natural Language Processing (EMNLP) (2020)
13. Yao, B., et al.: It is AI's turn to ask humans a question: question-answer pair generation for children's story books. In: Proceedings of the 60th Annual Meeting of the Association for Computational Linguistics (2022)
14. Tian, D., Li, M., Ren, Q., Zhang, X., Han, S., Shen, Y.: Intelligent question answering method for construction safety hazard knowledge based on deep semantic mining. Autom. Constr. **145**, 104670 (2023)
15. Mohamed Hassan, H. A., Marengo, E., Nutt, W.: A BERT-based model for question answering on construction incident reports. In: Rosso, P., Basile, V., Martinez, R., Metais, E., Meziane, F. (eds.) Natural Language Processing and Information Systems. NLDB 2022, Lecture Notes in Computer Science, vol. 13286. Springer, Cham (2022)
16. Kierszbaum, S., Lapasset, L.: Applying distilled BERT for question answering on ASRS reports. In: New Trends in Civil Aviation (NTCA), Prague, Czech Republic, pp. 33–38 (2020)
17. Swayamdipta, S., Parikh, A.P., Kwiatkowski, T.: Multi-mention learning for reading comprehension with neural cascades. In: International Conference on Learning Representations (2018)
18. Min, S., Zhong, V., Zettlemoyer, L., Hajishirzi, H.: Multi-hop reading comprehension through question decomposition and rescoring. In: 57th Annual Meeting of the Association for Computational Linguistics (2019)
19. Gu, Y., et al.: Domain-specific language model pretraining for biomedical natural language processing. ACM Trans. Comput. Healthc. (HEALTH) **3**(1), 1–23 (2021)
20. Lee, J., et al.: BioBERT: a pre-trained biomedical language representation model for biomedical text mining. Bioinformatics **36**(4), 1234–1240 (2020)
21. Chen, D., Fisch, A., Weston, J., Bordes, A.: Reading Wikipedia to Answer Open- Domain Questions. arxiv.org/abs/1704.00051 (2017)

22. Devlin, J., Chang, M., Lee, K., Toutanova, K.: BERT: pre-training of deep bidirectional transformers for language understanding. In: Proceedings of the 2019 Conference of the North American Chapter of the Association for Computational Linguistics: Human Language Technologies (2019)
23. Kocisky, T., et al.: The NarrativeQA reading comprehension challenge. Trans. Assoc. Comput. Linguist. **6**, 317–328 (2018).
24. Rajpurkar, P., Zhang, J., Lopyrev, K., Liang, P.: SQuAD: 100,000+ questions for machine comprehension of text. In: Proceedings of the 2016 Conference on Empirical Methods in Natural Language Processing. Austin, Texas. Association for Computational Linguistics, pp. 2383–2392 (2016)
25. Liu, Y., et al.: RoBERTa: a robustly optimized BERT pretraining approach. In: International Conference on Learning Representations (2020)
26. Lan, Z., Chen, M., Goodman, S., Gimpel, K., Sharma, P., Soricut, R.: ALBERT: a lite BERT for self-supervised learning of language representations. In: International Conference on Learning Representations (2020)
27. Joshi, M., Chen, D., Liu, Y., Weld, D.S., Zettlemoyer, L., Levy, O.: SpanBERT: improving pre-training by representing and predicting spans. Trans. Assoc. Comput. Linguist. **8**, 64–77 (2020)

Author Index

T. Shreekumar et al. (Eds.): ISCComm 2023, CCIS 2231, pp. 375–377, 2025.
https://doi.org/10.1007/978-3-031-75605-4

SPRINGER NATURE

GPSR Compliance

The European Union's (EU) General Product Safety Regulation (GPSR) is a set of rules that requires consumer products to be safe and our obligations to ensure this.

If you have any concerns about our products, you can contact us on ProductSafety@springernature.com

In case Publisher is established outside the EU, the EU authorized representative is:

Springer Nature Customer Service Center GmbH
Europaplatz 3
69115 Heidelberg, Germany

The manufacturer's authorised representative in the EU is Springer Nature Customer Service Centre GmbH, Europaplatz 3, 69115 Heidelberg, Germany. If you have any concerns regarding our products, please contact ProductSafety@springernature.com

Printed and bound by CPI Group (UK) Ltd, Croydon, CR0 4YY
30/09/2026
02225597-0017